Buddhism and Science

COLUMBIA SERIES IN SCIENCE AND RELIGION

Columbia Series in Science and Religion

The Columbia Series in Science and Religion consists of peer-reviewed scholarly and general interest titles that probe salient issues relating to science and religion. The series is a forum for the examination of issues that lie at the boundary of these two complementary ways of comprehending the world and our place in it. By examining the intersections between one or more of the sciences and one or more religions, the CSSR hopes to stimulate dialogue and encourage understanding.

Buddhism & Science

BREAKING NEW GROUND

B. Alan Wallace, editor

COLUMBIA UNIVERSITY PRESS / NEW YORK

Columbia University Press
Publishers Since 1893
New York   Chichester, West Sussex
Copyright © 2003 Columbia University Press
All rights reserved
Library of Congress Cataloging-in-Publication Data

Buddhism and science : breaking new ground / B. Alan Wallace, editor.
   p. cm.
  Includes bibliographical references and index.
  ISBN 0-231-12334-5 (cloth : alk. paper) — ISBN 0-231-12335-3 (pbk. : alk. paper)
   1. Buddhism and science. I. Wallace, B. Alan.

BQ4570.S3 B836 2003
294.3′375—dc21

2002031502

Columbia University Press books are printed
on permanent and durable acid-free paper.

Printed in the United States of America
c 10 9 8 7 6 5 4 3 2 1
p 10 9 8 7 6 5 4 3 2 1

Dedicated to the memory of Francisco J. Varela (1946–2001)

Contents

Contributors

WILLIAM L. AMES

William L. Ames has an M.S. in physics from the California Institute of Technology and a Ph.D. in Buddhist studies from the University of Washington. He has taught a course on physics and Buddhism at California Institute of Integral Studies. He is a coauthor of a paper on black holes and is the author of several papers on various aspects of the Madhyamaka school of Mahāyāna Buddhist thought. He is currently a librarian at Fisher Library at John F. Kennedy University in Orinda, California and is engaged in translating and publishing a number of chapters from an Indian Madhyamaka text, Bhāvaviveka's *Prajñāpradīpa*.

MICHEL BITBOL

Michel Bitbol is directeur de recherche at the CNRS, at the Centre de Recherches en Epistemologie Appliquee (CREA), which depends on the Ecole Polytechnique, Paris. And he also teaches at the University Paris, and he is a permanent visiting member of Linacre College (Oxford). He was born in 1954, and obtained successively an M.D., a Ph.D. in physics, and a habilitation in philosophy in Paris. From 1980 to 1989 he worked as a researcher in biophysics. Then from 1990 on he specialized in the philosophy of physics. His main publications include *Schrödinger's philosophy of quantum mechanics* (Kluwer, 1996), *Mecanique quantique, une introduction philosophique* (Flammarion, 1996), and *Physique et philosophie de l'esprit* (Flammarion, 2000).

JOSÉ IGNACIO CABEZÓN

José Ignacio Cabezón is the XIV Dalai Lama professor of Tibetan Buddhism and cultural studies at the University of California, Santa Barbara. He did his undergraduate work at Caltech, with an emphasis in physics, and his doctoral studies at the University of Wisconsin-Madison, in Buddhist studies. He was a Buddhist monk for ten years; six of them he spent engaged in the traditional curriculum of studies at Sera monastery in South India. He is the author/editor/translator of several scholarly books and articles, among them *A Dose of Empti-*

ness, Buddhism and Language, Buddhism, Sexuality, and Gender, and *Tibetan Literature* (coedited with R. Jackson), and his most recent book, *Scholasticism: Cross-Cultural and Comparative Perspectives.*

HIS HOLINESS THE FOURTEENTH DALAI LAMA

His Holiness the Fourteenth Dalai Lama is the spiritual and temporal leader of Tibet and the 1989 recipient of the Nobel Peace Prize. An international advocate of nonviolence and premier representative of Buddhism in the modern world, he has a long-standing interest in science. Among his numerous encounters with scientists of the caliber of David Bohm and John Bell, since 1987, he has participated in a series of biannual conferences on Buddhism and science known as the Mind and Life conferences. Edited proceedings from these conferences have been published under the titles *Gentle Bridges: Conversations with the Dalai Lama on the Sciences of Mind* (Shambhala, 1992), *Sleeping, Dreaming, and Dying: An Exploration of Consciousness with the Dalai Lama* (Wisdom, 1997), *Healing Emotions: Conversations with the Dalai Lama on Mindfulness, Emotions, and Health* (Shambhala, 1997), and *Consciousness at the Crossroads: Conversations with the Dalai Lama on Brainscience and Buddhism* (Snow Lion, 1999).

NATALIE DEPRAZ

Natalie Depraz earned her Ph.D. in philosophy in 1993, focusing on the writings of Husserl and the phenomenology of intersubjectivity. Since 1997 she has been directrice de programme au Collège International de Philosophie (Paris), and since 2000 she has been maître de conférences en philosophie à l'Université de la Sorbonne (Paris IV). Her publications include *Transcendance et incarnation. Le statut de l'intersubjectivité comme altérité à soi chez Husserl* (Paris: Vrin, 1995), *Alterity and Facticity: New Perspectives on Husserl*, edited with D. Zahavi (Dordrecht: Kluwer 1998), *Lucidité du corps. De l'empirisme transcendantal en phénoménologie* (Dordrecht: Kluwer, 2001), and *On Becoming Aware: An Experiential Pragmatics* (with Fr. Varela and P. Vermersch) (Amsterdam: Benjamin, Amsterdam, forthcoming). She is the journal editor of *Alter, revue de phénoménologie* (Paris, 1993 onward) and *Phenomenology and the Cognitive Sciences* (with S. Gallagher and F. Varela; first issue to appear with Kluwer in January 2002).

DAVID RITZ FINKELSTEIN

David Ritz Finkelstein is working on a theory of nature as a neural network, a kind of quantum relativistic brain, in which nature operates as a quantum pattern of annihilations and creations. This research touches on quantum space-time, quantum logics, quantum set theory, elementary particles, and quantum gravity, and appears in journals like the *Physical Review, Classical and Quantum Gravity*, and *International Journal of Theoretical Physics*, which he edits, and in his recent book, *Quantum Relativity*. He was a student at Stuyvesant High School and City College in New York City and then at MIT. He now lives in Atlanta with his wife Shlomit and his daughter Aria, and works and studies at Georgia Tech with inspiring students of his own, such as James Baugh, Sukanya Chakrabarti, Andrej Galiautdinov, J. Michael Gibbs (former student), William Kallfelz, and Zhong Tang. His main scientific correspondent is Heinrich Saller of the Heisenberg Institute, Munich.

DAVID GALIN

David Galin earned his medical degree in 1961 at Albert Einstein College of Medicine, New York. He presently holds the position of associate professor in residence in the Department of Psychiatry, and he is the director of the Neurodevelopment Laboratory of the Langley Porter Neuropsychiatric Institute at the University of California School of Medicine, San Francisco. His research background includes four decades of neuro- and psychophysiology in animals and humans, and he has authored widely cited publications on the two halves of the brain, their differences and integration. He has also conducted research on dyslexia and on neuropsychological aspects of psychiatry. His current interests include religious experience from a neuropsychological perspective, theories of consciousness and the self, and rehabilitating the concept of spirit for the nonreligious and the scientifically minded.

PIET HUT

Piet Hut is professor in the School of Natural Sciences at the Institute for Advanced Study in Princeton. While his main research area is theoretical astrophysics, he is frequently involved in multidisciplinary collaborations, from geology, paleontology, and cognitive science to particle physics and computer science. He is currently involved in a Tokyo-based project aimed at developing a special-purpose computer for simulations in stellar dynamics, with a speed of 1 Petaflops. He recently wrote a textbook, *The Gravitational Million-Body Problem*, coauthored with Douglas Heggie (Cambridge University Press, 2002). He is a member of the Husserl Circle and a founding member of the Kira Institute, which has organized summer schools on ways of knowing.

THUPTEN JINPA

Thupten Jinpa was educated according to the traditional Tibetan Buddhist academic system and received his Geshe Lharam degree (the Tibetan equivalent to a doctorate in divinity) from the Shartse College of Ganden Monastic University in south India. He taught Buddhist epistemology and metaphysics at Ganden for five years. He later received his B.A. honors in Western philosophy and his Ph.D. in Religious Studies from Cambridge University. Since 1985, he has been the principal English interpreter for H. H. the Dalai Lama. He has translated and edited several books by the Dalai Lama, including *Good Heart: The Dalai Lama Explores the Heart of Christianity* (Rider, 1996). His own works include *Songs of Spiritual Experience: Tibetan Buddhist Poems of Insight and Awakening* (Shambhala, 2000) and *Tsongkha-pa's Philosophy of Emptiness* (forthcoming from Curzon). From 1996 to 1999, he was the Margaret Smith Research Fellow in Eastern Religions at Girton College, Cambridge University. He is currently the president of the Institute of Tibetan Classics, Montreal, Canada, where he works also as the chief editor for the Classics of Tibet Series, to be developed and published by the institute.

STEPHEN LABERGE

Stephen LaBerge earned his Ph.D. in 1980 in psychophysiology at Stanford University. He is a recipient of the Woodrow Wilson and Dean's Fellowships and has been an NIH Postdoctoral Fellow at Stanford University. He is presently a research associate in the Department of Psychology at Stanford University and he is the director of the Lucidity Institute. He serves on the editorial board of *Dreaming, Sleep and Hypnosis,* and *Consciousness and Cognition* and he is the author of several books and many articles on dreaming and consciousness.

VICTOR MANSFIELD

Victor Mansfield is professor of physics and astronomy at Colgate University, where he also teaches popular courses containing Tibetan Buddhism and Jungian psychology. He developed his keen interest in Eastern thought and depth psychology while earning his Ph.D. in theoretical astrophysics at Cornell University. He has published widely in theoretical astrophysics and interdisciplinary studies including the book *Synchronicity, Science, and Soul-Making.* To view some of his publications see www.lightlink.com/vic. A student of Buddhism for over twenty-five years, he has practiced and studied with spiritual leaders in the U.S., Europe, and India.

MATTHIEU RICARD

Matthieu Ricard has been a Buddhist monk for twenty years at Shechen Monastery in Nepal and is the French interpreter for the Dalai Lama. Born in France in 1946, he earned a Ph.D. in cell genetics at the Institut Pasteur under Nobel laureate François Jacob. He wrote a widely reviewed book on *Animal Migrations* (Hill and Wang, 1969). He first visited India in 1967, where he began studying and practicing Tibetan Buddhism. He has lived in the Himalayan region since 1972, training for many years under the personal guidance of Dilgo Khyentse Rinpoche, one of the Dalai Lama's teachers. He is coauthor of *The Monk and the Philosopher* (Schocken, 1999), a book of dialogues with his father, the French agnostic philosopher Jean-François Revel, which has been translated into twenty-one languages, as well as *Journey to Enlightenment,* a photo book on the life of Khyentse Rinpoche (Aperture, 1996) and numerous translations of Tibetan texts, among which is *The Life of Shabkar* (State University of New York Press, 1994). Recently he has published a dialogue with the astronomer Trinh Xuan Thuan on science and Buddhism, which is published in English as *The Quantum and the Lotus* (Crown, 2001).

FRANCISCO J. VARELA

The late Francisco J. Varela, cofounder of the Mind and Life Institute, received his Ph.D. in biology from Harvard University in 1970. His interests centered on the biological mechanisms of cognition and consciousness, and he contributed over two hundred articles to scientific journals on these matters. He also wrote or edited fifteen books, including *The Embodied Mind* (MIT, 1992), and more recently *Naturalizing Phenomenology* (Stanford University Press, 1999), and *The View from Within: First-person Methods in the Study of Consciousness* (London: Imprint Academic, 1999). Until his untimely death in May 2001, he was Foundation de France Professor of cognitive science and epistemology at Ecole Polytechnique, director of research at the Centre Nationale de la Recherche Scientifique (CNRS), and head of the Neurodynamics Unit at LENA (Laboratory of Cognitive Neurosciences and Brain Imagining) at the Salpetrière Hospital, Paris.

WILLIAM S. WALDRON

William Waldron teaches South Asian religions at Middlebury College in Middlebury, Vermont. He received a B.A. in South Asian studies and a Ph.D. in Buddhist studies from the University of Wisconsin-Madison, after spending many years living and studying in India, Nepal, and Japan. His research area is South Asian Buddhism, with strong interests in comparative and cross-cultural philosophies of mind. He has written articles and a forthcoming book on the *ālayavijñāna*, a theory of the cognitive unconscious in Yogācāra Buddhism.

B. ALAN WALLACE

Trained for many years as a monk in Buddhist monasteries in India and Switzerland, Alan Wallace has taught Buddhist theory and practice in Europe and America since 1976 and has served as interpreter for numerous Tibetan scholars and contemplatives, including H. H. the Dalai Lama. After graduating summa cum laude from Amherst College, where he studied physics and the philosophy of science, he earned his M.A. and Ph.D. in religious studies at Stanford University, where his research focused on contemplative ways of training the attention. He has served as lecturer at the Centre for Higher Tibetan Studies (Switzerland), the American Institute of Buddhist Studies, UCLA, and the University of California, Santa Barbara. Presently an independent scholar and contemplative, he lectures and teaches meditation throughout Europe and America. He has edited, translated, authored, and contributed to more than thirty books on Tibetan Buddhism, medicine, language, and culture, and the interface between science and religion. His published works include *Choosing Reality: A Buddhist View of Physics and the Mind* (Snow Lion, 1996), *The Bridge of Quiescence: Experiencing Buddhist Meditation* (Open Court, 1998), and *The Taboo of Subjectivity: Toward a New Science of Consciousness* (Oxford, 2000).

ANTON ZEILINGER

Anton Zeilinger, after being educated in the classical humanist (ancient Greek and Latin) tradition, studied physics at the University of Vienna. He held employment and visiting positions at many institutions worldwide including the Technical University of Vienna, the Massachusetts Institute of Technology, University of Melbourne, Technical University Munich, College de France and Merton College

(Oxford). He is presently professor of physics at the Institut für Experimental-physik at the University of Vienna. His main researching interests are the foundations of quantum mechanics, where, over the years, he, together with his team, has realized a number of key fundamental experiments, most recently quantum teleportation.

Preface

Since 1987, I have been involved in a series of cross-cultural scientific dialogues organized by the Mind and Life Institute (www.mindandlife.org) between H. H. the Dalai Lama and eminent Western scientists. This book, which was inspired by a suggestion to me by H. H. the Dalai Lama after one of those meetings, consists of a collection of essays on specific topics that have been examined within the fields of Buddhism and the cognitive and physical sciences. With the interface of Buddhism and the natural sciences, the contributors not only examine the fruits of inquiry from the East and the West but also shed light on the underlying assumptions of these disparate worldviews. In this way we have attempted to break down some of the barriers that have inhibited fruitful dialogue regarding Buddhism and science, anticipating that each discipline may bring fresh understanding and insightful challenges to the assumptions and methodologies of the other.

Following this preface, I have written an introduction to the volume as a whole, where I express my own views (not necessarily shared by the other contributors to this volume) concerning some of the problems in any comparative study of Buddhism and modern science. I have also written short prefaces to each of the essays in parts 1–3. Part 1 of this anthology includes two historical overviews of the engagements between Buddhism and modern science. Part 2 focuses on points of intersection between Buddhism and the cognitive sciences. Among all the natural sciences this is the field that most directly pertains to Buddhism, which holds the understanding of the mind and its relation to the rest of the world to be of paramount importance.

Because of the many common interests between Buddhism and the cognitive sciences, this interface has been a common theme in most of the

Mind and Life conferences with H. H. the Dalai Lama and various groups of scientists since 1987. These conferences were initiated by Adam Engle and Francisco J. Varela and continue on in the present, and I have joined Thupten Jinpa as an interpreter for all but one of these meetings. With Francisco's untimely death in May 2001, shortly following the ninth Mind and Life conference, we have lost a cherished friend and respected colleague. He is dearly missed, and this volume is dedicated to his memory.

Part 3 focuses on Buddhism and the physical sciences. Since physics has been the paradigm for the natural sciences as a whole, dialogues between Buddhism and science cannot avoid discussions of this mode of inquiry into the physical universe. Moreover, twentieth-century physics—most notably quantum mechanics—has raised profound epistemological and ontological issues that challenge many of the traditional assumptions underlying science as a whole. Some of these topics, such as the relation between subject and object, lend themselves to dialogue with Buddhist philosophy, especially the Madhyamaka view, which is discussed in several essays in this volume. This volume concludes with an essay by physicist Piet Hut, who brings the challenges of this type of interdisciplinary inquiry into the context of daily life.

I believe that this anthology of essays will be of interest not only to Western scientists, Buddhologists, and scholars of religion but also to a much broader range of readers interested in East-West dialogue and the interface between science and religion as a whole. It is our hope that this work will illuminate multiple ways of exploring the nature of human identity, the mind, and the universe at large and thereby lead to greater well-being for all humanity.

I would like to thank H. H. the Dalai Lama for his continuing inspiration for such cross-cultural interdisciplinary dialogues and collaboration, the John E. Fetzer Foundation and Richard Gere Foundation for their support of this project, all the scholars, contemplatives, and scientists who contributed essays to this volume, Taline Goorjian for preparing the index, and Jonathan Slutsky, Holly Hodder, Robin Smith, Alessandro Angelini, and Susan Pensak of Columbia University Press for bringing this project to completion.

Buddhism and Science

B. Alan Wallace

Introduction: Buddhism and Science—Breaking Down the Barriers

The publication of a volume of essays on Buddhism and science presupposes that these two fields are commensurable and that the interface between Buddhist theories and practices and scientific theories and modes of inquiry can somehow be fruitful. But serious objections to this presupposition can be raised from the outset, so I would like to introduce this work by presenting arguments against such a coupling of Buddhism and science together with my responses to those arguments. The first idea to be considered is the view that religion and science are autonomous, their domains of concern mutually exclusive, so they really have little, if anything, to say to each other. I shall respond to this assertion by first analyzing whether Buddhism can properly be categorized according to modern Western notions of religion, then I shall describe specific elements within Buddhism that may be deemed scientific. I shall then distinguish between empirical science itself and the metaphysical dogma of scientific materialism that is often conflated with it. Next I shall address objections raised by proponents of postmodernism, to the effect that Buddhism and science are cultural specific and hence fundamentally incomparable. Finally, I shall present suggestions for a dialogic approach to the study of Buddhism and science that may enrich both fields and consequently broaden our understanding of the subjective and objective domains of the natural world.

ARE RELIGION AND SCIENCE AUTONOMOUS?

Most mainstream religious thinkers and many scientists share the view of religion and science as independent and autonomous rather than conflicting realms, with each discipline having its own domain and methods that can be justified on its own terms. One of the most prominent scientists to promote this view is paleontologist Stephen Jay Gould. In his book *Rocks of Ages: Science and Religion in the Fullness of Life* Gould argues that religion and science are logically distinct and fully separate in terms of their styles of inquiry and goals. But, rather than suggesting that they are irrelevant to each other, he emphasizes the need to integrate insights from both in order to build a rich and full view of life (Gould 1999:29). One of Gould's central ideas is that the domains of religion and science consist of "non-overlapping magisteria." In his view the magisterium of science includes the empirical realm, and it addresses the questions of what the universe is composed of and how it works. The magisterium of religion, on the other hand, consists of the realm of human purposes, meaning, and value. His solution for the apparent conflicts between religion and science is to maintain that the two should coexist in a spirit of respectful noninterference. Religious texts, therefore, should not be read as scientific texts, and the claims of scientists should not be used to disprove the basis of religious belief (93).

In a similar vein, theologian Langdon Gilkey declares that religion addresses questions concerning the meaning and purpose of life, our ultimate origins and destiny, and the experiences of our inner life. Science, in contrast, seeks to explain objective, public, repeatable data with theories that are logically coherent and experimentally adequate, presenting quantitative predictions that can be tested experimentally (Gilkey 1985:108–116).

Not all scientists (or religious believers), however, go along with the amicable assertion of the nonoverlapping domains of religion and science. Zoologist Richard Dawkins, for example, poignantly argues that religious beliefs are not outside the domain of science and there are consequently irreconcilable differences between religion and science. Since religions do make claims about the nature of existence, and do not confine themselves solely to questions of meaning and values, religious beliefs and dogmas should be subjected to scientific criticism (Dawkins 1999:62–64).

Sociobiologist Edward O. Wilson takes a somewhat more equivocal position regarding the relation between religion and science. He first defines science as the "*organized, systematic enterprise that gathers knowledge about the world and condenses the knowledge into testable laws and principles*" (Wilson 1998:58). This, he claims is the accumulation of humanity's organized, objective knowledge and is the first medium devised able to unite people everywhere in common understanding. Religion he defines as "the ensemble of mythic narratives that explain the origin of a people, their destiny, and why they are obliged to subscribe to particular rituals and moral codes" (247).

At first glance Wilson seems to follow Gould's premise of nonoverlapping magisteria. He suggests that the proper role of religion is to codify and put into enduring, poetic form the highest values of humanity consistent with empirical knowledge. And the responsibility of science is to test relentlessly every assumption about the human condition, thereby eventually uncovering the bedrock of the moral and religious sentiments. This appears to be an affirmation of the familiar fact/value split between science and religion. He even bemoans the tragedy it would be if the United States, for example, were to abandon its sacral traditions. For example, even the most secular Americans would be ill advised, he admonishes, to expunge "under God" from the American Pledge of Allegiance; oaths should continue to be taken with hand on the Bible; everyone should bow their heads in communal respect as ministers and rabbis bless civil ceremonies with prayer.

In apparent agreement with Dawkins, Wilson's scheme of consilience, or the grand unification of knowledge, requires that all religious truth claims be subjected to empirical testing using the methods of objective scientific inquiry. Thus the existence of God is a problem for astrophysics, and the nature of the human mind and soul is to be determined at the juncture of biology and psychology, in terms of nerve cells, neurotransmitters, hormone surges, and recurrent neural networks. Only in this way, he declares, can humanity discover which cellular events compose the mind (241, 99–100). In this way religious narratives about the nature of the universe and humanity will be replaced by scientific theories that possess "more content and grandeur than all religious cosmologies combined" (265). In the final analysis, he claims, belief in the gods and belief in biology are not factually compatible. "As a result those who hunger for both intellectual and religious truth will never acquire both in full measure" (262). The implication of his view seems to be that society should maintain the rituals suggestive of belief in a fictitious God without attributing any more reality to God than is granted to Santa Claus or the Easter Bunny.

All the above views have profound implications for the interface between Buddhism and science, assuming that Buddhism can be categorically regarded as simply a religion. Let us now turn to the important question of whether this common classification is justified.

IS BUDDHISM SIMPLY A RELIGION?

Whether we categorize Buddhism as a religion depends, of course, on our definition of the word *religion*. According to religionist Van Harvey, we deem a system of belief and practice to be religious if it expresses a dominant interest in certain universal and elemental features of human existence as those features bear on the human desire for liberation and authentic existence (Harvey 1981:chapter 8). So defined, Buddhism may indeed be viewed as a religion. On the other hand, according to the above definition suggested by Edward Wilson, Buddhism does not as a whole neatly fit the criteria of a religion. Its core is the Four Noble Truths: the truths of suffering, the sources of suffering, the cessation of suffering together with its source, and the path to such cessation. This has little to do with any "mythic narratives that explain the origin of a people, their destiny, and why they are obliged to subscribe to particular rituals and moral codes."

Wilson's assumptions and speculations about religion, expressed in his book *Consilience*, are evidently based almost entirely on the Judeo-Christian tradition, while ignoring the specific features of all other religious traditions throughout the world. Considering the wealth of information we now have in European languages about all the major religious traditions of the world, this oversight seems unjustified. But this form of ethnocentricity is unfortunately quite common, even in the contemporary academic study of religion. Religionist Richard King cites the long tradition in Western Orientalist scholarship to uncritically bring Christian assumptions about the nature of religion to bear on Buddhism. Academic Buddhologists have traditionally regarded Asian Buddhists as "native informants," and involvement with them, he writes,

> was usually subordinated to the insights to be gained from careful reading of the "canonical" works of ancient Buddhists. This was seen as the most effective way to discern the true essence of Buddhism. The consequence of this trend was that "pure" or "authentic Buddhism" became located not in the experiences, lives or actions of living Buddhists in Asia but rather in the university libraries and archives of Europe—specifically in the edited manuscripts

and translations carried out under the aegis of Western Orientalists. (King 1999:150)

To understand Buddhism on its own terms, it is imperative that we in the West recognize the cultural specificity of our own terms *religion, philosophy,* and *science* and not assume from the outset that Buddhism will somehow naturally conform to our linguistic categories and ideological assumptions. Buddhism clearly includes profoundly religious elements, as outlined by Harvey, as well as strong philosophical themes and reasoning from its inception. Most important for the theme of this volume, it has also, from its very origins, established rigorous methods for experientially exploring the personal and impersonal phenomena that make up the natural world. Such techniques, many of which are designated by the English term *meditation,* frequently entail careful observation followed by rational analysis. In short, there are elements of Buddhist theory and practice that may be deemed scientific, but in flatly classifying Buddhism as a religion both its philosophical and scientific features are simply overlooked.

Such ideological hegemony crops up frequently in the writings of Western Buddhologists to the present day. For example, Buddhologist Luis Gómez characterizes Buddhist doctrine simply as a "religious ideology," which stands in opposition, he suggests, to any form of rational public discourse (Gómez 1999:369). As indicated in many of the records of Buddha's own teachings, as well as the eminent history of Buddhist dialectics and public debate in India and Tibet in particular, many Buddhist theories are obviously expressions of rational public discourse. The notion that only nonreligious people have some kind of a monopoly on such discourse is simply untenable. Moreover, Buddhists have a long tradition of studying the mind and presenting rational descriptions of its functions and ways of healing its afflictions and developing wholesome mental behavior. Yet Gómez expresses his bafflement upon encountering the term *Buddhist psychology* on the grounds that such a word is no more justified than *Christian chemistry.* By drawing a spurious parallel between Buddhist psychology and Christian chemistry, Gómez dismisses the rapidly increasing number of essays and books on Buddhist psychology and its relation to modern psychology written by psychologists and Buddhist scholars alike. No one, on the other hand, is seriously suggesting that there is such a thing as Christian chemistry. His reasoning here seems to be that Buddhists lack the tools and methods of modern Western psychology, so whatever methods for studying the mind and whatever conclusions they may have drawn cannot be deemed psychological. This same line of reasoning is the one used for excluding Buddhist philosophy from virtually all academic departments of philosophy in Europe and America: if Buddhists don't philosophize following the same rules as Western philosophers, they don't philosophize at all. But if we should follow this line of reasoning ad absurdam, since Buddhism does not even affirm the existence of a divine Creator who rules the universe, punishes sinners, and rewards the faithful, like the "genuine" religion of Christianity, it can't even be counted as a religion. It simply falls through the cracks and counts for nothing at all.

It is true that Buddhism fails to fit neatly into any of our categories of religion, philosophy, and science, for the simple reason that it did not develop in the West, where these concepts originated and evolved. Buddhism offers something fresh and in some ways unprecedented to our civilization, and one of its major contributions is its wide range of techniques for exploring and transforming the mind through firsthand experience. But many scholars of religion, including Buddhologists, appear incapable of imagining that the Buddhist tradition may have developed ways of knowing that have not already been developed in the West. Gómez, for instance, refers to Buddhist meditations as forms of "ritualized behavior" that are "rehearsed" in the hopes that "conforming" to such conduct will transform oneself and others in favorable ways (368). Buddhologist Roger R. Jackson similarly portrays Buddhist meditation as a type of ritual act (Jackson 1999:231). While such characterizations are certainly valid for some types of Buddhist meditation, they are profoundly misleading for the practices of meditative quiescence (*śamatha*) and contemplative insight (*vipaśyanā*), which are the two core modes of Buddhist meditative training. Techniques of meditative quiescence entail the rigorous cultivation of attentional stability and vividness, methods having a strong bearing on William James's psychological theories of attention (Wallace 1998, 1999a). And the wide range of Buddhist techniques for the cultivation of contemplative insight, based upon the prior training in refining the attention, also bear great relevance to modern theories of clinical and cognitive psychology.

What many Buddhologists seem to do is stuff Buddhism into familiar files, such as "religion" or "philosophy," without attending closely to the ways in which it does not fit our Western categories. Having comfortably classified Buddhism in ways that do not challenge any of their preconcep-

tions about religion, philosophy, or science, they conceive of the great proponents of traditional Buddhism over the ages in their own image: as scholars who spent their time reading other people's books and writing their
own books about other people's books. While contemporary Buddhologists
may be validly regarded as professional scholars of Buddhism, few consider
themselves, or are considered by others, as professional contemplatives
within the Buddhist tradition. This contrast between scholarly profession-
alism and contemplative inexperience has introduced a glaring bias into
modern academic Buddhist scholarship (Wallace 1999b).

A flagrant example of this trend occurs in the writings of religionist Paul
Griffiths. In his extensive writings on the nature and goals of Buddhist
meditation, Griffiths candidly acknowledges that in terms of his own
methodology he does not even begin to address whether or not there actually are or were virtuoso Buddhist practitioners who claim to be able to enter the meditative state called "the attainment of cessation" (*nirodhasamā-
patti*), which is a primary goal of Buddhist meditation. On the basis of his
text-critical analysis of the attainment of this meditative state, he concludes
that it is analogous to "some kind of profound cataleptic trance, the kind of
condition manifested by some psychotic patients and by long-term coma
patients" (Griffiths 1986:11). Having drawn this conclusion, he does not
speculate on why Buddhist contemplatives would undergo long years of
training in philosophy (Wallace 1980), ethical discipline, attentional refinement, and experiential, contemplative inquiry just to achieve a state that
could more readily be achieved through a swift blow to the head with a
heavy, blunt instrument. This is the type of absurd conclusion that emerges
from the Orientalist approach taken by so many Buddhologists since the inception of this Western academic discipline (Almond 1988; Wallace 1999c).
What is most odd about this approach is that it is somehow deemed by its
advocates to be scientific. On the contrary, modern science emerged in part
as a rebellion against just this kind of nonempirical, dogmatic, scholastic
mode of inquiry!

Regardless of the commonly unscientific study of Buddhism in Western
academia, the question remains whether elements of Buddhism itself may
be considered scientific in some meaningful sense of the term. Wilson accurately describes science as the "*organized, systematic enterprise that gathers
knowledge about the world and condenses the knowledge into testable laws and
principles*" (58). This enterprise is centrally concerned with the networks of
cause and effect across adjacent levels of organization, and it is character-

ized by an empiricism that stands in stark contrast to the transcendentalism that lies at the core of so much philosophy and theology.

Returning to the core theme of Buddhist theory and practice—the nature and causal origins of suffering, the possibility of freedom, and the causes that lead to such freedom—we see that Buddhism too is centrally concerned with causality within human experience. In this sense it is a form of naturalism, not transcendentalism. Buddhism, like science, presents itself as a body of systematic knowledge about the natural world, and it posits a wide array of testable hypotheses and theories concerning the nature of the mind and its relation to the physical environment. These theories have allegedly been tested and experientially confirmed numerous times over the past twenty-five hundred years, by means of duplicable meditative techniques (Wallace 2000:103–118). In this sense, too, Buddhism may be better characterized as a form of empiricism rather than transcendentalism. This is not to deny, of course, the diversity of views among Buddhists about the nature and significance of specific contemplative insights, rather, over the history of science, in each generation, its theories and discoveries have also been open to varying interpretations. A major difference between science and Buddhism is that scientists largely exclude subjective experience from the natural world and attribute causal efficacy only to physical phenomena. Buddhism, in contrast, takes subjective mental phenomena at least as seriously as objective physical phenomena and posits a wide range of interdependent causal connections between them.

To take a specific example, to a much greater extent than modern psychology Buddhism presents rigorous means of investigating the necessary and sufficient causes of suffering and happiness. It is intent not only on counteracting suffering once it has arisen but on identifying and counteracting the causes of suffering before it arises. All conditioned phenomena arise from multiple causes, and the central theme of Buddhism is to identify especially the *inner* causes of joy and sorrow, for they have been found to be more crucial than *outer*, physical causes. This is perhaps the most scientific aspect of Buddhism, and it addresses issues in the realm of human experience and consciousness itself that have been largely overlooked by modern science. Surely it is unscientific to declare that humans are prone to dissatisfaction simply because of "human nature," as is commonly assumed in modern psychology!

Buddhist insights into the nature of the mind and consciousness are presented as genuine discoveries in the scientific sense of the term: they can be replicated by any competent researcher with sufficient prior training. But are the means by which these alleged discoveries have been made truly rigorous? They are certainly not quantitative, nor do they lead to the formulation of mathematical laws. But the criteria of rigor in one field, such as the exploration of objective physical processes, may be superfluous or inapplicable in another, such as the exploration of subjective experience and its relation to the environment. Whether a method is deemed rigorous or not depends on what one is trying to achieve. In some contexts precise quantitative measurement is crucial to rigorous observation, in others it is impossible or irrelevant. In such cases new criteria of rigor need to be devised in relation to the specific epistemic and pragmatic goals of the research.

One distinction commonly made between science and the contemplative traditions of the world is that science entails collective knowledge, whereas contemplative insights are always private and cannot be shared. As Edward Wilson, points out, "One of the strictures of the scientific ethos is that a discovery does not exist until it is safely reviewed and in print" (59). What he means, of course, is that a discovery is not accepted within a scientific community until it has been published and acknowledged. This valid assertion cannot refute the obvious fact that a genuine discovery actually takes place prior to its publication! And even after it is published a scientific discovery can normally be validated only by a relatively small number of experts within a specific field of research. Other scientists and the general public will, for the most part, accept the discovery on the basis of their faith in the experts. This situation is not so different from discoveries made by Buddhist contemplatives. The discoveries are made in terms of their own firsthand experience. They may then be reported either verbally or in print, and their claims are subject to peer review by their fellow contemplatives, who may debate the merits or defects of the reported findings. Critiques by anyone other than professional contemplatives are taken no more seriously than critiques of scientific theories by nonscientists.

The assertion that Buddhism includes scientific elements by no means overlooks or dismisses the many explicitly religious elements within this tradition. As Stephen Jay Gould says of religion, Buddhism is very much concerned with human purposes, meaning, and value. But, like science, it is also concerned with understanding the realms of sensory and mental experience, and it addresses the questions of what the universe, including both objective and subjective phenomena, is composed of and how it works. In accordance with Langdon Gilkey's portrayal of religion, Buddhism does address questions concerning the meaning and purpose of life, our ultimate origins and destiny, and the experiences of our inner life. But the mere fact that Buddhism includes elements of religion is not sufficient for singularly categorizing it as a religion, any more than it can be classified on the whole as a science. To study this discipline objectively requires our loosening the grip on familiar conceptual categories and preparing to confront something radically unfamiliar that may challenge our deepest assumptions. In the process we may review the status of science itself, in relation to the metaphysical axioms on which it is based.

EMPIRICAL SCIENCE AND THE DOGMA OF SCIENTIFIC MATERIALISM

In this presentation of the salient features of science and scientific materialism, I shall refer frequently to the writing of Edward Wilson, who is both a distinguished scientist as well as an articulate, self-avowed proponent of scientific materialism and scientism. The extent to which I dialogue with Wilson in this introduction may seem incommensurate with the fact that he has so little to say about Buddhism. But scientific materialism as he so well presents it is widely accepted by many scientists and nonscientists, and it is this dogma in general that presents formidable obstacles to any meaningful collaboration between Buddhism and science. Wilson points out five diagnostic features of science that distinguish it from other modes of inquiry (53). 1. The first feature is repeatability, which is characteristic of experiments in which the phenomena under study can be controlled. But this is rarely possible in sciences such as astronomy, where the Scientific Revolution began, and geology, so it is not true of all the natural sciences. 2. The second feature is economy, namely, the abstraction of knowledge in the simplest and most aesthetically pleasing way possible. While simplicity may be an objectively measurable quality, beauty is not. At this point Wilson rightly points to an obviously subjective element in the formulation of scientific theories. 3. The third feature is mensuration, which is to say that science focuses on things that can be measured using universally accepted scales. While quantitative measurements are perfectly appropriate for objective, physical phenomena, their application is less feasible for the scientific study of subjective mental phenomena. This leaves science with two options: (a) either to exclude subjective experience from the domain of science, or (b) to acknowledge that the concept of rigorous measurement must be reappraised when studying the mind. 4. The fourth feature is heuristics, which means that the best science stimulates further discovery. The implication, of course, is that if a scientific dogma inhibits further discovery it should be expelled from scientific thinking. 5. The final feature cited by Wilson is consilience, meaning that the scientific explanations that survive are those that can be connected and proved consistent with one another. If consilience is truly to unify all aspects of the natural world, as he envisions, it must include the empirical study of subjective experience as well as objective phenomena, but Wilson offers no strategy for unifying both these elements of the natural world.

Like most other scientific materialists, Wilson conflates empirical science with the metaphysical assumptions of scientific materialism, but I shall now argue that the latter is actually a type of dogma that has long impeded discoveries especially pertaining to the mind and consciousness. By the term *dogma* I mean a coherent, universally applied worldview consisting of a collection of beliefs and attitudes that call for a person's intellectual and emotional allegiance. A dogma, therefore, has a power over individuals and communities that is far greater than the power of mere facts and fact-related theories. Indeed, a dogma may prevail despite the most obvious contrary evidence, and commitment to a dogma may grow all the more zealous when obstacles are met. Let us now turn to some of the fundamental assumptions of the dogma of scientific materialism.

Objectivism   As a metaphysical dictate the principle of scientific objectivism requires one to disregard that which is individual, private, uncontrolled, unique, and anomalous. Even though such events occur frequently in the natural world, they are not included in the scientific picture of reality and so are not regarded as real. Thus, with a single metaphysical stroke of the pen, subjective experience is written out of nature and consigned to the status of an epiphenomenon or illusion. The affirmation of scientific objectivism also implies a commitment to the view that, in Wilson's words, "outside our heads there is freestanding reality. . . . Inside our heads is a reconstitution of reality based on sensory input and the self-assembly of concepts" (60–61). The proper task of scientists, he claims, is to correctly align the subjective representation of reality inside our heads with the objective external world. What this implies, then, is that the objective world *lies beyond* the subjective world of appearances, including all the evidence from our senses, which exists only in our heads. Wilson correctly acknowledges that there is no objective yardstick on which to mark the degree of correspondence between the objective world and our subjective representations (59). Therefore, as much as he tries to promote empiricism, in opposition to transcendentalism, he is in fact a transcendentalist with respect to the very existence of the objective world and our knowledge of it. For the "real world," which is the domain of science as he understands it, transcends all empirical data and can be known only indirectly, by way of the representations inside our heads. While admitting that contemporary science has no criteria of objective truth, Wilson places his faith in future discoveries in the brain sciences, which he hopes will reveal the physical bases of thought processes and thereby reveal the nature of the mind itself (Wilson 1998:60, 64).

Reductionism   Wilson places great store in the principle of reductionism, which he describes as the cutting edge of science that breaks nature apart into its natural constituents. As a research strategy, reductionism has proved its usefulness countless times in the history of science. But this guideline must be used wisely. In the brain sciences, for example, if one focuses one's attention on the operations of individual subatomic particles, atoms, molecules, cells, or even entire ganglia of neurons, this excessively narrow vision can obstruct insight into the global processes occurring in diverse regions of the brain. Moreover, if one focuses solely on objective brain functions and ignores subjective mental events, this mode of reductionism prevents one from discovering mind-brain correlates. All one learns about is the brain, which, by itself, reveals no objective evidence for the existence of consciousness or subjective experience of any kind! Thus, if one dogmatically assumes that the mind is composed of nothing more than brain functions, as Wilson does, this fixation ensures that the mind, as it is experienced firsthand, will remain a mystery. And if all our representations of the objective world exist only in our minds, this type of reductionism also leaves the objective world a transcendent mystery. In other words, Wilson's commitment to ontological reductionism seems to undermine his whole ideal of consilience.

Monism   Scientific materialists appear to be in wholehearted agreement that the entire universe fundamentally consists of one kind of stuff, and that is matter. Any divergence from this view, they claim, brings us inevitably back to an antiquated, discredited form of Cartesian dualism, which posits the existence of two primary kinds of substances: mind and matter. But why should we limit our imaginations to these two options alone? Consider the ontological status of numbers, including real numbers such as the gravitational constant and Planck's constant, as well as imaginary and complex numbers, mathematical laws, space, time, ideas, sensory and dream imagery, and consciousness. Why should we believe that all such phenomena really consist of one type of stuff? Why could the natural world not be comprised of a wide range of material and immaterial phenomena? Only the dogmatic principle of monism prevents us from considering other possibilities.

Physicalism   The research instruments of science, since the time of Galileo, have been designed to measure physical phenomena only. Thus if other types of phenomena exist, they must lie outside the domain of science as it has developed thus far. Advocates of the metaphysical principle of physicalism, however, have concluded that only those phenomena that can be detected with the tools of science actually exist. The universe is believed to consist solely of matter and its emergent properties. To understand this principle, it is crucial to recognize that the matter in question is not the familiar stuff that we bump into in everyday experience. A rock held in the hand, for instance, is experienced as having a certain color, texture, and weight. But all those qualities are *secondary attributes* that exist, according to Wilson, not in the objective world but as representations inside our heads. The matter that is the fundamental stuff of the objective universe, according to scientific materialism, exists independently of all such secondary attributes that arise only in relation to a conscious subject. The real properties of matter are its inherent *primary attributes* that exist independently of all modes of detection.

What does science know today about the nature of matter? Physicists agree that matter consists of atoms, which in turn are made up of elementary particles such as electrons and protons. There are then further speculations concerning quarks, superstrings, and so on with regard to the component parts of elementary particles. But the actual nature of these fundamental building blocks of the universe is somewhat shrouded in mystery. Some physicists argue that atoms are emergent properties of space or space-time. But which space are they referring to? There are actually countless possible spaces with their own geometries, each of which is equally valid and self-consistent. Others maintain that atoms are not *things* at all but are better viewed as sets of relationships (Wallace 1996:55).

Even if matter is regarded as some independent stuff existing independently in the objective universe, its mass and spatial and temporal dimensions are not fixed or absolute but, according to relativity theory, depend upon the inertial frame of reference from which they are measured. And, in terms of quantum mechanics, it appears increasingly dubious whether the elementary particles of matter have any discrete location independent of all systems of measurement. Ever since the origins of quantum mechanics experts have expressed diverse views, ranging from the assertion that elementary particles exist independently as real, distinct entities to the view that there is no objectively existing quantum realm at all (Herbert 1985)! As physics continues to progress, the primary status of matter appears to be on the decline. As physicist Steven Weinberg recently commented, "In the physicist's recipe for the world, the list of ingredients no longer includes particles. Matter thus loses its central role in physics. All that is left are principles of symmetry" (Cole 1999).

Upon confronting such startling lack of consensus about the nature and primacy of matter, the physicalist may take refuge in the notion of energy and its conservation as the primary stuff of the universe. But, once again, one is bound for disappointment, for, according to physicist Richard Feynman, the conservation of energy is a mathematical principle, not a description of a mechanism or anything concrete. He then goes on to acknowledge, "It is important to realize that in physics today we have no knowledge of what energy *is*" (Feynman, Leighton, and Sands 1963:4–2).

For scientific materialists such as Edward Wilson, signs of the existence and primacy of matter are to be found everywhere, even though those signs are all indirect (existing, as they do, as mere mental representations). Although matter is never detected as an independently existent stuff in the objective world, it is assumed to be the origin and basis of all that we experience. As to its actual nature, there have always been many competing views, and the number of hypotheses does not appear to be on the decline. Upon reflection, it seems that matter presently fills the role for the materialist that God has traditionally filled for the theist. And the diverse speculative theories of "materialogians" provide little support for the belief that such mysterious stuff can support the ontological burden of the entire universe of subjective and objective phenomena.

The Closure Principle   Whether or not immaterial phenomena exist, advocates of the closure principle maintain that they never exert any influences within the physical universe. That is, the universe is closed off from nonphysical causation. Wilson alludes to this assumption when he writes, "The central idea of the consilience world view is that all tangible phenomena, from the birth of stars to the workings of social institutions, are based on material processes that are ultimately reducible, however long and tortuous the sequences, to the laws of physics" (266). The implications of this principle, of course, are enormous, both in terms of the boundaries of scientific knowledge and the nature of human existence. One field in which it is especially pertinent is the study of evolution. Wilson points out that "biological capacity evolves until it maximize the fitness of organisms for the niches they fill, and not a squiggle more" (48), and that such evolution occurs solely due to the laws of physics, with no immaterial influences. And yet he goes on to admit that one cannot comprehend progressively the formation of cells by understanding electrons or atoms. Rather, once one has understood the cell, one can work backward to understand it in terms of more basic elements. In other words, there's an asymmetry of knowledge here: the explanations of the physical sciences are necessary but not sufficient for understanding biological processes (Wilson 1998:68).

Actually, if one insists that all biological and psychological processes are ultimately reducible to the laws of physics, many facets of human existence, including those commonly deemed the most meaningful, remain inexplicable. Since natural selection does not anticipate future needs, how did it prepare the human mind for civilization before civilization existed? How did the human mind evolve symbolic language, which was necessary for igniting the exponentiation of cultural evolution? If the brain is nothing more than a machine assembled not to understand itself but to survive, as Wilson claims, how is it that humans have the capacity to develop such a sophisticated brain science? Finally, if, as Richard Dawkins maintains, and Wilson agrees, the human brain and sensory system evolved as a biological apparatus to preserve and multiply human genes (Wilson 1998:52), how is it that we humans have the capacity to experience universal love and concern for the welfare of the human race as a whole? As Dawkins admits, such facts "simply do not make evolutionary sense" (Dawkins 1978:2). What exactly doesn't make sense? The capacity of the human mind to develop symbolic language, the human yearning and ability to pursue truth, whether by means of religion, philosophy, or science, and the human capacity for unconditional love and compassion? Shall we say these don't make sense? Or shall we abandon the metaphysical assumption that they all inexplicably evolved due to natural selection—in accordance with the closure principle—even though this assertion violates a central principle of evolution?

The assertion of the closure principle also has great ramifications for the question of free will. If there is such a thing as freedom of the will, there must be someone to exert that freedom, to make free decisions. But when he addresses the question of the relation between the brain and the self, Wilson writes, "Who or what within the brain monitors all this activity? No one. Nothing. The scenarios are not seen by some other part of the brain. They just *are*" (119). If there is no individual identity or self apart from brain function, the question of free will seems moot. But Wilson doesn't leave it there. The hidden, cerebral preparation of mental activity, he claims, gives the *illusion* of free will, and humans need this illusion for our survival. The fact that it is an illusion, he assures his readers, is protected by the un-graspable complexity of the material influences on the brain. In short, ac-cording to scientific materialism our very survival depends in part upon the maintenance of the illusion of free will. But if this is true, and scientific ma-terialism has shown us that we do not even exist as individuals who make real choices, it would follow that the proliferation of scientific materialism is undermining our very chances of survival as a race. For once an illusion is unveiled—for example, when a child is told that Santa Claus doesn't real-ly exist—its ability to influence the course of our lives is impaired.

The impact of the closure principle doesn't stop even there. Humans are concerned not only with survival and procreation but with the pursuit of meaning and happiness. But meaning, Wilson writes, "is the linkage among the neural networks created by the spreading excitation that enlarges im-agery and engages emotion" (115). Concerning the pursuit of happiness, he acknowledges that millions seek it and "feel otherwise lost, adrift in a life without ultimate meaning," but he suspects that it will eventually be ex-plained as "brain circuitry and deep, genetic history" (260–261). He offers no clue as to how humans might actually experience happiness, for this is one more facet of human existence that does not make "evolutionary sense."

The picture, thus far, that scientific materialism gives us of the nature of human existence appears bleak at best. Each of us, it maintains, is an organ-ic robot, dominated by our brains, which are biologically programmed to preserve and multiply human genes. Human identity, therefore is an illu-sion, as is free will, and the pursuit and experience of meaning and happi-ness finally boils down to neural activity operating under the impersonal laws of physics. But Wilson conceals the implications of this dismal vision of reality when he writes that scientists "have begun to probe the foundations of human nature, revealing what people intrinsically most need, and why. We are entering a new era of existentialism—giving complete autonomy to the individual" (297). To my mind, the strangest aspect of this view of human existence is anyone would want to embrace it, when neither empirical facts nor rational arguments compel one to do so.

THE RELIGIOUS STATUS OF SCIENTIFIC MATERIALISM

The preceding overview of the central tenets of scientific materialism clearly reveals its status as a dogma that far transcends the domains of empirical science. But, in the twentieth century in particular, it took on the status of a religion. In Wilson's presentation of scientific materialism he has a great deal to say about the realm of human purposes, meaning, and value, which Stephen Jay Gould argues is the sole domain of religion. Wilson also writes at length on the meaning and purpose of life, our ultimate origins and destiny, and the experiences of our inner life, which Langdon Gilkey reserves for religion. In scientific materialism the boundaries between science and religion dissolve, and a new religion is presented as a substitute for all traditional religions. The sacred object of its reverence, awe, and devotion is not God or spiritual enlightenment but the material universe, which exists transcendently, "outside out heads." In other words, scientific materialism appears to be a modern kind of nature religion, which has innumerable precedents in the preliterate history of humanity (Goodenough 1998; Wallace 2000:30–39).

Edward Wilson, however, would have us believe just the opposite. All traditional religions, he claims, are in fact hereditary, "urged into birth through biases in mental development encoded in the genes" (257). There is little in his work to suggest that he is aware of the extent to which the tenets of his own creed of scientific materialism are rooted in the theological premises of the Judeo-Christian tradition (Wallace 2000:41–56). In his view the human mind evolved to believe in gods, but it did not evolve to believe in biology. The empirical basis for this assertion is that throughout recorded history there is evidence humans have believed in a God or gods, whereas biology has emerged only within the past few centuries. But there is no evidence of humans believing in gods before the human mind developed symbolic language, and that capacity remains unexplained in terms of the principles of natural selection. So if the use of symbolic language cannot be explained in terms of evolution, as something encoded in the genes, the same must be true of religion as well. Wilson overlooks this fact, and while he claims to be an advocate of empiricism, pitting it against transcendentalism, when it comes to the origins of biology, he presents this as a mystery that cannot be explained in terms of the natural laws of evolution! While he inaccurately tries to naturalize religion by attributing it to evolution, he transcendentalizes biology by setting it above the laws of evolution. Here is a clear-cut move to sanctify his own field of scientific expertise, which he claims, time and again, holds the key to the deepest questions of our existence.

Lest there be any doubt about Wilson's elevation of scientific materialism to the status of a new nature religion, he elaborates on this point with great clarity. "If the sacred narrative cannot be in the form of a religious cosmology," he declares,

> it will be taken from the material history of the universe and the human species. That trend is in no way debasing. The true evolutionary epic, retold as poetry, is as intrinsically ennobling as any religious epic. Material reality discovered by science already possesses more content and grandeur than all religious cosmologies combined. (265)

He is advocating not some notion of nonoverlapping magisteria for science and religion but an unambiguous conversion to a new creed, reflecting his own early conversion from Christian fundamentalism to scientism. Such conversion, he admits, "cannot be learned by pure logic; for the present only a leap of faith will take you from one to the other" (238). The future validation of this leap of faith, he encourages his readers, will eventually be reached through the accumulation of objective evidence acquired by scientists, with biologists leading the way. Thus the ultimate validation of this creed rests on the authority of future biologists, who will take on the role of messiahs to redeem humanity from ignorance and delusion.

Wilson acknowledges that he is an advocate not only of scientific materialism but of scientism (11), which appears to be the fundamentalist branch of this nature religion. Like other religious fundamentalists throughout the world, Wilson claims that his belief system is the *sole* way to understand reality and it holds the keys for solving all humanity's problems. Without the instruments and accumulated knowledge of the natural sciences, he writes, humans are trapped in a cognitive prison. They are like intelligent fish born in a deep, shadowed pool. . . . They invent ingenious speculations and myths about the origin of the confining waters, of the sun and the sky and the stars above, and the meaning of their own existence. But they are wrong, always wrong, because the world is too remote from ordinary experience to be merely imagined. (45)

Without belittling in any way the extraordinary achievements of science in shedding light on the objective world of physical phenomena, it must be pointed out that it has left us mostly in the dark regarding the subjective world of mental phenomena. As philosopher John Searle acknowledges, "In spite of our modern arrogance about how much we know, in spite of the assurance and universality of our science, where the mind is concerned we are characteristically confused and in disagreement" (Searle 1994:247). Ignoring this oversight, Wilson declares that prior to the rise of science there was only "the debris of millennia, including all the myths and false cosmologies that encumber humanity's self-image" (61). But if humanity devotes itself to the sole leadership of science, "we will in time close in on objective truth. While this happens, ignorance-based metaphysics will back away step by step, like a vampire before the lifted cross" (62).

In short, apart from science, Wilson declares, "Nothing else ever worked, no exercise of myth, revelation, art, trance, or any other conceivable means; and notwithstanding the emotional satisfaction it gives, mysticism, the strongest prescientific probe into the unknown, has yielded zero" (46). Given his distinguished career as a practicing scientist, one might hope that his conclusion about a topic as significant as the deepest modes of religious experience would be based on compelling empirical evidence. Unfortunately, in his evangelical zeal, Wilson throws to the winds any attempt to study this subject objectively, rigorously, or thoroughly. In his sham attempt at comparative religious scholarship, he claims, "Within the great religions . . . enlightenment . . . is expressed by the Hindu samadhi, Buddhist Zen satori, Sufi fana, Taoist wu-wei, and Pentacostal Christian rebirth. Something like it is also experienced by hallucinating preliterate shamans" (260). This facile conclusion is the sole reference in his book that he is even *aware* of the existence of non-Western religious traditions. But judging by his uncritical way of tossing them all together and all but equating them with hallucinations of preliterate shamans, religious scholars might prefer that he ignored them completely.

While scientific inquiry is characterized by careful observation, rigorous analysis, and open-mindedness that allows one to question even one's most cherished assumptions, these exemplary qualities seem to be flagrantly missing in Wilson's advocacy of scientific materialism. Even though the scientific tradition includes both elements—of empirical science and dogmatic scientific materialism presented as a nature religion—it would be misguided to label science as a whole a religion. But Buddhism, too, includes elements of rigorous experiential inquiry and rational analysis as well as explicitly religious elements. If it is misleading to categorize science as a religion—despite the common conflation of science with scientific materialism—it is equally misleading to categorize Buddhism as a religion to the exclusion of its scientific and philosophical elements. Buddhist science, if we acknowledge that such may exist, is certainly no substitute for modern science. Buddhism has no sophisticated theory of the brain or methods for exploring it, nor has it devised any body of objective knowledge comparable to modern physics, chemistry, and biology. On the other hand, modern science has left us humanity in the dark as to the nature and potentials of consciousness, subjective experience and its relation to the objective world, and the pursuit of a life of meaning and fulfillment. Once science is freed from the ideological shackles of scientific materialism, its modes of open-minded inquiry may well complement those of Buddhism and other ancient contemplative traditions.

THE DOGMA OF POSTMODERNISM

While scientific materialism dominates much of the thinking in the natural sciences, and has made deep inroads in capturing the imagination of the public at large, postmodernism continues to exert a considerable influence in the social sciences and humanities. Despite some deep ideological differences between them, many intellectuals have adopted both views, with the tenets of postmodernism laid over those of scientific materialism, like an eiderdown comforter laid over a granite mattress.

A fundamental tenet of postmodernism that can be launched against any dialogue or collaboration between Buddhism and science is the principle of *cultural particularism*, which asserts that different societies are culturally unique, incommensurable, and hence fundamentally unknowable by outsiders (Patton and Rav 2000:7). This would imply that the various schools of Buddhism are culturally unique to the Asian societies in which they developed, therefore their theories and methods of inquiry cannot be compared to those of science. While it is certainly true that different societies are culturally unique, and an outsider's knowledge of a society will never be identical to that of an insider, to absolutize this principle is to undermine any pursuit of cross-cultural or interdisciplinary understanding. When this line of reasoning is extrapolated to its logical conclusion, it implies that none of us can really understand anyone else, nor should we try, for each person is unique and fundamentally unknowable by others. In other words, this principle provides a recipe for the breakdown of empathy and dialogue, which in turn leads to the erosion of human civilization itself.

Undeterred by the implications of their stance, postmodernists emphasize differences over similarities whenever two fields, disciplines, or assertions are compared. Differences they assume to be somehow "real" and "objective," whereas similarities are deemed "imaginary" and thought to exist only in the subjective mind of the beholder (Patton 2000:157). This view goes hand in hand with the theme of the "social construction of reality," which is widely ignored or ridiculed in the sciences but still fashionable in the humanities. When presented with the suggestion that any type of religious experience reveals an aspect of reality, or the suggestion that there may be common insights between, say, Buddhism and science, postmodernists commonly respond with alarm or derision. Their metaphysical assumptions simply do not allow for such an occurrence *as a matter of principle.*

Having declared that the worldviews of "the other" are fundamentally unknowable, many postmodernists adopt the elitist Foucauldian premise that what a religious tradition says of itself is not what it is really about. Specifically, religions commonly claim to portray the relation of the human to ultimate reality. But Foucault claims, on the contrary, that this is not what religions are *really* about. What they are actually about is relationships of power: who has it, how do they get it, and how do they wield it? It is absurd, of course, to deny that power has a significant role in the formulation of religious doctrines. But the reductionistic claim that this is *all there is* to religion implies that the outsider, specifically the postmodernist, can know "others" better than they can know themselves. When applied to the study of Asian religions, this form of methodological condescension has been labeled "Orientalism," but it persists despite numerous, cogent critiques.

Postmodernism does not, of course, limit itself to a critique of religion. Its broader claim is that it is impossible to determine finally the "truth" of *any* particular worldview or vision, whether traditional or modern. This conclusion can be seen to stem from a sober reflection upon the transience of so many theological and scientific assertions that have been held virtually sacrosanct at one time or other, only to be later proven false. But this postmodernist assertion is itself a truth claim, and few of its advocates display any doubt whatsoever about having finally determined the truth of this matter. The insistence on the lack of absolute truth in any worldview other than postmodernism appears to be one of the fundamental articles of faith of this dogma, which indicates its close similarity (dare we say the word?) with scientism and other forms of fundamentalism. In short, the very assertion that *no one* can determine with complete certainty the validity of any truth claim implies that the one who makes this assertion has absolute knowledge about the limits of knowledge of *everyone else*! Whatever the merits of postmodernism, modesty is not its long suit.

Another recurrent theme in postmodernist literature is the primacy of aesthetics. Not only philosophy but science itself is viewed more as an art form than a rigorous pursuit of objective knowledge. Needless to say, this hypothesis has little appeal in the scientific community, nor has it persuaded the media, the general population, or governments, who consistently allot greater funding and support to scientific research than the fine arts and humanities.

This same theme is prevalent in the postmodernist evaluation of religion. In light of the differences between the doctrines of the world's religions and other worldviews, Roger Jackson writes, "The choice, in short, is an aesthetic one, for that may be the only sort of choice that, in a postmodern setting, remains open" (Jackson 1999:238). Those who are not under the sway of the metaphysical injunctions of postmodernism happily have a variety of choices as to how to choose among alternative worldviews. But postmodernists, according to Jackson, have only one option, meaning no choices at all. Shall we apply this same rule to all choices among competing theories and hypotheses? When presented with diverse scientific theories, philosophical ideas, ethical principles, and religious views, shall we choose only those that we somehow find aesthetically pleasing? Once again, postmodernism appears to present a recipe for the collapse of all intellectual and empirical rigor in the pursuit of understanding. As religionist Kimberley C. Patton cogently concludes, as a result of following the dictates of postmodernism, "we end up . . . only talking about ourselves and our own prejudices, victims of a kind of narcissistic epigraphy that poses as methodological sophistication" (Patton 2000:166).

In all traditional accounts of the Buddha's teachings and later Buddhist writings on contemplative practice, philosophy, ethics, and so on it is patently obvious that numerous truth claims are made concerning a wide range of subjective and objective phenomena. As mentioned before, Buddhism begins with the Four Noble Truths, which consist of one truth claim after another. But despite the overwhelming evidence of this obvious fact, postmodernists still try to conceal this with their appeal to aesthetics. Once again, they, as outsiders, assume to have knowledge of Buddhism that overrides and refutes the traditional views of its own advocates. Numerous Buddhist contemplatives have made the astonishing claim, allegedly based on their own experiences, that humans *can* meditatively train their minds to such a degree that they can experientially discover the reality of individual experience following death and prior to conception. And they make many other extraordinary truth claims about the nature and capacities of human existence, including the possibility of realizing the ground of being and achieving enlightenment. But postmodernists refute *in principle* the possibility of any such knowledge. Like the clerics who challenged Galileo's discoveries through his telescope, they claim to know beforehand what can and cannot be known through the aided senses. In the case of scientific research, the physical senses have been enhanced and extended with the use of technology. In the case of Buddhism and other contemplative traditions, the faculty of mental perception has been allegedly enhanced and extended through the cultivation of extraordinary states of meditative concentration and techniques for cultivating insight. But most contemporary Buddhologists are more prone to scholastic rumination than they are to the discipline of devoting years to rigorous contemplative training. Postmodernism, in the final analysis, provides the modern intellectual with the easiest way possible of coping with the diversity of scientific, philosophical, and religious worldviews. No precise intellectual analysis is required and no rigorous experiential investigation is needed. Instead, simply regard all Buddhist truth claims not as propositions to be proved or refuted but simply as metaphors or images that help to form the imaginative and affective landscape in which Buddhists live and move and have their being (Jackson 1999:231). Jackson refers to this strategy as cutting Buddhism back to its "bare doctrinal bones" (236). But in his rejection of the validity of the Four Noble Truths, past and future lives, and the attainment of enlightenment, he actually serves up an eviscerated filet of Buddhism that is free of all the bones of contention between Buddhism, scientific materialism, and postmodernism.

Oddly enough, this postmodernist interpretation of Buddhism does not compel Jackson to jettison Buddhism altogether. Rather, having adopted his aesthetic approach to Buddhism, he claims he could still praise enlightened beings for qualities he doubts they, or anyone, literally could possess. He could vow to liberate sentient beings in future lives he doubts they would experience. And he could contemplate as primordially pure a mind he is not convinced is more than a by-product of the brain (237). All this is eerily similar to Edward Wilson's encouragement to believe in free will, to continue in ritual verbal references to God, to show deference toward the Bible and communal respect as ministers and rabbis bless civil ceremonies with prayer—*even though one is confident that all of this has no basis in reality.* According to Wilson's scientific materialism and Jackson's postmodernism, our lives will be more meaningful and fulfilling if we *pretend* to believe in illusions that we assume to be false. During the twentieth century, when advocates of scientific materialism, under the banner of communism, were systematically slaughtering tens of thousands of religious believers, destroying monasteries, temples, and churches, burning religious books, and forcefully banning all behavior suggestive of religious practice, it is hard to imagine that such make-believe religion would have provided any solace or support for the victims of this ideological warfare and genocide. Such a contrived approach to religion is viable only when little or nothing is at stake. As soon as the balloon of such pretense is struck with the sharp edge of an existential crisis, it pops.

From a Buddhist perspective scientific materialism falls to the extreme of metaphysical realism, which claims knowledge of absolutely objective realities. As Edward Wilson acknowledges, there is no objective yardstick on which to mark the degree of correspondence between the objective world and our scientific representations of it. All scientific knowledge is mediated by specific modes of observation, experimentation, and analysis. And the hypothetical objective world that allegedly exists independently of all such mediation, is, as Kant pointed out long ago, forever beyond our ken. A Buddhist evaluation of postmodernism, on the other hand, suggests that it falls to the opposite extreme of nihilism and solipsism.

The tenets of scientific materialism have guided scientific research in many useful ways and have provided the theoretical grounding for many technological advances. Postmodernism has enriched human understanding by pointing out the culturally embedded nature of both scientific and religious theories. After postmodernism the role of cultural context can no longer be ignored or marginalized, as has so often been done in both scientific and religious writings. In short, both scientific materialism and postmodernism have proven their pragmatic usefulness in a variety of ways. But when they take on the role of a dogma and claim to present a realistic way of viewing reality as a whole, they are disastrous both for the individual and society at large.

THE WAY OF DIALOGUE AND COLLABORATION

There is nothing trivial about the differences in views among the world's religions or between science and Buddhism in particular. But understanding the relations among these modes of inquiry is not facilitated by reducing religious beliefs to genetic programming or by reducing the differences among diverse worldviews to their cultural situatedness. There is no scientific evidence to support the notion, for example, that the origin of Christian doctrine of the Trinity or Buddhist doctrine of the three embodiments of the Buddha can be explained in terms of genetic programming. And it is equally preposterous to explain away the origins of these theories solely in terms of the sociopolitical climates of Israel or India two thousand years ago. Christianity, Buddhism, and other world religions have become global phenomena; they are embedded everywhere, as is science.

How then might one grapple with the differences among the truth claims of diverse religions and science? One alternative is to take a position of cultural relativism regarding others' beliefs—asserting that they are *valid and useful for their adherents*—while maintaining an absolutist stance for one's own beliefs—insisting that they are uniquely valid in the sense of depicting reality as it truly is. One advantage of this asymmetrical perspective regarding one's own and others' views is that it enables one to appreciate the diversity of worldviews corresponding to the predilections and intellectual capacities of different cultures and individuals. In effect, one regards others' religious beliefs as if they were medicines appropriate for their specific spiritual needs and inclinations. This makes sense if one believes the primary function of religions is to help people overcome vices, cultivate virtues, and find happiness. From this perspective one might still maintain that some views of specific religions are more profound or authentic than others, but one would reject the notion that there is one religion that is the best for everyone, just as there is no one medicine that is the best for all.

The above perspective is, in effect, taking a realist stance regarding one's own religion, and an instrumentalist, or even utilitarian, stance regarding others' religions. Traditional Buddhists who adopt this perspective, for instance, still think non-Buddhists are subject to the effects of mental afflictions and to the karmic consequences of their deeds in future lives. And they believe that scientific materialists still experience a continuity of individual consciousness after their own death, however firmly they might deny this possibility. Likewise, Buddhists believe the aggregates of the body and mind are impermanent, subject to suffering, and are devoid of an unchanging, unitary, independent self. Those who disagree with these defining themes of Buddhist doctrine are thought to be wrong, for these are considered by Buddhists to be universal truths.

Buddhism does not define itself as a religion or as a science, and traditionally it has made no distinction between religious truths and scientific truths. H. H. the Dalai Lama, who has taking a leading role in dialogues between Buddhism and science, has repeatedly claimed that if compelling scientific evidence refutes any Buddhist assertion, Buddhists should abandon their own discredited assertion. This attitude stems, presumably, from the Buddhist belief that sentient beings are fundamentally subject to suffering because of ignorance and delusion, and the way to freedom is by coming to know reality as it is. Thus, if scientific research illuminates errors in Buddhist doctrine, Buddhists should be grateful for such assistance in their own pursuit of truth. In other words, the Dalai Lama is flatly rejecting the notion of nonoverlapping magisteria between Buddhism and science. And he equally rejects the postmodernist notion that Buddhist assertions are not subject to verification or refutation but rather consist simply of metaphors that are to be appraised for their aesthetic appeal alone.

Buddhism, of course, is not the only religious tradition to make truth claims that are of great importance to its adherents. Thus, a false dichotomy is maintained when one assumes a stance of cultural relativism regarding others' religious assertions while taking a realist stance regarding scientific assertions or one's own religious beliefs. Moreover, especially when it comes to the nature of human existence, and specifically the nature of the mind, consciousness, and the human soul, the notion of nonoverlapping magisteria between religion and science is simply untenable. I would therefore suggest that a uniform ontological stance, combined with a wide range of modes of investigation and analysis, should be adopted in the evaluation of truth claims from all religions and branches of science.

The way forward, I would argue, is through mutually respectful dialogue and collaboration in both empirical and theoretical research. This en-

tails reaching out across disciplines and cultures to increase mutual understanding of areas of common interests. In terms of the interface between
Buddhism and science, we must be self-conscious of the assumptions we
bring to Buddhist studies, while entertaining the possibility of learning
about the world *from* Buddhism, as opposed to studying this tradition
merely as a means to learn *about* Buddhism. The aspects of Buddhism that
are most inviting for such interdisciplinary inquiry are those that are accessible to empirical and analytical inquiry. Moreover, such research will take
fully into account the experiences of Buddhist practitioners, of the present
and the past, and not focus on texts alone. In this way Buddhism may be
viewed as a form of "natural philosophy" (the label for early European science), challenging us to ask the deepest possible questions (as in religion)
by means of rigorous logical analysis (as in philosophy) and empirical investigation (as in science). This way of grappling with Buddhist truth
claims seeks not only an objective appraisal of the textual *doctrines* of Buddhism but also its claims of experiential *insights*. And the objective appraisal of the latter may require testing these assertions by engaging in the Buddhist practices oneself, just as one might test a scientific theory by running
experiments oneself.

   In her insightful essay, "Dialogue and Method: Reconstructing the Study
of Religion," religionist Diana L. Eck comments that a hallmark of Orientalism was the accumulation of knowledge about the colonized "other," without listening to the voice of the other. Moving beyond Orientalism, she
points out, means entering the methodological terrain of dialogue, a way of
working in which the situatedness of both our own voices and the voices of
those we study become integral to the process of understanding (Eck
2000:140). The scientific engagement with Buddhism can shed a fresh light
on our own subjectivity, our own language, and our own categories, for example, of religion, science, and philosophy. By recognizing the unique contexts of both Buddhism and science, all participants in such dialogue may at
least begin to escape from the tendency to unwittingly attribute a privileged
status to our own preconceptions. Surely this is sufficient reason to engage
in such a cross-cultural and interdisciplinary pursuit of understanding.

References

Almond, P. C. 1988. *The British Discovery of Buddhism*. Cambridge: Cambridge
   University Press.

Cole, K. C. 1999. "In Patterns, Not Particles, Physicists Trust." *Los Angeles Times*, March 4, 1999.

Dawkins, R. 1978. *The Selfish Gene*. New York: Oxford University Press.

———— 1999. "You Can't Have It Both Ways: Irreconcilable Differences." *Skeptical Inquirer*, July/August.

Eck, D. L. 2000. "Dialogue and Method: Reconstructing the Study of Religion." In K. C. Patton and B. C. Rav, eds., *A Magic Still Dwells: Comparative Religion in the Postmodern Age*, pp. 131–149. Berkeley: University of California Press.

Feynman, R., R. B. Leighton, and M. Sands. 1963. *The Feynman Lectures on Physics*. Reading, Mass.: Addison-Wesley.

Gilkey, L. 1985. *Creationism on Trial*. Minneapolis: Winston.

Gómez, L. 1999. "Measuring the Immeasurable: Reflections on Unreasonable Reasoning." In Roger R. Jackson and John Makransky, eds., *Buddhist Theology: Critical reflections by Contemporary Buddhist Scholars*, pp. 367–385. Surrey: Curzon.

Goodenough, U. 1998. *The Sacred Depths of Nature*. New York: Oxford University Press.

Gould, S. J. 1999. *Rocks of Ages: Science and Religion in the Fullness of Life*. New York: Ballantine.

Griffiths, P. J. 1986. *On Being Mindless: Buddhist Meditation and the Mind-Body Problem*. La Salle: Open Court.

Harvey, V. 1981. *The Historian and the Believer*. Philadelphia: Westminster.

Herbert, N. 1985. *Quantum Reality: Beyond the New Physics*. Garden City, N.Y.: Anchor/Doubleday.

Jackson, R. R. 1999. "In Search of a Postmodern Middle." In Roger R. Jackson and John Makransky, eds., *Buddhist Theology: Critical Reflections by Contemporary Buddhist Scholars*, pp. 215–246. Surrey: Curzon.

King, R. 1999. *Orientalism and Religion: Postcolonial Theory, India, and "The Mystic East."* London: Routledge.

Patton, K. C. 2000. "Juggling Torches: Why We Still Need Comparative Religion." In K. C. Patton and B. C. Rav, eds., *A Magic Still Dwells: Comparative Religion in the Postmodern Age*, pp. 153–171. Berkeley: University of California Press.

Patton, K. C. and B. C. Rav. 2000. "Introduction." In K. C. Patton and B. C. Rav, eds., *A Magic Still Dwells: Comparative Religion in the Postmodern Age*, pp. 1–19. Berkeley: University of California Press.

Searle, J. R. 1994. *The Rediscovery of the Mind*. Cambridge: MIT Press.

Wallace, B. A. 1980. *The Life and Teachings of Geshé Rabten*. London: George Allen and Unwin.

———— 1996. *Choosing Reality: A Buddhist View of Physics and the Mind*. Ithaca: Snow Lion.

———— 1998. *The Bridge of Quiescence: Experiencing Tibetan Buddhist Meditation*. Chicago: Open Court.

———— 1999a. "The Buddhist Tradition of *Samatha*: Methods for Refining and Examining Consciousness." *Journal of Consciousness Studies* 6(2–3): 175–187.

———— 1999b. "Three Dimensions of Buddhist Studies." In Roger R. Jackson and John Makransky, eds., *Buddhist Theology: Critical reflections by Contemporary Buddhist Scholars*, pp. 61–77. Surrey: Curzon.

———— 1999c. "The Dialectic Between Religious Belief and Contemplative Knowledge in Tibetan Buddhism." In Roger R. Jackson and John Makransky, eds., *Buddhist Theology: Critical Reflections by Contemporary Buddhist Scholars*, pp. 203–214. Surrey: Curzon.

———— 2000. *The Taboo of Subjectivity: Toward a New Science of Consciousness.* New York: Oxford University Press.

Wilson, E. O. 1998. *Consilience: The Unity of Knowledge.* New York: Knopf.

PART 1
Historical Context

In this opening essay, Buddhologist José Ignacio Cabezón presents an illuminating overview of the historical interface between Buddhism and science, emphasizing the structural and typological facets of this interrelationship. In first addressing the study of Buddhism and Buddhists as objects of scientific inquiry, he raises important ethical concerns. In what sense, for instance, can Buddhist meditators safely be regarded as "informed and consenting subjects" in such research? And to what degree might positive scientific findings inadvertently legitimize certain Buddhist orders or sects, thereby exacerbating social tensions that already exist within Buddhist communities?

As he questions the status of Buddhism as an object of inquiry in the history of science, Cabezón critiques familiar diagnoses that allegedly account for the rise of science in Europe and its failure to emerge in Asia. He then proceeds to analyze the relation between Buddhism and science in terms of three models: conflict/ambivalence, compatibility/identity, and complementarity. In the third of these models, arguably the most interesting and provocative, he points out that Buddhism and science are presented as being able together to contribute epistemically to a more complete way of knowing a common subject. Whether the difference between them is identified principally in terms of content, method, or goal, the limitations of each is thought to be overcome by bringing them together in a harmonious way. But he goes on to point out the dangers of reifying metaphors for Buddhism and science in ways that silence their voices and insights into matters where they may be incompatible. In conclusion, Cabezón expresses his thoughts concerning the future of the conversation between Buddhism and science in ways that may well lead to deeper and richer dialogue and collaboration.

José Ignacio Cabezón

Buddhism and Science: On the Nature of the Dialogue

PREAMBLE

It is the purpose of this essay to consider some of the ways in which Buddhism and science have engaged each other: to take stock of the historical interaction of these two spheres, and to suggest, by way of conclusion, some directions for future engagement. Some caveats are in order at the outset, however. Although I use the terms *Buddhism* and *science* throughout, I am not unaware of the problems involved with the use of such generalities. Both Buddhism and science are of course highly internally differentiated categories. At times I will resort to evoking some of that internal structure (e.g., when I discuss the biological theory of evolution or Indian Buddhist views of matter). But I do not apologize for the fact that on other occasions I am painting a picture in broad strokes.

First of all, part of my goal in this essay is to consider the ways in which scientists and scholars of Buddhism have *themselves* depicted their mutual interaction. It is clear that, especially in their earliest encounters, that engagement was rhetorically constructed as one between Buddhism and science generally. To the extent that my remarks are a historical characterization of that encounter, then, it is fitting that I resort to the categories the participants in that encounter themselves utilize. Second, even more dis-

tinct categories, like *evolution* and *Tibetan Buddhism*, are themselves generalities. The point of course is that there is no escaping generality. A brush
stroke is always broad by comparison to one that is finer. Finally, and more
pragmatically, there is always a place for generalities as long as one remains
mindful of the fact that in the process of resorting to them one is sacrificing
detail.

Although I will be dealing here with many different time periods, this
essay does not purport to be anything even remotely close to a complete
historical overview of the interaction of Buddhism and science. If anything,
I am more interested in characterizing this interaction in structural and typological, rather than in historical, terms, although even here the reader
will find, as I have, the complexity of the real historical engagement of these
two spheres evades even this form of categorization. This being said, it is my
hope that this essay will be provocative, if only as a starting point for others
who, like me, would seek to make some explanatory sense of the complex
interactions of Buddhism and science and some normative suggestions for
their future intercourse.

BUDDHISM AS THE OBJECT OF SCIENCE

The earliest encounters between Buddhism and science cast Buddhism
not as the partner of science in a dialogue, but as the object of scientific inquiry. It was the Enlightenment penchant for modeling the humanities after the natural sciences that led to the rise of the "science of religion." As
Buddhism came within the purview of this new "science," there emerged,
on the one hand, the rise of Buddhist philology, perceived as the application
of systematic scientific principles to the study of Buddhist texts, and, on the
other, the rise of the social scientific study of Buddhist cultures. In this way
Buddhist texts and societies became fodder for the "scientific inquiries" of
figures like Max Mueller and Max Weber. That today we find *Religionwis-
senschaft* as a movement much less neutral, much less the disinterested and
objective analysis of pure fact, and much more theory and theologically
laden than did the founding fathers of the discipline does not belie the fact
that for these late nineteenth-century scholars Buddhism was to be ap-
proached as the object of such a science.

More important, this early rhetoric, a rhetoric that cast Buddhism as the
subject matter of scientific inquiry, has in many ways set the tone for one
important strand in the encounter between Buddhism and science, even to

the present day. Such a mode of interaction is presumed in a good deal of current sociological and anthropological work being done in regard to Buddhist cultures and societies. And it is of course the dominant model in much of the psychological and neuroscientific work currently being undertaken in regard to Buddhist meditation.

At the risk of digression, I consider this form of encounter—the objectification of Buddhism, and especially Buddhists, by science as part of a scientific research program—to be sufficiently important to warrant some further remarks at this point. There is, of course, an inherent danger in the scientific objectification of subjects in an experimental setting. The peril lies in the possibility that those being tested come to be considered *mere* objects and thus dehumanized. Such a problem becomes especially acute when subjects are separated from researcher not only by professional but also by cultural distance. One way to lessen the negative effects of scientific experimental objectification—a tack taken by a group of researchers, to which I myself belong, studying the effects of meditation in a group of Tibetan monks[1]—is to involve the subjects, to whatever extent possible, in the actual planning and execution of experiments, that is, to acknowledge them as intellectual equals and thus to give them a voice as colleagues. But even when this is done, casting Buddhists' collective or individual behaviors, or their bodies, in the role of examined object is an enterprise fraught with ethical perils. My purpose in bringing this up is not to suggest that this form of encounter is to be avoided but only that it requires a great deal of forethought.

My own work as part of the research group mentioned above has made it clear to me that there are ethical issues involved in the cognitive scientific testing of Tibetan meditators that go beyond those traditionally covered and disposed of as part of traditional human subjects research screening.[2] Some of these are the result of cultural/religious factors that are unique to a Buddhist (and especially a Tibetan Buddhist) setting. For example, in a tradition where the wishes of the spiritual master are considered almost sacrosanct, how much freedom of choice do potential subjects really have if they know, or even believe, that their teacher or mentor is in favor of the program of testing? Other issues concern the procedures and effects of testing. Does mere participation in such experiments have a negative impact on a retreatant's meditative life? What responsibility does a researcher have to share the results of experiments with subjects? How might such results— both positive and negative—be interpreted by subjects, and what effects might such interpretations have on their personal practice? Even apart from the obvious interruption to a retreatant's isolation, can we say with certainty that, as regards the actual testing, apparently noninvasive procedures like EEGs will have no negative effects on the subtle physiology that a monk relies upon and manipulates in the more advanced forms of meditation, to take just one example of a real-life concern that emerged from our conversation with monks? And even assuming that one could be certain of this, which I think arrogant, is it not possible that a mediator's subsequent practice might be negatively affected by the *mere belief* that such testing might have negative consequences? How do we weigh the value of the knowledge gained from testing against the possible negative effects it might have on the subject being tested?

In addition, there are a host of ethical questions that have to do with the effects of research, not on the individual but on the society at large. When research is conducted on a specific sample of adepts from a particular religious tradition or from a particular subschool within a religious tradition—say, monks from a particular school of Buddhism—might "positive" results be socially interpreted as validating the meditative techniques or expertise of adepts of that school over others? Might this exacerbate already existing forms of interreligious rivalry or intrareligious sectarianism? Given that science is such an extremely powerful legitimizing force, how can such research avoid serving as a scientific imprimatur to social tensions that already exist between cultures or within a single society, or, worse, create new tensions? Can anonymity and confidentiality prevent this? What degree of anonymity, if any, should be required to assure that this does not occur?

Still other issues arise from the different presuppositions of Western science and Buddhism as worldviews. How can an intellectual rapprochement be achieved between the prevailing philosophical view of cognitive scientists, most of whom hold to strict mind-brain identity, and Buddhists, who, believing that the mind is nonmaterial, are dualists? It is true that one could envision such a rapprochement taking place on a ground where each side took a more nuanced position—the scientist eschewing materialist reductionism and the Buddhist granting the possibility that mental states could have physical, and therefore measurable, correlates. However, such conversations, and reconciliation of views, have rarely taken place prior to actual experimentation, and this once again raises ethical issues. Where a subject's ignorance of the nature and presuppositions of an experimental design can lead to fear, how much agreement must exist as a prerequisite for conducting research? What responsibility does a scientific researcher have to establish such consensus prior to testing?

Again, I will reiterate that my purpose here is not to argue against the objectification of Buddhism, or of Buddhists, by science, but simply to note that when this becomes a dominant mode of interaction between the two spheres there arise ethical issues to be sorted out, ethical issues that are in many ways more weighty than those arising when the mode of interaction is one of conversation.[3]

BUDDHISM AS OBJECT OF INQUIRY IN THE HISTORY OF SCIENCE

Not surprisingly, historians of science have long been concerned with etiological questions. In particular, they have sought to identify those factors which, by their presence in European culture, and their absence elsewhere, have brought about the rise of science. In this regard, Asian cultures have served for them as a kind of control. Concerned to establish the historical circumstances that led to the emergence of science in Euro-Christian society, it becomes natural to construct the rest of the world, including Buddhist Asia, as the barren and infertile site out of which science failed to arise.[4] It goes without saying that this kind of historiography begs to be subjected to the kind of literary critical analysis that weaves out the rhetorical construction of the other and its relationship to questions of race, gender, power, empire and its demise, but this of course is impossible here.[5] Suffice it to say that in much of this writing, and despite the subsequent antagonism between science and Christianity, it is precisely the Christian worldview that comes to be characterized as the causal sine qua non to the rise of science. Hence, Jaki (1974, 1985) believes that he explains Western Europe as the unique historical locus of the rise of science by identifying those factors within its Christian worldview that permitted its emergence: a monotheism that, in contradistinction to pantheism, led to a deanimized view of nature, a notion of the will of God as ascertainable and compatible with the existence of physical laws, and a noncyclical view of time, epitomized by the event of the Incarnation. Jaki contrasts the Indian notion of cyclical time, as instantiated, for example, in the doctrine of world-cycles, or *yugas*, to the notion of linear time, which, because of its focus on the "uniqueness of events," serves as a foundation for the kind of empirical observation that is necessary to the emergence of science.

Jaki's work is by no means ill-informed about the views of the various religions he examines, nor does it lack nuance. Ultimately, however, it remains unconvincing as an explanation, not because of errors in details but for various other reasons. For one, it can be pointed out that India and China *did* give rise to forms of empirically derived sciences that can be recognized as such even in Western terms, that is, that can be recognized as *science* without altering the semantic range of the word to take into account its understanding in these different cultures.[6] Moreover, neither cyclicality, nor pan- and polytheisms seem to have acted as deterrents to the acceptance of science in these various cultures since its movement east, something that would be expected were Jaki's thesis true.[7] On the contrary, there is an ever increasing, albeit mostly naive and misguided, literature in these various cultures that suggests that science is completely compatible with, and at times that its findings are even prefigured in, their respective religious traditions. The point I am making is not, of course, that this literature is true, but that its existence is difficult to explain given Jaki's thesis.

My chief objection to the work of Jaki and others like him, however, lies less in the realm of contingency than it does in the very logic of the form of historiography that it represents. For one, it places too much emphasis on ideology as *the* factor that determines or impedes the rise of science in a given culture. In so doing, it pays too little attention to material, sociological, and political forces that are at the very least as important as explanatory tools. Moreover, even if the importance of ideology were to be granted, there are questions as regards Jaki's categories. Notions like cyclical time, and even polytheism, are in large part vague and can be read out of or into a wide variety of traditions.[8] But, in addition to employing nebulous categories, studies like Jaki's ultimately fail because in the end they are speculative, for, to be convincing, they require controls, and the irreversibility of history makes controls impossible. Ultimately, all that we may be able to say of the rise of science in the West is that it is in some very complex way the result of many different and unique causes that resist generalization,[9] not the least of which is the bittersweet genius of individuals like Bacon, Descartes, Galileo, and Newton, something to which Jaki seems to pay little attention.

Apart from the historiographical literature that takes Buddhist and other Asian religious ideologies as contrasting vehicles for explaining the rise of science in the West, there is of course a literature that represents a more direct encounter between Buddhism and science. In his Gifford Lectures, Ian

Barbour (1990:3–30) suggests a fourfold typology for considering the interactions of religion and science: conflict, independence, dialogue, and integration. While illuminating, especially because of its explanatory power in
regard to the interaction between science and Christianity, I opt here for a
slightly modified schema that I consider more useful in characterizing the
interaction of science and Buddhism: conflict/ambivalence, identity/simi-
larity, and complementarity.

CONFLICT/AMBIVALENCE

I first consider conflict as a mode of engagement not, as is the case with
the other modes, because of its prevalence, but because of its relative absence. That conflict has existed between Buddhism and science can hardly
be denied. But conflict, where it has existed, is often attenuated. What is
perhaps more common than out-and-out antagonism is either mutual disregard, or, when the two spheres have interacted, a kind of ambivalence.

Of course, to come to conclusions about the presence or lack of conflict
between Buddhism and science from the Buddhist side requires not only a
complete historical survey but demands that we treat different geographical
regions of the Buddhist world individually as well. In Asia, where a rhetoric
of conflict exists, this will undoubtedly be enmeshed with (though in my
view not reducible to) Asian Buddhist opposition to European colonialism,
which was, at least in the early phase of contact, the vehicle for the introduction of science into most Asian cultures.

That Asian Buddhist views concerning science will be found to be ambivalent can, I think, be gleaned, at least anecdotally, from the case of Tibet.
We find that before the 1959 takeover of the country by Chinese forces, there
existed widespread skepticism on the part of the politically powerful Buddhist monastic elite about Western influence generally. Science and technology, if not viewed as coextensive with, were nonetheless perceived by
many conservative monastic officials as an essential part of, the Western
ideology that threatened to undermine the Buddhist worldview of the
country.[10] In spite of this, there were individuals, like the thirteenth Dalai
Lama and the regent sTag brag, who made valiant, though ultimately un-
successful, attempts at modernizing the country by, among other things,
educating young Tibetans in the theoretical and applied sciences and in
mathematics. There were also prominent intellectuals who, through their
exposure to Western culture, themselves developed a considerable personal

interest in science, among the most famous of these, the present, fourteenth Dalai Lama, and the well-traveled and controversial monk-scholar Gendün Chöphel, who lived in the first half of this century.[11]

More recently, there has been widespread skepticism among meditator monks as regards the exploding interest in the neuroscientific study of meditative states and the long-term effects of meditative practice. This skepticism, already alluded to above, is both theoretical and practical. The theoretical suspicion stems from the fundamental doubt (a) that nonphysical states of mind, like compassion and single-pointed concentration, can be measured using physical means and (b) that even if this were possible it would be a valuable thing to do. The practical skepticism expresses itself in these monks' suspicion that such tests could have deleterious effects on their health and practice.[12]

From these brief remarks it can be gleaned that from the time of Tibet's first major contacts with the industrialized West there has existed an ambivalence toward Western science, being seen by some as a threat to Buddhism and by others as an essential and positive part of the program to modernize the country. To my knowledge, however, there was never an elaborated and sustained critique that focused on Western science to the exclusion of other aspects of the Western intellectual tradition, even by those who viewed science and technological advancement in a negative light. It is of course the case that no one Asian Buddhist culture can be taken as paradigmatic of Buddhist Asia as a whole, but it would be surprising to me if the pattern of ambivalence toward science that we find in pre-1959 Tibet were not repeated in many other Asian settings.

If Buddhist attitudes toward science can be tentatively characterized as ambivalent, scientists' interest in Buddhism, at least up to the last few decades, can be viewed as practically nonexistent. Some Buddhists, writing from the late 1950s to the present day, see the relative lack of conflict between Buddhism and science in history as an indication of the fundamental harmony between these two spheres. But if Buddhism has not been singled out for attack by Western scientists and their philosophical backers, if the interaction between Buddhism and science has not been as polemical and antagonistic as that between Christianity and science, it is most likely and simply because Buddhism was not seen as a competitor in the arena where, from the seventeenth through the early twentieth centuries, the war for the intellectual hegemony of the West was being waged. Buddhism was not targeted for attack simply because it was, for all intents and purposes, absent as a serious intellectual option, and not because, as implied by some erstwhile scholars, it is, either in content or method, more in tune with science. It is then history, or, perhaps more accurately, the colonialist devaluing of the culturally other, that spares Buddhism the pains of having to pass through modernity kicking and screaming, for while Buddhism, in the not-so-subtle logic of colonialist superiority, was a curiosity, until very recently it could by no stretch of the imagination be considered an intellectual equal worthy of being engaged as a partner in conversation.

Today there is certainly to be found in the literature what might be considered anti-Buddhist sentiments on the part of scientists, but this is usually not directed specifically at Buddhism, being instead a part of more general antireligious sentiments that are founded on naive mechanistic materialism.[13] Likewise, there is a good deal of skepticism on the part of Buddhists as regards, for example, some of the technological fruits of science, but this critique falls short of a full-blown repudiation of science.[14] Although most contemporary Buddhists and scientists continue simply to disregard each other, in the last few decades there has begun to emerge a real conversation between the two, and this is best characterized not in terms of either conflict or ambivalence but in terms of the modes of interaction to be discussed next.

COMPATIBILITY/IDENTITY

In his study of the spread of Buddhism in the United States, Thomas Tweed (1992) argues that the first phase of the dissemination of Buddhism in North America (1844–1857) was characterized by a deemphasizing of Buddhism's distinctiveness, and by a concomitant accentuation of the similarities between Buddhism and Christianity, especially Catholicism. This is followed, says Tweed, when more of the details about Buddhism are known, by a period in which Buddhism's *otherness* is emphasized, especially its otherness vis-à-vis Christianity. Tweed's work is of interest to me here because it suggests that when two very different cultures or traditions meet the first reaction is to treat the culturally other in terms of the culturally familiar. This, I will claim, is what happens when Buddhism and science first make contact.

Of course there are degrees of similarity, from vague compatibility to selfsameness. At one end of the spectrum, the end that begins to meld into the next mode of interaction, which I call complementarity, is the claim that Buddhism and science are similar: that they share common concerns, that they reach similar conclusions, that they have similar aims or utilize analogous methods. At the other is the stronger claim that Buddhism *is* science: that the objects of investigation, the results, aims, and methods of the two are identical. But wherever different Buddhists or scientists come out in this spectrum of views, the mode under discussion here emphasizes similarity rather than difference. Although by no means a complete survey, let me cite some examples of this mode of interaction between the two spheres.

One of the earliest sources for exploring the relationship between Buddhism and science is Henry Steele Olcott's *Buddhist Catechism*, first published in the late 1880s or early 1890s.[15] Olcott states that Buddhism is "in reconciliation with science," that there is "an agreement between Buddhism and science as to the root idea" (1889:30, 33). The reason for this is basically twofold. Olcott believes that both Buddhism and science teach evolutionism, "that man is the result of a law of development, from an imperfect lower, to a higher and perfect condition" (30). Further, "both Buddhism and science teach that all beings are alike subject to universal law" (33), which he relates both to the "law" of karma, and to the "law of motion" (51, 57) that brings the universe into existence. Thus, Olcott implies, Buddhism and science are in agreement because they subscribe to the view that there are natural laws governing the development of both persons and the world. But Olcott is also ambivalent about science. Although he states that Buddhism encourages the teaching of science (55), just two pages later he tells the reader that Buddhism is not "a chart of science" but rather "chiefly a pure moral philosophy" since Buddhism (unlike science?) believes that it is "unprofitable to waste time in speculating as to the origin of things" (57–58).

A slightly later *Buddhist Catechism*, written originally in German in 1888 by a German lay Buddhist under the pen name of Subhadra Bhikshu (Subhadra 1970), follows Olcott's lead, albeit cautiously, in speaking of Buddhism's belief that "in the universe there reigns strict conformity to law" (34, 48), which it relates to the law of cause and effect (40, 77). Subhadra is more explicit, however, and perhaps more "separatist" and triumphalist, than Olcott in his depiction of the relationship of Buddhism to science:

> Buddhism does not intend to teach natural science; it does not concern itself with the outward condition of things, but with their inner being, and therefore stands neither in a hostile nor a dependent relation in regard to science.

The educated Buddhist occupies a perfectly unprejudiced position concerning natural science; he examines its results and accepts, uninfluenced by religious scruples, such of its teachings as appear to him correct. . . . The Buddhist knows that science, like all earthly things, is changeable, progresses continually, and can teach many useful and great things now-a-days which were unknown at the time of the Buddha; but that on the other hand nothing can be discovered, no matter how far scientific research may progress, which could contradict the words of the Buddha. Science teaches us to find our way in . . . the material world. . . . But the eternal truth which the Buddha proclaimed leads to consummation and deliverance. He who has completely apprehended and thoroughly grasped the Four Noble Truths can do without science. While the most extensive scientific knowledge still belongs to ignorance (*avijjā*) from the point of view of the highest wisdom as it does not lead to deliverance. (93–94)

Subhadra evinces a real ambivalence concerning the relationship of Buddhism and science. On the one hand, there seems to be a compatibility between the two (Buddhists are free to accept the findings of science). But this is because Buddhism is neutral as regards the findings of science, since it is concerned with the metaphysics (the "inner being") rather than the physics of things. Hence, in Subhadra, perhaps for the first time, we find the beginnings of the view that Buddhism and science are complementary. But even if they are complementary in terms of their subject matter (that is, in terms of the things they examine—Buddhism/inner world, science/outer world), the knowledge of Buddhist doctrine trumps that of science. It is Buddhism that is ultimately worth knowing, so that someone who comes to the end of the Buddhist path has no further need of science. Indeed, says Subhadra, science is in actuality a form of ignorance since it is incapable of delivering human beings from suffering. Thus, in the end, Subhadra's is a mixed view that cuts across the three general types being discussed here, since it contains elements of conflict/ambivalence, compatibility, and complementarity.

Another early record of Buddhist views concerning science, a record that stresses the similarity of the two, is the proceedings of the World Parliament of Religions, held in Chicago as part of the Columbian Exposition of 1893. It was as part of that meeting that the Sri Lankan Buddhist leader Anagarika Dharmapala "launched into a favorite theme of the nineteenth-century Buddhist reformers: that it was Buddhism, not Christianity, that could heal the breach between Science and religion" (Fields 1981:126).[16] Stressing the fact that Buddhism repudiated the notion of a creator, Dharmapala claimed that in Buddhism there was no need for explanations that went beyond that of science, there being no need for miracles or faith.[17] As part of that same convocation the great Japanese Zen Master Soyen Shaku drew at least implicit parallels between the law of karma and the laws of science. Paul Carus, an influential American editor and publisher, a monistic rationalist, and himself already a strong advocate of the view that Buddhism was in a unique position among the world's religions to be reconciled with science, was greatly influenced by the World Parliament and especially by its Buddhist representatives, a fact attested to by his subsequent efforts on behalf of Buddhism in the United States and his writings on Buddhism. These included his widely read *Gospel of the Buddha*, in which, in the words of D. T. Suzuki, he "combined the spirit of science and philosophy."[18] Tweed (1992:23) points out that this trend to see Buddhism, and more specifically the Buddhist repudiation of God and soul, as compatible with Western science was a view prevalent among Buddhist sympathizers in the United States in the last half of the nineteenth century.

Taking the views of the figures mentioned here together with others of the same period, like Thomas B. Wilson and Dyer Daniel Lum, we can glean something of the general tenor of the rhetoric of the compatibility of Buddhism and science in the mid to late Victorian era. Influenced by the prevailing rationalism, empiricism, and free-thinking views of the Enlightenment, these men saw in Buddhism a lack of credal dogmatism that they believed was in marked contrast to the tenets of traditional Christianity. Buddhism was therefore like, or concordant with, science precisely because it partook of those elements that, lacking in Christianity, made a reconciliation between the latter and science impossible. The perceived Buddhist emphasis on "the authority of the individual" and its critical spirit were seen as analogous to the methods of science, and the "universal law" of karmic causation was perceived by many as harmonious with science's search for causes, especially for impersonal causal laws that were independent of the will of a deity. Given that, especially in the wake of Kant, ethics was seen as the core of religion, it should also be noted that many of these thinkers saw the Buddhist path as offering a scientific, that is, systematic and empirically verifiable, method for the moral perfection of the individual.

Some late nineteenth-century North American Buddhists seem to have been content simply to point out what they saw as the similarities between science and Buddhism. They believed Buddhism to be more versatile in its ability to respond to the findings of the (then) "new sciences" (especially evolutionism and psychology). Some, while considering Buddhism and science as distinct spheres, saw between them an unspecified but "close intellectual bond." Others believed that when science had brought about the final demise of Christianity, Buddhism would stand alone as the only plausible alternative religious view. Others would go further. For example, Paul Carus grew to believe that Buddhism *was* the "Religion of Science," the religion that would make "scientific truth itself . . . the last guide of a religious conception of mankind."[19] In these various views we see reflected the entire range of opinions of this second mode of interaction: from vague similarity to total identity. It is important to consider these early views because they will in many ways set the tone for the rhetoric of compatibility that is to follow.

In the late 1950s the English-trained Sri Lankan scholar K. N. Jayatilleke wrote what has become one of the most influential essays on the relationship between Buddhism and science.[20] "Buddhism and the Scientific Revolution" reiterates many of the themes of the nineteenth-century rhetoric of compatibility, although now in a more sophisticated way. Jayatilleke's basic thesis is that the rift between religion and science would not have occurred had the scientific revolution "taken place in the context of Early Buddhism." This is not only because Buddhism "accords with the findings of science" (similarity in content and conclusions) but also because Buddhism "emphasizes the importance of a scientific outlook," in that "its specific dogmas are said to be capable of verification" (similarity in method). As is evident from his other writings, most notable his *Early Buddhist Theory of Knowledge*, Jayatilleke, heavily influenced by the prevailing tendencies in British philosophy, sought to portray the Buddhism of the earliest Pāli texts as "an analytical approach, combined with an empiricism" (1980:276). In the essay under discussion Jayatilleke points out similarities between the Buddhist and scientific conceptions of the cosmos and between Buddhism and psychology. While acknowledging that Buddhism offers no theory of biological evolution, he sees the former as compatible with the latter, a view put forward more recently and in much greater detail by Robin Cooper.[21] But Jayatilleke is concerned more with showing similarities in the *methods* of Buddhism and science than in demonstrating parallels as regards *content* or *conclusions.* Both science and Buddhism, he claims, are committed to critically (and not dogmatically) establishing the existence of universal laws, laws that can be verified through "personal experience." While acknowledging, for example, that rebirth and karma may not seem very scientific to the modern mind, he stresses that these "laws" were accepted by early Buddhists not dogmatically but only after they had been personally verified. Underlying this search for laws is the common belief in universal causation: that everything can be explained "without the need for teleological explanations or divine intervention."[22]

Despite these parallels, Jayatilleke does not hold the stronger view of the identity of Buddhism and science. He makes this abundantly clear when he states that it is not his intention to show that "Buddhism teaches modern science." This does not mean that such a view is dead. A century after the Victorians, and over two decades after Jayatilleke, we find the rhetoric of identity still very much alive, and there is arguably no clearer statement of it than in the work of Gerald Du Pre (1984). For Du Pre, just as the Madhyamaka is the philosophy of Buddhism par excellence, so too is it the philosophy of science (105). Scientific psychology, he tells us, was not founded in nineteenth-century Germany by Wundt, but "two thousand five hundred years earlier, in India, by Prince Siddhartha" (110). Du Pre finds many "amazing" instances of modern scientific findings being prefigured in the Buddhist (mostly Pāli) texts. Finally, meditation, he states "is scientific examination itself!" (1984:141).

As naive and unsophisticated as this view may seem to many of us today, it is by no means a relic of bygone ages, for even today we find scholars given to the hyperbolic claim that *Buddhism is science*, and Buddha the quintessential scientist. Guenther (1984), while claiming to be utilizing the concepts and terminology of science only metaphorically and heuristically (4, 56)—in his case as a way to make the Buddhist rDzogs chen material more accessible to a Western audience[23]—at times betrays his underlying belief in some real and substantial compatibility between the two spheres, even as regards the size of elementary particles and the size of the universe (98–99). Far more prevalent and more subtle, though nonetheless a vestige of the rhetoric of identity described here, is the view that Buddhism is its own unique type of science: an interior science, a mind science. But this takes us into a new mode of the engagement of Buddhism and science, one that I call complementarity.

COMPLEMENTARITY

Conflict/ambivalence as a mode of interaction between Buddhism and science presumes radical and irreconcilable differences between the two spheres. In a parallel fashion the rhetoric of compatibility presumes a fundamental similarity between (and in its extreme form, the identity of) Buddhism and science. Complementarity as a mode of engagement lies somewhere between the first two modes: negotiating both similarities and differences.

Just as conflict and compatibility as modes of engagement were seen to be of different types, depending upon the kind and degree of difference or similarity they stressed, so also with complementarity. Where similarity in method, and difference in the object of study, is maintained, there emerges the rhetoric of Buddhism as an interior science, or a science of the mind. Here Buddhism—now seen as an empirical and verifiable technology, but a technology *of the spirit*—is portrayed as complementing science whose object of study is the exterior world of matter. While different in *what* they analyze, they are claimed to be similar in *how* they go about the analysis.[24] The challenge then is seen as one of building on that common methodology (the similar *how*) to extend the purview of each sphere (the disparate *what*). In this model science stands to gain, for example, by being pushed to consider mind or consciousness nonmechanistically or by having to confront extraordinary inner mental states that are not normally within the purview of its investigations.[25] Buddhists stand to profit by gaining access to new facts concerning the material world (body and cosmos)—facts that have lain outside of traditional Buddhist speculation due to technological limitations. This, in any case, is the rhetoric of this way of envisioning the complementary nature of Buddhism and science.

Another form of complementarity stresses *difference in method* and *similarity in content*. Here both Buddhism and science are seen as engaging the selfsame object (whether matter or consciousness or both), but do so utilizing different modes of analysis. Science is seen as operating rationally, conceptually, and analytically. Buddhism, it is claimed, engages its object experientially, using nonconceptual modes of intuitive understanding that emerge as the result of the practice of meditation. Science yields *factual* knowledge that is useful in practical and mundane tasks; Buddhism is the purveyor of *transformative* knowledge that brings about positive personal and social change. In studies that exemplify this mode of engagement it is claimed that, even if Buddhists and scientists continue to utilize their own unique methods of gaining knowledge in their respective traditions, *as human beings* they can (and should) learn to access reality using the entire range of epistemic possibilities. Thus the complementarity of Buddhism and science here lies in the ability of each sphere to contribute epistemically to a more complete way of knowing a common object.

Obviously, what I have just outlined are ideal-typical forms of complementarity. These modes of engagement in the real world will be more complex and often intertwined. Regardless of what types and degrees of similarity are stressed, however, the various forms of the rhetoric of complementarity all have in common the tendency to see *in the differences* between Buddhism and science the basis for a dichotomizing of the field or mode of knowledge that makes of each a part that, when united, creates an even greater and more worthwhile whole. It is this holism that distinguishes complementarity from mere compatibility. While the intrinsic value of each of the two parts—Buddhist and scientific modes of knowledge—may not be denied, the implication of course is that in isolation each is also lacking, if only because each by itself fails to realize its full potential as part of—that is, that in isolation each fails to realize its full contribution to—the whole.[26]

Complementarity as a mode of engagement operates, then, according to a structuralist logic that uses the perceived differences between Buddhism and science to construct them as the binary parts of the greater whole. Hence, science is concerned with the exterior world, Buddhism with the interior one. Science deals with matter, Buddhism with mind.[27] Science is the hardware, Buddhism the software.[28] Science is rationalist, Buddhism experiential.[29] Science is quantitative, Buddhism qualitative.[30] Science is conventional, Buddhism contemplative.[31] Science advances us materially, Buddhism spiritually.[32] But whether the difference is identified principally in terms of content, of method, or of goal, the perceived problem—diagnosed in terms of the overemphasizing of one of the two elements[33]—is overcome by a balance that is achieved when the two parts are brought together harmoniously. Unlike conflict/ambivalence as a mode, the logic of complementarity eschews the kind of triumphalism in which one of the two spheres emerges as victorious over the other. Unlike identity/compatibility as a mode, by holding firmly to the notion of irreconcilable differences it refuses to allow either Buddhism or science to be reduced to the other.

It should also be mentioned that the logic of complementarity being described here is ultimately axiological insofar as it presumes and promotes certain values, e.g., a *balance* of external/rationalist/material and internal/experiential/mental development. In brief, in the mode of complementarity Buddhism and science, though distinct, and though individually valuable, each contribute to the creation of a whole that is of even greater worth.

One of the clearest examples of the rhetoric of complementarity is found in a work that, though not strictly devoted to the dialogue between *Buddhism* and science, has been extremely influential in the subsequent conversation between the two. The work is Fritjof Capra's *The Tao of Physics* (1984 [1976]). Despite its many limitations—not the least of which is the fact that it conflates many of the "Eastern" traditions, treating them as if their most fundamental doctrinal claims were the same—the implications of Capra's work are to be felt even today. As the yin/yang symbol on the cover admonishes, *The Tao of Physics* touts the complementarity of physics and "Eastern mysticism." Capra's work stresses the similarity—and even the "complete harmony"—of the conclusions of modern physics and the "mystical" (especially Hindu, Buddhist, and Taoist) traditions, and thus at times reads more like a work belonging to the second mode of engagement, that of compatibility or identity. But in the epilogue to *The Tao of Physics* Capra makes it clear that his view is one of complementarity:

> I see science and mysticism as two complementary manifestations of the human mind; of its rational and intuitive faculties. . . . The two approaches are entirely different and involve far more than a certain view of the physical world. However, they are complementary, as we have learned to say in physics. Neither is comprehended in the other, nor can either of them be reduced to the other; but both of them are necessary, supplementing one another for a fuller understanding of the world. . . . Science does not need mysticism and mysticism does not need science, but men and women need both. Mystical experience is necessary to understand the deepest nature of things, and science is essential for modern life. What we need, therefore, is not a synthesis, but a dynamic interplay between mystical intuition and scientific analysis. (297)

The dialogue between Buddhism and science has come a long way since Capra's work, but the underlying logic of complementarity can still be found in a great many studies even today. Vic Mansfield, a physicist with a longstanding interest in Buddhism, writes in a recent essay:

I hope to show that understanding a little about time in modern physics helps us more deeply appreciate some of the most profound ideas in Buddhism. Furthermore, I will also suggest that some appreciation of Middle Way Buddhist ideas could aid in the development of physics. Thus a nontrivial synergy between these two very different disciplines is possible, one that results in deeper understanding and more compassionate action.

Mansfield's synergistic model of the interaction of Buddhism and science is prevalent today; it evokes a specific kind of complementarity. It is not that Buddhism and science use similar methods or reach identical conclusions but that each has implications for the other, implications that, when seriously considered, can yield greater insight in each of the two spheres.

The dialogue between Buddhism and physics is by no means in its infancy,[34] but it cannot compare to the state of the dialogue between Buddhism and the mind sciences. Let us turn to a discussion of some of the recent work in this latter field to see the model of complementarity at work. In a recent book (Goleman and Thurman 1991:7–8, 59, 73) Robert Thurman suggests that, whereas the West has been concerned principally with the exploration of the material universe, Buddhism has been concerned with developing a refined "inner science," and that, whereas Western science has been concerned with the hardware of the brain, the Tibetan mind sciences provide us with the software for understanding and modifying the mind. In the end, however, it is not clear whether Thurman's is a form of complementarity or a kind of triumphalism, for in both the work just mentioned and in a more recent book (Thurman 1998:275 et passim) it eventually becomes clear that for Thurman the "whole" to which both Buddhism and science contribute is none other than *Buddhist omniscience*, something that Thurman clearly believes can be accessed through the Buddhist path alone. (Cf. the discussion of Subhadra above.)

Daniel Goleman (Goleman and Thurman 1991), though recognizing that there are similarities *in content* between Western and Buddhist psychology, believes these to be "surface" similarities and, instead, sees the complementarity *in their methods* to be the locus of their most fruitful interaction:

> The telling difference is in the methods, the means to mental health that exist in each system. . . . By and large, psychotherapy focuses on the *content* of consciousness. It does not attempt the more radical transformation posited in

the Tibetan Buddhist approach, which focuses on the *process* of consciousness. . . . When you put the two psychologies together, you get a more complete spectrum of human development. (100–101)

This "full spectrum" model is a view that has been put forward most clearly and forcefully by Wilber, Engler, and Brown (1986). *Transformations of Consciousness*, a collection of essays by these three (and other) authors, reprinted mostly from the *Journal of Transpersonal Psychology*, puts forward the view that there is a basic underlying similarity to the structure of developmental (meditative) stages in various contemplative religious traditions.[35] Buddhism plays a major role in their analysis. Science (specifically psychology) has provided us with a map of the "conventional" stages of human development. Buddhism extends those stages by providing a further structuring of human development beyond the conventional realm. Only when the conventional and contemplative templates are taken together (and in that order) do we have a full picture (the full spectrum view) of human development. Despite these authors' (and especially Brown's) very detailed attempt to argue for a common structure to the meditative path across cultures and religious traditions—a view that, even if not a perennialism of goal, is a perennialism of method (220)—I remain unconvinced about such a common structure, just as I remain unconvinced that the various "conventional" psychological schemes of human development can be harmonized into a unified and consistent template that would be acceptable to each of the various *Western* theorists on whose work these three authors draw. Equally problematic is the tendency in this study to ascribe to the contemplative traditions a lack of thematization concerning the *conventional* workings of the mind and the nature of the conventional self. Be that as it may, *Transformations of Consciousness* presents us with one of the clearest examples of the logic of complementarity. In their words,

> If it is true that the conventional schools have much to learn from the contemplative schools (especially about possibly higher development), it is equally true—and we believe as urgent—that the contemplative schools surrender their isolation and apparent self-sufficiency and open themselves to the vital and important lessons of contemporary psychology and psychiatry. (8).

Thus in *Transformations* science and Buddhism complement each other by contributing, respectively, models of the conventional and contemplative stages of human development, which, when taken together, provide a complete picture of what it means to be fully human. Taken in isolation, each provides us with only a partial picture of human development.

A different form of complementarity is found in Christopher deCharms's *Two Views of the Mind* (1998), a work that explores the complementary perspectives on the mind offered by Buddhist Abhidharma and neuroscience. Although the two perspectives, he claims, are very different— "The Tibetan approach to mind is largely descriptive, explaining by illustration and metaphor, whereas Western science is predominantly mechanistic, explaining in terms of simple material forces acting on small constituent parts" (50)[36]—deCharms believes that the two spheres can contribute to each other, but not by justifying each other's conclusions. Many studies that tout the complementarity of Buddhism and science focus on the ways in which the former can contribute to the latter. DeCharms's work, for all its limitations, at least gives equal time to the other side of the equation. He believes that science can contribute to Buddhism because the former has at its disposal methods for achieving consensus through verifiability. DeCharms portrays Buddhism (and the contemplative religious traditions generally) as fractured in regard to the details of their views of the mind. "By using the methods of science, methods based on commonly verifiable observations, it might be possible to start to find a similar kind of consensus regarding debated points within Buddhism, or even debated points within traditions" (48). Hence, science, the paragon of consistency in deCharms's view, will act as the adjudicator of—or, perhaps more accurately, the model for adjudicating—intra-Buddhist and interreligious doctrinal disputes. Whatever one might think of such a view (and the work of Kuhn and Feyerabend alone may be enough to seal its fate), at least it evinces the virtue of offering some concrete suggestion as to how Buddhism stands to gain in its encounter with science. But if Buddhism can profit, so, of course, can science. Buddhism can contribute to neuroscience insofar as "the observational methods of Buddhism are what the present [Western] science of mind largely lacks in systematic form" Their main value to Western thinking may be that they are both subjective and systematic to a level of detail that current Western systems of observation have not yet reached" (1998:46).

A similar view can already be found in the 1991 work of Varela, Thompson, and Rosch, *The Embodied Mind*, a study that also sees in Buddhism a dialogue partner for cognitive science, one that can infuse the latter with a perspective that it is presently lacking: "a direct, hands-on, pragmatic approach to experience." Both similarities in content (e.g., both Buddhism and cognitive science claim the self to be "fragmented, divided or nonunified") and similarities in method (both Buddhism and science are pragmatic, systematic, and disciplined techniques for dealing with the phenomenal world and with human experience, respectively) are important to the authors of this work. Buddhism's sophisticated phenomenology of experience allows it to confront science on equal terms, as it were, and ultimately allows it to serve as a corrective to science, which, up to now, has given "the spontaneous and more reflective dimensions of human experience . . . little more than a cursory, matter of fact treatment" (1991:xviii). Buddhism thus complements science by providing it with the means to reclaim a sense of the embodiment of experience, something that it has lost. But, in addition to this, Buddhism, and especially the Buddhist philosophical perspective known as Madhyamaka, serves as another kind of corrective, a corrective to the "nihilism"—both in the cognitive sciences and in society generally—that is the natural response to groundlessness in general, and to the breakdown of a belief in a unified self in particular. Like Mansfield, then, Varela, Thompson, and Rosch believe that Buddhism is more than just intellectually useful to science. It is also an ethical corrective to the problems that face us as human beings today, problems that have in large part been either caused or exacerbated by science.

This notion of Buddhism as dual corrective—both intellectual and ethical—is also at the core of Alan Wallace's earlier study, *Choosing Reality* (1996). The first part of this work is unique in taking as its subject matter not Buddhism and science itself but Buddhism and the philosophy of science. Wallace is concerned, first of all, to establish the naïveté of the view that science (and his focus is principally on physics) is metaphysically neutral. He next shows how two metaphysical views—"realism" and "instrumentalism"—have acted as philosophical underpinnings for science and argues that they have failed in that regard. Wallace then suggests that the philosophy of the Madhyamaka school may be a more suitable alternative.[37] In addition to being intellectually complementary in the way just described, Buddhism, according to Wallace, can also complement science ethically, by reinvigorating it with the spirit of ethical responsibility, "the longing to be of service to others," a spirit that is found in science at its best but has too often been lost as science became disassociated from religion and philosophy.

Finally, let me mention, if just briefly, and by way of concluding this discussion of the dual-correctionist form of complementarity, the work of Jeremy Hayward (1984, 1987), who was one of the first scholars to take the discussion in this direction. Much of what has already been said as regards the complementary nature of Buddhism and science in the work of Varela, Thompson, and Rosch and of Wallace is also true of the work of Hayward. Where Hayward perhaps differs is in his more detailed and historical thematization of the role that modern science has played in creating personal and social alienation. For him Buddhism complements science by counteracting/correcting this tendency of modernist science to create a rift between nature and its observer, and between matter/body and mind. In this way Buddhism acts as a force for reenchanting the world.

Even a cursory survey of the interaction of Buddhism and science, which is all that the present essay purports to be, would be remiss in failing to mention the role that the present Dalai Lama has played in the ongoing dialogue. His interest in the dialogue between Buddhism and science is evidenced by wide-ranging engagements: from conversations with physicist David Bohm (Weber 1986:231–242) to the series of Mind and Life conferences that, in turn, have dealt with topics as diverse as the mind sciences (Hayward and Varela 1992), emotions and health (Goleman 1997), sleep, dreaming, and dying (Varela 1997), compassion and altruism, physics, and, most recently, destructive emotions. On one level, the Mind and Life conferences, by consciously striving to choose topics and participants carefully, have brought the level of the dialogue between Buddhism and science to new heights, both in the level of sophistication, and in terms of public exposure. At the same time, the Mind and Life conferences are conversations between Western scientists and the Dalai Lama, almost exclusively, as the sole Buddhist interlocutor. Although these dialogues have become accessible to Western audiences through the publication of proceedings, they are known in the Tibetan world, again, almost exclusively, through the efforts of the Dalai Lama himself. No one figure, even one as extraordinary as the Dalai Lama, can carry the burden of being the sole Buddhist representative in Buddhism's engagement with science, nor, for that matter, of being the sole representative of that dialogue before an entire culture. If that dialogue is to flourish, then, it will require greater participation and commitment on the part of the Buddhist scholarly world, both Asian and Western.

THE FUTURE OF A CONVERSATION

Even in this brief and impressionistic account of the interaction of Buddhism and science, it would be difficult not to notice change and even progress. A dialogue that began in an idiom of broad generalities—"Buddhism," "science," "universal laws" and so forth—has shifted to a more concrete conversation that is increasingly cognizant of, and more informed about, the complex internal texture of these two spheres. Today the partners in the dialogue are not simply *Buddhists* and *scientists* but Japanese Rinzai Zen Buddhists and cognitive neuroscientists or Tibetan dGe lugs pa Buddhists and elementary particle physicists. Today we are more likely to see the partners in such a dialogue discussing not simply "cosmology" but the Buddhist and Western cosmological understandings of the structure of space in the early universe; not just "the mind" but the probable outcomes of PET-scan studies of dying meditators. This more specialized and sophisticated dialogue is partially the result of the development of the various fields themselves. We simply know more about human physiology, human behavior, the workings of the brain, the structure of matter, and the evolution of the universe than we did a few decades ago. Concomitantly, Western scholarship in Buddhist studies has progressed from the musings of armchair scholars to the serious study of texts in their original languages, to the exploration of the interface of text and culture in a variety of Asian settings. Specialization—both in science and in Buddhist studies—obviously has its downside, but it also has its advantages. In the absence of such specialization it is difficult to imagine that the dialogue between Buddhism and science could have reached the level of sophistication it has today.

But specialization alone is not enough to explain the changes that have occurred in the interaction between Buddhism and science. Increased accessibility of information about these two traditions has been just as important. Scientists now have access to resources for the study of Buddhism, both textual and human, that were simply unavailable a generation ago, and vice versa. Shifts in the intellectual ethos of the scientific West have also positively impacted the dialogue. Especially in recent years, we have witnessed a slowly decreasing resistance to the idea that mind may have a substantial role to play in medicine, neuroscience, and even physics. (Hayward 1987). Finally, there are sociological factors that have contributed to the changing nature of the conversation between Buddhism and science, not the least of which is the spread of Buddhism to the West. Despite a rhetoric that sometimes conceals this fact, many, perhaps even most, of the scientists active in the dialogue today are practicing Buddhists. Nor can the importance of the interest in science shown by a charismatic Buddhist leader like

the Dalai Lama be underestimated as a factor in the current, and more refined, state of the dialogue.

That Buddhism and science are today engaged in a dialogue that is more
nuanced and mature is not of course to deny that more naive ways of envisioning their mutual engagement still exist. There are still individuals—
Buddhists, scientists, or both—who find a rhetoric of identity or conflict ir-
resistible. What is less obvious, perhaps, is that even some models of
complementarity can act as impediments to mature dialogue. Dialogue can
be stunted by a dichotomizing logic of strict complementarity that is taken
too literally, and applied too strongly, by a structuralist logic of binary opposition creating impermeable categories that cease to operate as
metaphors and come to be believed as real. Let us consider for a moment
the rhetoric that casts Buddhism as an inner science of the mind, and science as the investigation of matter. As a fluid metaphor this mode of complementarity certainly has its uses. It can be illuminating, and can enhance
our understanding of the interaction of Buddhism and science. If taken too
literally, however, this model would seem to imply that Buddhism has the
spiritual realm as its *sole* concern, and that it is therefore unconcerned with
the analysis of matter. Nothing could be farther from the truth, of course.
Buddhist scholars have elaborated complex theories of matter. They have
debated the relative merits of competing theories and have speculated extensively about the relationship of matter to other phenomena. Where the
metaphor of Buddhism as an inner science is reified, it precludes a dialogue
between Buddhism and science on the nature of matter; it prevents a serious engagement between Tibetan and Western medicine on the physical
causes of illness. In short, it acts to silence Buddhism when it attempts to
address any issue related to the material world. Similarly, a strict structural-
ism of this sort, by granting to Buddhism a monopoly on the life of the
mind, dismisses in an ad hoc fashion the possibility that, say, psychoanalysis or cognitive science could have anything valuable to contribute on the
nature of mental processes.

Of course, there is something to be gained by both Buddhists and scientists in creating and maintaining such a strict segregation of the two
spheres, or so it would seem. Such a tactic, as Wayne Proudfoot (1985) has
shown in a different context, serves as a kind of protectionist strategy: the
kingdom is divided between the two factions for the sake of peace. In so doing, each side assures for itself total control over its own sphere of influence.
Buddhism need fear no challenge from science when it comes to the realm

of experience, and science can rest peacefully in its knowledge that Buddhism poses no threat when it comes to the external material world. Peace appears to be the reward, but it is a false peace, one based on convenience and not on truth, for the *truth* is that both Buddhism and science are highly complex, totalizing worldviews that defy the literalist and strict structuralist attempts to delimit them.

Protectionist strategies of the type just mentioned, whether conscious or unconscious, seem to me to be at work in many of the real-life dialogical encounters between Buddhists and scientists. How so? The tendency manifests itself in an unwillingness to go beyond the merely informative, in an intellectual laziness that too readily accepts differences, and that, when conscious, justifies such acquiescence in the name of complementarity, or its offspring, the romantic idealization of the other: Buddhist: "How can they be wrong about matter . . . they have the machines," or scientist: "Who are we to dispute the fact that they can attain such states? . . . They have been doing this for generations." Of course, the subtext here is, "Don't challenge us about the mind, and we won't challenge you about the brain," or, "Don't challenge us about the structure of matter, and we won't challenge you about meditation." The dialogue, when conducted in this vein, quickly degenerates into show-and-tell, as the critical faculty enters a state of suspended animation. That an informative stage is a prerequisite to dialogue is not of course at issue here. My point is simply that dialogue is not coterminous with show-and-tell, that it must go beyond it.

As I have stated, it is my belief that in many cases the failure of dialogue to do so is a result of the tendency on the part of the participants to take a literalist view of the dichotomies constructed in models of complementarity. Eventually, however, such dichotomies will have to be seen as the metaphors they are, and, when this happens, Buddhists and scientists will have to face the truly contentious issues—issues that "belong" exclusively neither to Buddhism nor to science, because they belong to both: Is there mind separate from brain? Is the human personality consciously mutable? If so, to what extent? Are there elementary particles that are fundamental in the sense that they have no constituent parts? Is sentient life separate from matter possible?

I do not mean to imply, of course, that there has not been *any* grappling over these issues, only that there has not been enough. This is certainly understandable. On the one hand, confronting such issues will undoubtedly be the cause of some discomfort, since intellectual antagonism will be unavoidable. On the other, the logic of strict and literalistic complementarity is seductive: scientists are the masters of matter, Buddhists of the mind: let us walk the road to inner and outer progress together! Or again, scientists provide the hardware, Buddhists the software, let us create the great computer together! The problem, of course, is that such segregationist metaphors are, like all metaphors, artificial. They are artificial because scientific claims impinge, and sometimes impinge *negatively*, upon Buddhist ones, and vice versa.

Let us consider just one example by way of illustration. The extent to which the Madhyamaka doctrine of emptiness is invoked in models that advocate the complementarity of Buddhism and science will not have gone unnoticed. In such contexts, the Madhyamaka is frequently characterized as a philosophical and/or religious system, one that neither affects nor is affected by conclusions concerning the physical world: a good example of the religion-mind-experience versus science-matter-fact dichotomy of strict complementarity. The problem, of course, is that this disregards an important though little known fact: that the Madhyamaka theory of emptiness[38] requires the existence of an external world, and it implies, additionally, that there can be no elementary particles fundamental in the sense that they have no constituent parts.[39] Why this is so—that is, the details of these arguments—need not concern us here. It is sufficient to note that the truth of the doctrine of emptiness—one of the most important and unique tenet of Mahāyāna Buddhism—depends upon a fact of the physical universe. If it can be shown that there are partless particles—particles that are no further divisible, that are not made up of anything more elementary—then the theory of emptiness as formulated in the classical Indian and Tibetan sources, and therefore Mahāyāna Buddhism as we know it, would be false. The point, of course, is that the realm of physics and metaphysics cannot be so easily segregated, that Buddhism is not *simply* a "philosophical view" or a "mind science," and that the truth of Buddhist doctrine both implies and is dependent upon facts related to the material world.

What does all of this imply for the future of the dialogue between Buddhism and science? It implies, first of all, the necessity of realizing the metaphorical nature of the models we use to help us envision the interactions of Buddhism and science. Such a realization requires, in turn, an understanding of Buddhism and science (and their subdisciplines) as complete systems that resist dichotomizing: systems that can both support and challenge each other at a variety of different levels—no monopolies, no holds barred! Let me conclude by offering one very concrete suggestion for the future of the conversation between Buddhism and science, a suggestion that I believe can help us avoid the pitfalls that I have just described.

It is an astounding fact that of the various studies cited as part of my attempt to chart the interaction of Buddhism and science, almost all the serious studies in the last fifteen years—the period in which I believe we have witnessed the most sophisticated and interesting work—have been written *by scientists*. What has been missing, of course, is the voice of the Buddhist scholar, both Asian and Western.[40] Many of the scientists, or former scientists, who have published in this field, of course, have considerable background in Buddhism; many, in addition, have long-standing commitments to Buddhist practice, but none of them are trained scholars of Buddhism, nor have any, to my knowledge, mastered any of the languages of the Buddhist texts. If, as I have suggested, the strict dichotomizing and protectionist tendencies of a literalist model of complementarity is the next major hurdle to be overcome, and if this will require a broader and deeper understanding of science, and especially of Buddhism, as I think it does, then perhaps it is precisely the voice of the scholar of Buddhism in the dialogue that is the greatest desideratum at this point in time.[41] But perhaps my suggestion is, as of the printing of this book, already moot, since one of the virtues of this volume is precisely that it has brought trained scholars of Buddhism formally into the discussion.

Notes

1. The Training of the Mind research project, funded by the Fetzer Institute, was an exploratory first attempt to determine principally what measurable cognitive changes occurred as a result of intensive, prolonged meditation. The primary subjects were a group of Tibetan monks who had been in long-term retreat in the mountains above Dharamsala, India. The other members of the research team were Richard Davidson, Clifford Saron, Gregory Simpson, Francisco Varela, and Alan Wallace.
2. This would suggest that it is not only fitting but indeed necessary that there be addenda to traditional human subject research guidelines that are more situationally specific. In the case under discussion there seems to me to be an urgency in creating specific guidelines that address the issues that follow.
3. That even conversation is not an ethically neutral enterprise—that it involves economic and political power differentials with definite ethical consequences—is of course an observation that has been made most forcefully

by anthropologists. See Rabinow 1996; and Clifford and Marcus 1986. But, on the spectrum of urgency, dealing with ethical issues in the case of the one-sided scientific objectification of Buddhism seems to me clearly more pressing.

4. The metaphor of barrenness, with its obvious implications to procreation, is not my own. Jaki, for example, speaks of the "stillbirth" of science in non-European societies. See Jaki 1985:132.

5. That "science" in this form of historiography (and elsewhere) is simply assumed to be coterminous with science as it developed in the West not only vitiates against considering science to be culturally situated but also against the possibility that it can be situated elsewhere than in the West. It acts, among other things, against the possibility of considering science as a *multicultural* phenomenon and hence of even entertaining the possibility that certain forms of traditional speculation and practices specific to, say, Buddhist societies are forms of science. Although I will use the term *science* in this essay to refer to the specific form of science that emerged in Europe, I am cognizant of the fact that this tends to reinforce an unhealthy Eurocentrism as regards the nature of science, much in the same way as using the term *man* to refer to the generic human acts to reinforce androcentrism. For an excellent discussion of these and related issues, see Harding 1998.

6. See, for example, Singhal 1972:153–188; and Needham 1981, which is a synopsis of his multivolume *Science and Civilization in China*. Needham's work suffers from the same kind of faulty generalizations about Indian and Buddhist views of time (that it is cyclical, idealist, subjectivist) as does the work of Jaki, but his work also shows, without question, that there was such a thing as *science* in classical China.

7. Science has encountered opposition in its spread to various Asian cultures, but I believe that this has been principally due to sociopolitical rather than ideological reasons.

8. How less cyclical is the doctrine of a coming foretold, and a second coming prophesied, and how less polytheistic is a trinitarian God that lords over a pantheon of angels and saints? As regards Indian and Buddhist notions of time, it can be pointed out that these are as varied and complex, and therefore as resistant to generalization, as any notion of time found in any other culture and religion. That, despite the cosmological doctrine of the repeated creation and destruction of the universe, there is, at least in many Buddhist texts, a definite sense of "linear time" is witnessed by the clear distinction made in these texts between past, present, and future (Poussin 1988:593) and by the notion of instantaneous and irreversible change (Poussin 1988:474). See also the considerable range of views in the debate over the nature of time in the Buddhist world in the early centuries of the Christian era, as reported in Vasubandhu's *Abhidharmakośa* (Poussin 1989:808–816).

9. I should hasten to add that I find equally problematic the claim that the reason for the relative nonemergence of science in Asia can be easily identified.

I find baseless, for example, Thurman's claim that the nonemergence of science in India represents a conscious choice. Goleman and Thurman 1991:57. A choice on the part of whom?

10. On the sabotaged attempts on the part of Tibetans to introduce a Western system of education that included science into Tibet in the 1920s and then again in the 1940s see Goldstein 1989:121, 421–426.

11. On the life and works of Gendün Chöphel, see Stoddard 1985 and Lopez 1996.

12. That the theoretical skepticism may, upon further examination, be found to be groundless, insofar as Buddhism has no qualms with the notion that mental states can have physical correlates, does not diminish my point here, which is meant as a report of monks' actual views, reports based on extensive interviews with long-term meditators as part of my work with the Training the Mind research group; see note 1. See the report of one such meditator in Lobsang Tenzing 1990.

13. See, for example, Stephan Hawking's dismissal of "eastern mysticism" as obscurantism, in Weber 1986:210.

14. See, for example, Hayes's 1996 critique of the Internet in *Cybersangha*; and the Ven. Hsuan hua's more thoroughgoing critique, one that almost borders on paranoia, "Electric Brains and Other Menaces," in the same venue. Though not openly conflictual or polemical, there is also to be found in contemporary Asian Buddhist writing a rhetoric that seeks to portray Buddhism as the *true* innovation, and science as having to catch up with it. See, for example, Zoysa 1998 and Yin Tak 1998.

15. The edition available to me (Olcott 1889), which is a reprint of the Theosophical Publication Society's edition, has no date, but bears the library classification number of 145.72 O43b 1889. The "certificate" in the preface is dated 1881, and Olcott's preface 1886, but it is unclear to me when the first edition was printed.

16. Interestingly, H. H. the Dalai Lama has recently suggested a modified version of this claim, namely, that it is Buddhism's position outside of both radical materialism and theistic religion, "as belonging to neither camp," that could allow it to serve as a bridge between science and (other) religions. See Golemen and Thurman 1991:13.

17. In a later (1902–1904) lecture tour of the United States, Dharmapala visited several technical schools, and, seeing the need for greater knowledge of science in Asia, was himself instrumental in establishing such a school in Sarnath, India. See Fields 1981:134.

18. On Carus, see Fields 1981:128, 141–143; and Tweed 1992:65–67, 103–105. Although ostensibly a representative collection of Buddhist scriptural passages, the texts collected in *The Gospel of the Buddha* in actuality reflected Carus's penchant for portraying Buddhism as a rational and science-compatible religion.

19. Cited in Tweed 1992:103.

20. The importance of this essay is evidenced by the fact that after its original

publication in the third issue of the Sri Lankan English-language journal the *Wheel*, it has, to my knowledge, been reprinted at least twice. Once in Nyanaponika 1971 and once in Kirtisinghe 1984.

21. Although making a case for the compatibility of Buddhism and evolution, Cooper also suggests ways in which evolutionary theory might be broadened under the influence of Buddhist insights, hence making his view fall, at least in part, under the mode of complementarity. For a very balanced review of Cooper's study, see Jones 1997.

22. For a similar contemporary claim, see Fenner 1995:10: "This (Buddhist karmic theory of the workings of habit formations in consciousness) is in distinction, for example, to theistic religions and transformational paths that introduce indeterminacy by the inherently inexplicable influence of grace."

23. Making the Buddhist material available to a wider audience is also Fenner's motivation for his System-Cybernetics reading of the Madhyamaka school of Buddhist thought. See Fenner 1995.

24. Hence, Diana Eck (in Goleman and Thurman 1991:106) states that "there is a common agenda and method in the fact that both the mind science of the Buddhist tradition and the exploration of the medical researchers are based on the traditions of experimentation. (Buddhism is) an experimental practice. It is not a form of religiousness that simply says, 'Believe this on faith.' It is an experimentally verified analysis of how the universe is."

25. Consider, for example, Francisco Varela's urging that subtle states of mind "merit respectful attention by anybody who claims to rely on empirical science." Varela 1997:216.

26. In the words of Wallace, "The meaningfulness of scientific and contemplative knowledge is therefore complementary. In the absence of either, the world is impoverished." Wallace 1996:205.

27. See Thurman in Goleman and Thurman 1991, chapter 4.

28. This is the dominant metaphor used by Thurman in Golemen and Thurman 1991:53–73.

29. See deCharms 1998.

30. See Wallace 1996:147.

31. See Wilber, Engler, and Brown 1986.

32. This is a theme that is to be found in much of the work of H. H. the Dalai Lama.

33. Capra 1984:xvi: "Our culture has consistently favored *yang*, or masculine, values over *yin*, or feminine, counterparts. We have favored self assertion over integration, analysis over synthesis, rational knowledge over intuitive wisdom, science over religion, competition over cooperation, expansion over conservation, and so on. This one-sided development has now reached a highly alarming stage; a crisis of social, ecological, moral and spiritual dimensions."

34. See, for example, Weber 1986. The last of the Mind and Life conferences

with the Dalai Lama was also dedicated to particle physics and cosmology (see below).

35. A similar, though by no means identical, project is Ornstein's much earlier *The Psychology of Consciousness* (1972), in which several pages are devoted to different forms of Zen meditation.

36. It is not at all clear to me that the distinctions are as clear-cut as deCharms makes them out to be, for surely there are materialist explanations of mind in "the Tibetan approach," nor is science allergic to the use of illustration and metaphor, as Ian Barbour and others have pointed out. More troubling, though, is deCharms's view that the Tibetan system subscribes to some kind of paradoxical logic that can live with contradiction (1998:26, 49), a view that I have taken issue with elsewhere; see Cabezón 1994.

37. Mansfield 2002 makes a similar point, arguing, in this instance, that Cartesian dualism has hindered the progress of science, and that the Madhyamaka is a more effective philosophical basis for science. This same point is reiterated, using a similar argument, in Mansfield 1995–1996.

38. Or more accurately, the Prāsaṅgika interpretation of the Madhyamaka theory of emptiness.

39. The existence of an external world is necessitated by the Prāsaṅgika's commitment to according with worldly conventions. The nonexistence of partless particles is a corollary of the fact that the theory of emptiness implies that all phenomena exist as mere imputations, as labels that require some basis (*gzhi*) on which to, as it were, adhere. In the case of material particles, such a basis can be none other than their parts. See Cabezón 1992:144, 149, 324–345.

40. The exception, from the Asian side, has of course been H. H. the Dalai Lama. The most prominent exception from the Western side has been Alan Wallace, though perhaps Peter Fenner and Herbert Guenther can also be named as exceptions in this regard.

41. One might say that expertise in the various sciences themselves is already well represented in the dialogue but that what has been missing from the side of science is a perspective with a broad overview of science, as exemplified in disciplines like the history, sociology, and philosophy of science. If I have a suggestion to make in regard to science, analogous to the one I put forward with respect to Buddhism, therefore, it is that representatives of these latter disciplines be brought into the discussion in a more consistent and self-conscious manner.

References

Austin, J. H. 1998. *Zen and the Brain: Toward an Understanding of Meditation and Consciousness.* Cambridge: MIT Press.

Barbour, I. 1990. *Religion in an Age of Science.* Gifford Lectures, vol. 1. San Francisco: Harper San Francisco.

Cabezón, J. I. 1992. *A Dose of Emptiness: An Annotated Translation of the sTong thun chen mo of Mkhas grub dGe legs dpal bzang.* Albany: State University of New York Press.

———— 1994. *Buddhism and Language: A Study of Indo-Tibetan Scholasticism.* Albany: State University of New York Press.

Capra, F. 1984 [1976]. *The Tao of Physics: An Exploration of the Parallels Between Modern Physics and Eastern Mysticism.* 2d rev. ed. Toronto: Bantam, 1984.

Clifford, J. and G. Marcus. 1986. *Writing Culture: The Poetics and Politics of Ethnography.* Berkeley: University of California Press.

Cooper, R. 1996. *The Evolving Mind: Buddhism, Biology, and Consciousness.* Birmingham: Windhorse.

DeCharms, C. 1998. *Two Views of the Mind: Abhidharma and Brain Science.* Ithaca: Snow Lion.

Du Pre, G. 1984. "The Buddhist Philosophy of Science." In B. P. Kirtisinghe, ed., *Buddhism and Science,* pp. 103–110. Delhi: Motilal Banarsidass.

———— 1984. "Buddhism and Psychology." In B. P. Kirtisinghe, ed., *Buddhism and Science,* pp. 111–118. Delhi: Motilal Banarsidass.

———— 1984. "Buddhism and Psychotherapy." In B. P. Kirtisinghe, ed., *Buddhism and Science,* pp. 97–102. Delhi: Motilal Banarsidass.

———— 1984. "Buddhism and Science." In B. P. Kirtisinghe, ed., *Buddhism and Science,* pp. 92–96. Delhi: Motilal Banarsidass.

———— 1984. "Science and the Skandhas." In B. P. Kirtisinghe, ed., *Buddhism and Science,* pp. 119–127. Delhi: Motilal Banarsidass.

———— 1984. "Science and the Way to Nirvāṇa." In B. P. Kirtisinghe, ed., *Buddhism and Science,* pp. 137–145. Delhi: Motilal Banarsidass.

———— 1984. "Science and the Wheel of Life." In B. P. Kirtisinghe, ed., *Buddhism and Science,* pp. 128–136. Delhi: Motilal Banarsidass.

———— 1984. "Scientific Buddhism." In B. P. Kirtisinghe, ed., *Buddhism and Science,* pp. 146–154. Delhi: Motilal Banarsidass.

Fenner, P. 1995. *Reasoning Into Reality: A System-Cybernetic Model and Therapeutic Interpretation of Buddhist Middle Path Analysis.* Boston: Wisdom.

Fields, R. 1981. *How the Swans Came to the Lake: A Narrative History of Buddhism in America.* Boston: Shambhala.

Goldstein, M. C. 1989. *A History of Modern Tibet: 1913–1951.* Berkeley: University of California Press.

Goleman, D. 1997. *Healing Emotions: Conversations with the Dalai Lama on Mindfulness, Emotions, and Health.* Boston: Shambhala.

Goleman, D. and R. A. F. Thurman. 1991. *Mind Science: An East-West Dialogue.* Boston: Wisdom.

Guenther, H. V. 1984. *Matrix of Mystery: Scientific and Humanistic Aspects of rDzogs-chen Thought.* Boston: Shambhala.

Harding, S. 1998. *Is Science Multi-Cultural: Postcolonialisms, Feminisms, and Epistemologies*. Bloomington: Indiana University Press.

Hayes, R. P. 1996. "The Perception of 'Karma-Free' Cyberzones." *Cybersangha*. www.hooked.net/csangha/hayessu95.htm.

Hayward, J. W. 1984. *Perceiving Ordinary Magic: Science and Intuitive Wisdom*. Boston: Shambhala.

————— 1987. *Shifting Worlds, Changing Minds: Where the Sciences and Buddhism Meet*. Boston: Shambhala.

Hayward, J. and F. Varela. 1992. *Gentle Bridges: Conversations with the Dalai Lama on the Sciences of Mind*. Boston: Shambhala.

Hsuan Hua, Ven. "Electric Brains and Other Menaces." *Cybersangha*. www.hooked.net/csangha/huasu95.htm.

Jaki, S. 1974. *Science and Creation: From Eternal Cycles to an Oscillating Universe*. Dordrecht: Kluwer.

————— 1985. "Science and Religion." In M. Eliade, ed., *The Encyclopedia of Religion* 13:121–133. New York: Macmillan.

Jayatilleke, K. N. 1971. "Buddhism and the Scientific Revolution." In Nyanaponika Mahathera, ed., *Pathways of Buddhist Thought: Essays from the Wheel*. London: George Allen and Unwin. pp. 92–100;

————— 1980 [1963]. *Early Buddhist Theory of Knowledge*. Delhi: Motilal Banarsidass.

Jones, C. B. 1997. "Review of Robin Cooper, The Evolving Mind." *Journal of Buddhist Ethics*, vol. 4. jbe.la.psu.edu.

Kirtisinghe, B. P., ed. 1984. *Buddhism and Science*. Delhi: Motilal Banarsidass.

Lobsang Tenzing. 1990. "Biography of a Contemporary Yogi." *Cho Yang* 3:102–111.

Lopez, D. S. Jr. 1996. "Polemical Literature (dGag lan)." In José I. Cabezón and Roger R. Jackson, eds., *Tibetan Literature: Studies in Genre*. Ithaca: Snow Lion.

Mansfield, V. 1995–1996. "Time in Madhyamika Buddhism and Modern Physics." *Pacific World: Journal of the Institute of Buddhist Studies*, vols. 11–12.

————— 2002. "Time and Impermanence in Middle Way Buddhism and Modern Physics." This volume.

Needham, J. 1981. *Science in Traditional China: A Comparative Perspective*. Cambridge: Harvard University Press.

Nyanaponika Mahathera, ed. 1971. *Pathways of Buddhist Thought: Essays from the Wheel*. London: George Allen and Unwin.

Olcott, H. S. 1889. *A Buddhist Catechism According to the Sinhalese Canon*. London: Allen, Scott.

Ornstein, R. E. 1972. *The Psychology of Consciousness*. San Francisco: Freeman.

Poussin, L. de la Vallée. 1988, 1989. *Abhidharmakośabhāṣyam: A Work of Vasubandhu*. Vols. 2–3. Trans. Leo M. Pruden. Berkeley: Asian Humanities.

Proudfoot, W. 1985. *Religious Experience*. Berkeley: University of California Press.

Rabinow, P. 1996. *Essays on the Anthropology of Reason*. Princeton: Princeton University Press.

Rolston, H. 1987. *Science and Religion: A Critical Survey*. New York: Random House.

Singhal, D. P. 1972. *India and World Civilizations*. Vol. 1. London: Sidgwick and Jackson.

Stoddard, H. 1985. *Le Mendicant d'Amdo*. Recherches sur la Haute Asie 9. Paris: Societé d'Ethnographie.

Subhadra, B. 1970 [1890]. *A Buddhist Catechism: An Introduction to the Teachings of the Buddha Gotama*. Trans. C. T. Strauss. Kandy: Buddhist Publication Society.

Thurman, R. A. F. 1998. *Inner Revolution: Life, Liberty, and the Pursuit of Real Happiness*. New York: Riverhead.

Tweed, T. A. 1992. *The American Encounter with Buddhism, 1844–1912: Victorian Culture and the Limits of Dissent*. Bloomington: Indiana University Press.

Varela, F. J. 1997. *Sleeping, Dreaming, and Dying: An Exploration of Consciousness with the Dalai Lama*. Boston: Wisdom.

Varela, F. J., E. Thompson and E. Rosch. 1991. *The Embodied Mind: Cognitive Science and Human Experience*. Cambridge: MIT Press.

Wallace, B. A. 1996. *Choosing Reality: A Buddhist View of Physics and the Mind*. Ithaca: Snow Lion.

Weber, R. 1986. *Dialogues with Scientists and Sages: The Search for Unity*. London and New York: Routledge and Kegan Paul.

Wilber, K., J. Engler, and D. P. Brown. 1986. *Transformations of Consciousness: Conventional and Contemplative Perspectives on Development*. Boston: Shambhala.

Yin Tak, Ven. 1998. "Buddhism and Science." tunglinkok.ca/9702/sources/sci.htm.

Zoysa, A. P. K. 1998. "Buddhism and Science: A Conversation." www.lanka.com/dhamma/science 1.htm.

In this essay Buddhologist Thupten Jinpa traces the recent history of the engagement between Tibetan Buddhism and modern science, drawing primarily from Tibetan sources. Among all Asian Buddhist civilizations, Tibet remained the most isolated from the West until the mid-twentieth century. So when Tibetans finally encountered modern science it was particularly alien to them, and their varied responses, as depicted in this essay, are particularly fascinating. Thupten Jinpa begins his essay with a discussion of the writings in the 1930s or early 1940s by the Tibetan Buddhist scholar Gendün Chöphel on the significance of science for Buddhist thinking. Gendün Chöphel's account of the relation between Buddhism and science is both historical and normative, as he counsels his countrymen on the importance of engaging constructively with scientific modes of inquiry and knowledge.

Thupten Jinpa then examines the long-standing interest of the present Dalai Lama in science and technology and his active engagement with scientists and promotion of science education for Tibetans. This discussion is followed by a presentation of three diverse conceptions of science among recent Tibetan Buddhist thinkers. One group of traditional Tibetan scholars views modern scientific thought as a rival philosophy, to be refuted on logical grounds wherever it conflicts with traditional Buddhist thought. This view corresponds closely to the first of three models presented by Cabezón, namely, that of conflict/ambivalence. According to Jinpa, a second group of Tibetan intellectuals views science as an ally to Buddhism, and such scholars are eager to see science validate Buddhist principles, while maintaining that Buddhism is, after all, superior to all mundane sciences. This view corresponds to the second model presented by Cabezón, namely, that of compatibility/identity. A third group of Tibetan Buddhist scholars, represented most prominently by H. H. the Dalai Lama, regards science as an equal partner to Buddhism. Advocates of this view, which corresponds to the model of complementarity discussed by Cabezón, assume that a critical engagement between the two disciplines could expand the horizons of common human knowledge, thus giving rise to a more comprehensive understanding of human existence and the world we inhabit. One important feature of this approach is its deep respect for the integrity of both Buddhism and science, so that there is no urge to reduce one to the other.

Jinpa concludes this insightful essay with his own views regarding the most fruitful engagement between science and classical Buddhist thought. He argues that, among all the religions of the world, Buddhism may be best suited for critical dialogue with science, because of its suspicion of any notions of absolutes, its insistence on belief based on understanding, its empiricist philosophical orientation, its minute analysis of the nature of mind and its various modalities, and its overwhelming emphasis on knowledge gained through personal experience.

Thupten Jinpa

Science As an Ally or a Rival Philosophy? Tibetan Buddhist Thinkers' Engagement with Modern Science

THE FIRST SOUNDINGS: GENDÜN CHÖPHEL (1903–1951)[1]

Around the end of his twelve-year travel through central Asia, India, and Sri Lanka, the Tibetan philosopher and historian Gendün Chöphel wrote a passionate oeuvre appealing to his fellow Tibetan thinkers to engage positively with modern science. The piece opens with the following lines: "Now, I shall, from the depth of my heart, make the following suggestion to my colleagues—who belong to the same spiritual community as myself—those who are objective and far-sighted."[2] This was back in the 1930s or early 1940s. What follows is a brief attempt to trace the trajectory of the path of Tibetan thinkers' response to this over a period of more than five decades.

Gendün Chöphel's oeuvre actually marks the conclusion of a sixteen-part journal, which was based on the author's observations, insights, and experiences accumulated during his travels through various countries, peoples, and cultures. From the tone of his journal it is evident that the author realized, not surprisingly from early on, that his travel experiences could potentially have a transforming impact on the Tibetan intellectual world. The knowledge that Tibet has, for whatever reasons, chosen to remain isolated from the outside world seems to have weighed heavily on Gendün Chöphel's mind.[3] Yet he was also quick to realize that sooner or later Tibet would be forced to confront the realities of the modern world. It is in this context that the author saw his work as that of a reconnaissance, that is, reporting back to his colleagues his personal observations and assessment of what he had witnessed. Gendün Chöphel's impressive academic background made him an ideal candidate. Like many of his Tibetan colleagues, his primary education was that of a rigorous philosophical training in classical Buddhist thought, especially epistemology, phenomenology, and metaphysics. In addition, he had a thorough grounding in classical Tibetan literary studies, which included among others poetry, linguistics, and grammar.

Gendün Chöphel's piece begins by acknowledging the basic fact of science's pervasive influence in the contemporary world. It reports that in many countries the critics of modern science, some learned and some foolish, were silenced as they ran out of any valid arguments against the scientific worldview. "Even the Indian Brahmins," the author observes, "who regard the literal truth of scriptures dearer even than their own life, were eventually compelled to accept [modern science]."[4] The author then relates the experience of Christianity's initial encounters with modern science. He observes that even with strong alliances with the ruling power such as the monarchy, and despite its sustained persecution of scientists, which included excommunication, imprisonment, and at times burning them alive, the Church was eventually compelled to accept the validity of science. In fact, the author writes, "They felt the need to articulate their own religious beliefs within the framework of the 'new knowledge' even when the ideas were [often] incompatible."[5] Gendün Chöphel concludes this section of his work with the observation that opposition to reason is most unfortunate. To underline this point to his Tibetan colleagues, the author cites the following quotation from the seventh-century Buddhist epistemologist Dharmakīrti, an Indian philosopher whose work is admired greatly in Tibet:

> The nature of things cannot be canceled
> Through means of falsity, even if attempted.
> The mind will [eventually] uphold that [truth].[6]

The author then relates the experience of a culture that is closer to home, namely, the Tibetan Buddhist culture of Mongolia. For Gendün Chöphel the example of Mongolia is deeply disturbing and confirms his view that blind faith to one's tradition can in fact be the seed of its own demise. He therefore suggests that negative opposition to science, on the one hand, and complete rejection of religion, on the other, are both equally extreme reactions to this new discipline. In contrast, Gendün Chöphel sees the Sri Lankan Buddhists' encounter with modern science as holding greater promise. He places important significance in the fact that there were important Western converts to Buddhism. The author singles out in particular the example of the German monk who became known as Nyanatiloka (Gendün Chöphel writes the name in its Sanskrit form as Trilokjñāna).[7] He accredits this monk with drawing a distinction between the "religion of reason" and the "religion of faith," Buddhism being the former.[8] From his remarks it is obvious that Gendün Chöphel deeply admired the Sri Lankan Buddhist renaissance, which historians today describe as Buddhist modernism.

So far as Gendün Chöphel's own views on the possible convergence between Buddhism and modern science are concerned, he makes explicit references to four areas. On the whole he seems to share the basic sentiments of the Sri Lankan Buddhist modernists. They see that modern science, especially with its development and application of powerful instruments, could provide valuable confirmation of many of the key insights of Buddhism. 1. For example, Gendün Chöphel sees the modern scientific understanding of matter as dynamic energy, possibly a reference to Einstein's equation $e = mc^2$, to provide powerful empirical confirmation of the fundamental Buddhist insight on the ever fluctuating, impermanent nature of things. He marvels at the invention of X-ray machines a decade earlier, the telephone machine that enables conversation between one person in India and the other as far away as China, and cinematography, which allows us to record moving images! For Gendün Chöphel these technologies give us what he calls the ultimate proofs of the matter-energy equation principle. This scientific principle also resonates, according to the Tibetan author, with Dharmakīrti's assertion that what we regard as "continua" and "composites" do not possess in actual fact any intrinsic objective reality. Rather, they are constructs of our minds.

2. Another area where Gendün Chöphel sees a profound convergence is the concept of relativity. He expresses amazement at the assumption that it had been only around five decades since it was understood there is no independent color spectrum called white except in relation to its contrasts. This idea of relativity and the rejection of any notion of absolutes that it entails are, Gendün Chöphel argues, at the heart of Nāgārjuna's (c. 150–250 C.E.) philosophy of the Middle Way.[9] 3. A further topic on which the author comments is the modern scientific insights into the deeply subjective and con-

tingent nature of our perception of the external world. According to this, what we perceive is, to a large extent, determined by our sensory faculties and the representations they produce. There is no such thing as seeing an object nakedly without any mediation. Gendün Chöphel points out that these ideas are very familiar to the Tibetan ear. 4. Finally, he makes a passing observation that the modern scientific neurological understanding of the human nervous system, which is based on empirical observation, shares surprising similarities with the human physiology explained in the Vajrayāna literature of the Highest Yoga class.[10]

Even though he sees such areas of possible convergence between Buddhist ideas and modern science, Gendün Chöphel warns against a dogmatic approach based on the false assumption that nothing found in the classical texts may be undermined. He argues that such an attitude may seem heroic but will not take us far. Interestingly, as an example of a possible challenge to established Tibetan Buddhist ideas, he cites the debate on whether or not trees are sentient. This is an old dispute between Buddhist and non-Buddhist epistemologists in ancient India. Gendün Chöphel suggests that some scientific experiments may be seen as providing greater weight to the non-Buddhist's claim that some trees are in fact sentient![11] For Gendün Chöphel, perhaps the greatest strength of modern science lies in its means of acquiring knowledge of the physical world. By this I am referring, of course, to modern science's predominant emphasis on empirical evidence as a basis for understanding. He makes this point in the following passage:

> The standpoints of this "new discipline" [science] are not formulated simply on the basis of refuting one person's views with those of another. With telescopes developed through new technology one can see [objects] up to a thousand *paktses*[12] as if they were on the palm of one's hand. Similarly, there are magnifying lenses that enable one to see small particles as if they were the size of a mountain, thus allowing one to observe their various attributes. These are empirical facts that are evident to all, against which, unless one closes one's eyes, there is simply no other way [but to accept the results].[13]

Gendün Chöphel concludes his observation with a powerful appeal. He requests his Tibetan colleagues not to misconstrue him as thinking that he is being gullible. He assures his readers that not only is his intelligence sharp but his concern for the survival of the Buddha's teachings is not less than that of his colleagues.[14] Thus he appeals to his colleagues not to waste time on contemplating how to refute his views. Rather, he warns, if they do not wish to see the tree of Buddha's teachings and its roots, the "inner science" (i.e., Buddhist philosophical thought), to be undermined, they should adopt a much wider perspective. They should, therefore, endeavor to cultivate the ability to recognize what is of primary and what is of secondary importance and adopt a standpoint of confidence and trust. And in this way, he suggests, Tibetan thinkers should strive in the means to ensure the continuation of the Buddha's teachings hand in hand with the ways of the new discipline called science.

It is extremely difficult at this stage in our historical knowledge to determine what impact, if any, Gendün Chöphel's oeuvre had on his Tibetan colleagues. To begin with, it is unclear whether his journals were ever published during his life. In fact, if we are to believe Horkhang's words, the Tibetan editor of the current edition of the journals, it appears that the entire sixteen-part series of the journal remained unpublished until after the author's death.[15] We do know, however, that Gendün Chöphel acquired a reputation for being rather radical in his thinking; some even went to the extent of accusing him of being a hardened skeptic! We also know that, after his return, Gendün Chöphel was mistreated by the Tibetan political establishment, which included his imprisonment on grounds of suspicion of Communist Party membership. This situation was further complicated by the thinker's own notorious personal life, with allegations of alcohol abuse and sexual perversions. All this obviously did not help in his mission to "awaken" his Tibetan colleagues to the new worldview. In fact, after his return from his landmark travels, the author seems to have forgotten about his enthusiastic appeal to his colleagues concerning the need to engage with modern science. On the whole, Gendün Chöphel appears to have devoted his time more to the modernization of the political system.[16] Whatever the case may be, it is a tragedy that before his words could exert any real effect the country became caught in a political upheaval that was to threaten the very survival of the Tibetan people and its culture.

THE SECOND SOUNDINGS: H. H. THE FOURTEENTH DALAI LAMA

The second Tibetan to play a critical role in this encounter between Tibetan Buddhism and modern science is the present Dalai Lama. We know from his autobiography that His Holiness early on developed a profound fascination for science and technology in particular. The young Dalai Lama's restoration of the old cinema projector, his love for telescopes, and his fascination with mending watches have today become popular knowledge thanks to Hollywood's two recent films on Tibet and the Dalai Lama's life.[17] Interestingly, the Dalai Lama's interest in modern science appears to have arisen independently of Gendün Chöphel's writings. In fact, the Dalai Lama read Gendün Chöphel's journals only when they were printed in India, more than four decades after they were written![18] Perhaps the turning points in the Dalai Lama's attitude toward modern science and technology may have come during his state visits to China in 1955 and to India in 1956–1957. The impact of powerful industries and modern telecommunication and transportation on the young Dalai Lama's mind cannot be overestimated. In any case, following his flight to India in 1959, and after the Tibetan refugee community became reasonably well-established in their chosen second home, namely, India, the Dalai Lama was once again able to pursue his interests in modern science. This time, however, he had access to practicing scientists, including some of the best scientific minds of our time. In particular, he initiated a series of informal discussions with the physicist David Bohm on quantum mechanics. And, since 1987, the Mind and Life conferences, which take place every two years at the Dalai Lama's residence, have given him an ideal forum to keep him abreast of the latest developments in the various scientific disciplines.[19]

The Dalai Lama hoped that, with the introduction of modern secular education in Tibetan society, and with the establishment of research and translation units in the education department, gradually publications on modern science might appear in Tibetan. This, he felt, could then open the way for a critical engagement between classical Tibetan scholars and modern science. Unfortunately this expectation proved to be misplaced. So in the early eighties the Dalai Lama began speaking in public, especially at major Buddhist monastic colleges, about the need to introduce studies of modern science and Western philosophy into the monastic curriculum. He suggested that this could lead to a mutually enriching dialogue between classical Buddhist philosophy and contemporary thought, including modern science. Interestingly, the Dalai Lama does not see the introduction of modern science within the classical curriculum as something radical. Rather, he sees this more as a matter of updating the curriculum. He argues that the study of physics updates the student's understanding of the nature of the physical universe, cosmology as updating classical Buddhist cosmology, biology as the study of life and consciousness, and psychology as updating the study of classical Abhidharma psychology and phenomenology.[20]

Like Gendün Chöphel, the Dalai Lama sees that the strength of modern science lies in its overwhelming reliance on empirical evidence. For a Buddhist thinker, he feels that any evidence that is grounded in empirical facts and can be experimentally demonstrated cannot be dismissed. Thus, the Dalai Lama argues, there is a fundamental convergence between what is called the *scientific method*, which consists of observation, reason, and experiment, and the Buddhist method of inquiry, which emphasizes the development of understanding derived through observation and critical analysis. By drawing attention to this key convergence of methodology, the Dalai Lama has warned Tibetan thinkers of the need for openness, especially with regard to any possible challenges scientific discoveries may pose to established ideas within the classical Buddhist worldview. In particular, the Dalai Lama has pointed out the need to discard many aspects of the Abhidharma cosmology, especially its ideas about the size and distance from earth of such celestial bodies as the sun and the moon. He has drawn attention to the weight of empirical evidence against the classical Abhidharma theory, evidence that is derived through experiments and the use of powerful scientific instruments.[21] By stressing the empirical nature of the evidence, the Dalai Lama is invoking a principle that is very dear to Tibetan Buddhist thinkers.[22]

For example, the Dalai Lama writes,

Suppose that something is definitely proven through scientific investigation, that a certain hypothesis is verified or a certain fact emerges as a result of scientific investigation. And suppose, furthermore, that that fact is incompatible with Buddhist theory. There is no doubt that we must accept the result of the scientific research.[23]

The Dalai Lama, however, offers an important caveat. He argues that it is critical to understand the scope and application of the scientific method. By invoking an important methodological principle, first developed fully as a crucial principle by Tsongkhapa (1357–1419), the Dalai Lama underlines the need to distinguish between what is negated through scientific method and what has been not observed through such a method. In other words, he reminds us not to conflate the two processes of *not finding* something and *finding its nonexistence*.[24] For example, through current scientific analysis so far we may have not found evidence for rebirth, but this does not imply by any means that science has somehow *negated* the existence of rebirth.

This idea of the need to accept the parameters of scientific understanding is related to the Buddhist division of objects of knowledge into three categories of *evident, slightly obscure,* and *extremely obscure* phenomena. Knowledge of things and events of the first class is derived primarily through direct perception and experience. In contrast, knowledge of the second class of phenomena is derived through inference based on some empirically observed facts. Many of the key insights of Buddhism, such as the ever fluctuating, impermanent nature of things, belong to this category of phenomena. However, the facts of the third category are thought to be accessible to our minds only on the basis of some third person's testimony. We have, for the time being, simply no direct access to knowledge of the third class of phenomena. However, the status of an object or a phenomenon as extremely obscure in this third sense may be simply relational and not an inherent attribute of the thing itself. For example, knowledge of certain customs of an ancient civilization may remain totally inaccessible to us at one point. Yet, gradually, as the result of archeological study, that knowledge can become more and more accessible to us. Still, our knowledge of that civilization will be based on inference. In contrast, for that civilization's members the characteristics of their customs were evident.

In delineating the scope of current scientific knowledge, it is unclear whether the Dalai Lama believes that 1. phenomena that currently remain outside this knowledge such as rebirth do so by the very nature of their existence or 2. that, as the current scientific paradigm changes, the scope of scientific analysis will expand, thus enabling such phenomena to fall within the parameters of what we call scientific investigation.

The Dalai Lama envisions dual benefits from this engagement between classical Tibetan Buddhist thought and contemporary thought, especially modern science. On the one hand, he believes that a dynamic encounter with scientific thought could help revitalize Buddhist analysis of the nature of objective reality and the mind. History shows that Buddhist philosophy gained tremendous insights from its long engagement with other systems of thought in ancient India. For scientists the Dalai Lama believes that engagement with Buddhist philosophy could provide new perspectives on their own various disciplines. For example, he sees potentials for mutual enrichment in the scientific investigation of the chemical and neurophysiological changes that occur in meditative states. This, he believes, could lead to a more comprehensive understanding of the nature and relationship between the body and mind.[25] Similarly, he feels that Madhyamaka Buddhist philosophical ideas about identity derived through dependent origination, despite the absence of any intrinsic reality of phenomena, may help deal with many of the conceptual challenges posed by quantum physics with regard to the question of reality.[26] Thus the Dalai Lama sees great potential for exploring areas of convergence and divergence between Buddhism and four key disciplines of modern science, namely, physics (especially particle physics), cosmology, neurobiology, and psychology. The proceedings of the past Mind and Life meetings stand as a testimony to the potentials that exist in engaged dialogues between Buddhist thought and these four fields of science.[27]

Unfortunately, so far no written work in Tibetan from the Dalai Lama has been published that articulates his views on the potential areas of engagement between Buddhist thought and science. It is also unfortunate that Gendün Chöphel, too, shunned any attempt to write a separate treatise on the fundamentals of scientific thought in Tibetan. He appears to have been deterred for two reasons. On the one hand, he felt that the task was likely to prove too demanding. He also felt that perhaps the time wasn't right for the appearance of such a work.[28] However, the Dalai Lama's repeated encouragement has today led to an intellectual climate among the younger generation of Tibetan scholars, especially within the academic monasteries, where a genuine thirst for basic scientific knowledge is strong today.

DIVERSE CONCEPTIONS OF SCIENCE

Among Tibetan thinkers who have taken up the call of Gendün Chöphel and the Dalai Lama, their diverse approaches to the issue of dialogue with science suggest three very different conceptions of science. On one side are those who primarily see modern scientific thought as representing a rival philosophy. In their view Buddhist thinkers should treat modern science in exactly the same manner as rival philosophical systems in ancient India. In other words, Buddhist philosophers must engage in debates with science and negate whatever can be demonstrated to be false. Needless to say, here argumentation, based on appeal to shared principles of logic and epistemological theories, is regarded as the primary means of validation or repudiation. Interestingly, one of the favorite issues they select for their criticism of scientific thought is what they see as modern science's materialistic theory of mind. Their arguments are, in actual fact, a reproduction of the very arguments Buddhist epistemologists employed more than a millennium ago to refute the Cārvāka's materialistic theory of mind![29] On the whole, members of this group tend to be scholars who do not read any contemporary European language and lack a real understanding of the fundamentals of modern science.[30] They also tend to conflate scientific theories and metaphysical assumptions that a particular scientific group may uphold, and much of their criticisms are actually directed against metaphysical rather than truly scientific concepts. Furthermore, their appreciation of the empirical dimension of modern scientific method of inquiry remains inadequate. These shortcomings seem to result primarily because of their lack of any access to scientific literature.

There is, however, a second group of Tibetan scholars who tend to view modern science as an ally. A primary example of this is Gendün Chöphel himself, who, as we observed earlier, wished to see some kind of Buddhist modernist movement in Tibet. This group also includes some Tibetan writers who either remained or grew up under Communist Chinese rule. Particular mention should be made of Chukye Samten. In the preface to his book on classical Tibetan logic, the author explicitly cites large segments of Gendün Chöphel's journal and identifies strongly with the sentiments expressed by Gendün Chöphel.[31] Like certain Sri Lankan Buddhist scholars, there is also an element of self-congratulation in that he cites what the author sees as the endorsement of Buddhism by well-known scientists. Gendün Chöphel too cites the statement attributed to a Sri Lankan monk that Buddhism will not only be able to keep pace with modern science but will also, in fact, outpace it, for Buddhism can go beyond the reaches of modern science.[32] Perhaps the most interesting part of Chukye Samten's engagement with science is his attempt to demonstrate the scientific nature of the key elements of the Tibetan classical discipline of *Düra* (lit. "Collected Topics"). He begins by first defining science as a discipline of reason developed in relation to our knowledge of both the material world and sentient beings, knowledge that is derived through observation, experience, and inference.[33] The main difference between Gendün Chöphel's attitude to science and later writers like Chukye Samten lies in their basic orientation. While Gendün Chöphel is interested primarily in opening his Tibetan colleagues' eyes so that they are prepared to confront the influence of modern science, Chukye Samten's approach is that of an apologist. This is understandable, given the radically different cultural and political conditions under which these two authors were writing.

Finally, we could identify a third group of Tibetan thinkers whose conception of science is that of an equal partner. In their view, although modern science and Buddhism may share many points of convergence, there is a clear awareness that certain topics within Buddhism may not fall within the scope of scientific analysis. An assumption exists that critical engagement between the two disciplines could expand the horizons of common human knowledge, thus giving rise to a more comprehensive understanding of ourselves and the world we inhabit. One important feature of this approach is its deep respect for the integrity of each discipline, so that there is no urge to reduce one to the other. Because of this, among the last group there is a greater interest to see the emergence of scientific literature in Tibetan rather than work comparing the two disciplines.[34] Those who belong to this group envision what could be called a critical engagement between classical Buddhist thought and modern scientific thought. The present Dalai Lama is a principal example.[35] Among the younger generation there have been students in all three main Gelu monastic universities, Sera, Drepung, and Ganden, albeit a handful, who have taken serious interest in modern science and Western philosophy. So far, this engagement with science by classical Tibetan scholars seems to be confined to the Gelu monastic centers of learning. This may be because historically the scholars of the Gelu monastic colleges tend to share deeper interests in epistemological and metaphysical issues. The author of this essay was one such student at Ganden, in south India. Not surprisingly, thinkers who share this perspective tend to be literate in English and therefore have access to materials outside the Tibetan language.

AS THE TIBETAN EXILE COMMUNITY becomes more exposed to the outside world, classical Tibetan thinkers become more aware of the pervasive influence of science in contemporary culture throughout the world. They may therefore become aware of the fact that many people in contemporary society derive their understanding of the world through science rather than religious belief. Furthermore, it will be realized that the social, cultural, and economic conditions that influence and shape individuals' lives in contemporary society naturally encourage a secularization process. History illustrates that advancement in scientific knowledge goes hand in hand with the modernization and secularization of a society. This is likely to be the case with the Tibetans as well. Today for the first time in history there is an entire generation of highly educated Tibetans whose primary educational background is not that of the classical monastic system. In the political realm the Dalai Lama himself is at the forefront of democratizing Tibetan society, thus secularizing the Tibetan government system. Gradually the impact of all of these changes will be felt.

After more than four decades into their diaspora, today the question is no longer the introduction of scientific knowledge to the Tibetans. There are already many young Tibetans with a conventional, secular educational background that includes the study of modern science. There are also individual Tibetans who are at home with even specialized fields of scientific studies. However, this situation has not really contributed to facilitating a critical engagement between modern science and classical Tibetan thought. Nor has it led to the emergence of any significant scientific literature in Tibetan. The challenge, therefore, is the introduction of scientific studies within the curriculum of Buddhist academic monasteries, where an established tradition of logical, speculative, and philosophical studies has existed for centuries. Until such time, modern science will fail to have any direct, significant impact on the classical Tibetan Buddhist worldview. As in many cultures today, the worldviews of the two disciplines will remain at best two parallel, and perhaps competing, descriptions of our own reality and the natural world.

The key to the successful introduction of scientific studies into the classical Tibetan monastic curriculum, however, is the emergence of scientific literature in Tibetan. Not only should there be textbooks on science and contemporary Western philosophy in Tibetan, more important, there should be books that bring fresh critical perspectives to some of the fundamental ideas of Buddhist thought. This latter class of literature is critical if we are to inspire serious interest in scientific thought among Tibetan thinkers. For example, an exposition of the basic concepts of the foundations of quantum mechanics, especially its idea of nonlocality, can be presented in a way that poses challenges to the Buddhist philosophical concept of causation. Similarly, some of the key disciplines of modern cognitive science, such as neurobiology, can be seen as challenging classical Buddhist epistemological theories of perception and cognition. It is also not difficult to envision how a detailed presentation of the premises of a materialist theory of mind written in classical Tibetan philosophical language will excite the monastic scholarly community. The emergence of such scientific and philosophical literature in Tibetan will no doubt revitalize and enrich philosophical discourse within Tibetan Buddhist thought. Only then can we expect to see a meaningful engagement with modern science by Tibetan Buddhist thinkers.

It may well be that of all religions Buddhism finds it easiest to engage in a critical dialogue with science. The following key features of Buddhism—its suspicion of any notion of absolutes, its insistence on belief based on understanding, its empiricist philosophical orientation, its minute analysis of the nature of mind and its various modalities, and its overwhelming emphasis on knowledge gained through personal experience—all make it easy for Buddhism to be in a dialogue with a system of thought that emphasizes empirical evidence as the key means of acquiring knowledge.

Notes

1. My dating is based on Horkhang Sonam Pelbar, who edited the *Writings of Gendün Chöphel*, 3 vols. (Tib.; Xinhua: Tibet's Old Texts Publishing House, 1990). Heather Stoddard, however, gives the dates as 1905–1951; see her *L'Mediant de L'Amdo* (Paris: Société D'Ethnologie, 1986). All publications in the Tibetan language are marked with the abbreviation Tib. and, unless otherwise stated, all translations from Tibetan are mine.
2. *Writings of Gendün Chöphel* 2:166.
3. It is difficult at this stage to fully appreciate the complex historical, social, political, and cultural conditions that may have led the Tibetans to isolate themselves from the rest of the world.
4. Ibid.
5. Ibid., p. 167.
6. *Pramāṇvārttikā*, chapter 1:213. Reprinted in typeset (Tib.; Mundgod: Drepung Loseling Library Society, 1987).
7. Nyanatiloka was a monk from Dodanduwa Island and the author of *Guide Through the Abhidhamma-pitaka*, published in 1938. He is known to have also translated the well-known Theravada text *Visuddhimagga* into German. See "Buddhist Studies in Recent Times," in B. V. Bapat, ed., *2,500 Years of Buddhism* (New Delhi: Ministry of Information and Broadcasting, Government of India, 1956), p. 373.
8. *Writings of Gendün Chöphel* 2:168.
9. Ibid., p. 170.
10. Ibid., p. 171.
11. Ibid., p. 172.
12. This is a unit of distance measurement known as *yojana* in Sanskrit.
13. *Writings of Gendün Chöphel* 2:167.
14. In explicitly stating his loyalty to the Buddhist faith, Gendün Chöphel may be attempting to preempt any potential negative reaction to his promotion of the study of science from the monastic establishment. Being a historian, the author must be keenly aware of the negative response to the thirteenth Dalai Lama's attempts at modernization in Tibet. For a detailed account of

this failed initiative, see Melvin Golstein, *A History of Modern Tibet, 1913–1951* (Berkeley: University of California, 1989), chapters 5 and 12.

15. See Horkhang's remarks in his foreword to *Writings of Gendün Chöphel*, vol. 1.

16. See, for example, Golstein, *A History of Modern Tibet*, pp. 452–463.

17. *Seven Years in Tibet* and *Kundun*, both released in 1998.

18. Personal conversation. Some sections of the journal appear to have first been published in India in the 1970s by the Higher Institute of Tibetan Studies, Sarnath, India.

19. These dialogues between the Dalai Lama and scientists from different fields take place every two years and are organized by the Mind and Life Institute, Boulder Creek, California.

20. Unpublished comments by the Dalai Lama.

21. See, for example, the Dalai Lama, *Policy of Kindness*, ed. Sydney Piburn (Ithaca: Snow Lion, 1990), p. 68.

22. For example, Zemey Rinpoche (1926–1996), a well-known scholar monk from Ganden Monastic University, wrote a short piece in the 1970s on the need to accept the modern cosmological view of earth as spherical. See Geshe Thupten Jinpa, ed., *Selected Writings of Kyabje Zemey Rinpoche* (Tib.; Mundgod: Tashi Gephel House, 1997), pp. 205–207. Again, like the Dalai Lama, Rinpoche places great weight on the empirical nature of the modern cosmological description.

23. Dalai Lama, *A Policy of Kindness*, p. 67.

24. Ibid., p. 69.

25. Dalai Lama, Herbert Benson, Robert A. F. Thurman, Howard E. Gardner, and Daniel Goleman. *Mind Science: An East-West Dialogue* (Boston: Wisdom, 1991), p. 18.

26. See, for example, the Dalai Lama's *The Four Noble Truths* (London: Thorsons, 1997), p. 102.

27. See J. W. Hayward and F. J. Varela, eds., *Gentle Bridges* (Boston: Shambhala, 1992), from Mind and Life 1; Daniel Goleman, ed., *Healing Emotions* (Boston: Shambhala, 1997), from Mind and Life 3; F. J. Varela, ed., *Sleeping, Dreaming, and Dying* (Boston: Wisdom, 1997), Mind and Life 4; and Z. Houshmand, R. Livinston, and B. A. Wallace, eds., *Consciousness at the Crossroads* (Ithaca: Snow Lion, 1999).

28. *Writings of Gendün Chöphel* 2:172.

29. See, for example, Geshe Lharampa Gashar Könchok Tsering, *A Mirror Reflecting the Nature of Mind* (Tib.; Mundgod: Drepung Loseling Library Press, 1983), pp. 67, 150. See also Geshe Yeshe Wangchuk, *Philosophical Tenets Attracting the Hearts of the Learned* (Tib.; Bylakuppe Sermey Monastery: Sermey, 1986), pp. 15–17.

30. By referring to these as groups I do not mean to suggest in any way that there exists some kind of self-conscious affiliation among the Tibetan thinkers who engage with modern science.

31. See Chukye Samten, *An Easy Road for the Intelligent on the Path of Reasoning* (Tib.; Xining: Qingai Minorities Press, 1996).
32. Cited in *Writings of Gendün Chöphel* 2:169.
33. Samten, *An Easy Road for the Intelligent*, p. 24.
34. For example, the Dalai Lama has suggested to this author that he write a book in Tibetan that would faithfully present the key arguments of the Western materialist philosopher's standpoint of the nature of mind.
35. The Drepung Loseling scholar Geshe Wangchen has written a book on the Buddhist and scientific understanding of the nature of life and sentience entitled *A Conversation on the Luminous and Knowing Mind* (Tib.; Mundgod: Drepung Loseling Library, 1991).

PART 2

Buddhism and the Cognitive Sciences

Among the many possible areas of fruitful interface between Buddhism and science, certainly none is more central than the nature of the mind and the possibility of positive mental transformation. These are the topics of this essay by His Holiness the Dalai Lama, who emphasizes the potential, mutual benefits of dialogue and collaboration between Buddhists and cognitive scientists. The Dalai Lama begins his presentation with a discussion of the relevance of the mind to each of the Four Noble Truths, which form the foundation of Buddhism as a whole. While the mind is certainly of primary importance in the Buddhist inquiry into the nature of reality, this tradition does not confine itself solely to subjective states of consciousness but is also very concerned with the nature of objective phenomena and their relation to consciousness. As the Dalai Lama points out, the very criterion for determining what does and does not exist is valid cognition, and the correct apprehension of the way things are is also pivotal for the cultivation of wholesome states of mind.

While Buddhism is commonly classified in the West as a religion, the Dalai Lama emphasizes here its concern with careful observation and rational analysis as opposed to reliance upon faith alone. Moreover, both Buddhism and science place great importance upon the need for objectivity, in the sense of freedom from subjective biases, in one's exploration of the natural world. In Buddhism the pursuit of knowledge is fundamentally pragmatic: one seeks knowledge in order to transform the mind and achieve freedom from suffering and its source for oneself and others. Such transformation, according to the Dalai Lama, can be brought about only by the mind, not by technology or other external modes of intervention. Moreover, the very possibility of positive mental transformation, he argues, is due to the impermanent nature of the mind and the fact that all wholesome and unwholesome mental states arise from prior causes and conditions. Thus the Buddhist understanding of the mind and the possibility of mental transformation are at root naturalistic: the mind itself is an integral part of nature, and changes in the mind take place because of identifiable prior causes.

His Holiness the Fourteenth Dalai Lama
Understanding and Transforming the Mind

UNDERSTANDING THE MIND

The topic of this presentation is the mind, the essential nature of which is luminosity and cognizance.[1] In fact I feel there will be great value in long-term dialogue and collaboration between Buddhists and neurobiologists, those who are studying the nature and functioning of the brain. In this regard, topics for collaborative research and discussion might include the relationship between the body and mind and the ways in which memory operates. Another topic is the manner in which habitual propensities in the mind manifest in experience. Up till now I have been able to participate in dialogues with various groups of cognitive scientists on a number of occasions, and I have found my understanding increasing with each such opportunity. Both neuroscientists and Buddhists may benefit from such collaboration. I have derived benefit from these conversations, and the neuroscientists themselves also appear to have gained some fresh perspectives and ideas as a result of these dialogues.

Now I would like to address the nature of the mind and related issues as they are understood within Tibetan Buddhism. As is well known, the root, or foundation, of the whole of the Buddhist teachings is known as the Four Noble Truths. Among the Four Noble Truths the First Noble Truth, the

Truth of Suffering, addresses the nature of suffering. The reason for this is because we are averse to suffering, and this subject is taught in terms of feelings. Among the three types of suffering, the first, called *blatant suffering*, is that very feeling of pain or suffering itself. Second, that which is called the *suffering of change* is in fact a tainted feeling of pleasure. The third form of suffering, known as the *ubiquitous suffering of conditioning*, pertains to the feeling of indifference, which is neither pleasure nor pain. Now all these three types of suffering pertain to feeling as it is directly related to consciousness. So the First Noble Truth, the Truth of Suffering, has a deep relevance to the nature of consciousness.

The Second Noble Truth, the Truth of the Origin of Suffering, pertains to mental afflictions and to karma, or the actions induced by mental afflictions. A number of Buddhist schools assert that some voluntary karmas are in fact of a material nature. But, on the whole, Buddhist theory asserts that the nature of karma is a mental factor pertaining to volition. Therefore, karma, being of the nature of volition, is of the nature of consciousness. And mental afflictions are certainly expressions of consciousness as well.

As for the Third Noble Truth, the Noble Truth of Cessation, although cessation itself is not consciousness, it is an attribute of consciousness. The Fourth Noble Truth, the Truth of the Path to Cessation, involves excellent qualities of the mind, or of consciousness, specifically those qualities that lead to liberation. In terms of the presentation of *saṃsāra*, the cycle of existence, and *nirvāṇa*, liberation, if the mind is not subdued, there is *saṃsāra*, and if the mind is subdued, there is *nirvāṇa*.

Given the tremendous importance of the mind, certain philosophical schools within Buddhism maintain that all phenomena are of the nature of the mind. They maintain that external objects—in the sense of phenomena that are totally independent of the mind—do not exist. But the most predominant philosophical school within Tibetan Buddhism does not take this position.[2] Rather, it says that physical, external entities, different in nature from the mind, do exist. In short, among Tibetan Buddhists there are some who deny the existence of eternal entities that are not of the nature of the mind, but, for the most part, Tibetan Buddhist philosophers do assert the existence of such external entities. There is a great deal of debate about this point.

Regarding the Buddhist classifications of the five psychophysical aggregates, the twelve sense-bases, and the eighteen constituents of reality, the mind is included among the twelve sense-bases and the eighteen constituents of reality.[3] Among the five psychophysical aggregates, the aggregates of feelings, recognition, and consciousness are all aspects of the mind. The aggregate of compositional factors includes both mental and nonmental phenomena. Thus, among the five aggregates, most are of the nature of consciousness. So if each of these aggregates could vote, those that are of the nature of the mind would win by a landslide! I should add that the fifth aggregate is the aggregate of form. So the five aggregates are form, feelings, recognition, compositional factors, and consciousness. Among those five, only one is completely nonmental, while feelings, recognition, and consciousness are of the nature of the mind, and compositional factors are of two sorts—some of the nature of consciousness and some not.

The Buddha said that if one trains the mind there is joy, and if the mind is undisciplined there is suffering. In this way the Buddha placed great emphasis on the mind. Thus, the basis that is to be purified is the mind. If it is trained, there is *nirvāṇa*, or liberation, and if it is not trained, one continues in the cycle of existence known as *saṣāra*. The principal things that must be purified are the contaminations of the mind, and these also are mental. That which purifies the mind are excellent qualities, or states, of the mind. The results of having purified the mind also consist of excellent qualities, or states, of the mind.

The fundamental criterion for determining what does and what does not exist hinges on whether or not something is apprehended by valid cognition. It is not sufficient for something to be merely cognized or merely to appear to the mind; rather, when the mind apprehends something, this cognition must be incapable of refutation. That is, when an object is apprehended by the mind, it must be incapable of being invalidated by some other sound knowledge. Thus the criterion for existence itself pertains to the mind, specifically to valid cognition. Therefore, some Westerners interested in Buddhism maintain that Buddhism is actually not a religion but a science of the mind. I think there are grounds for such a claim.

Now what is the nature of the mind? First, the Tibetan term for consciousness, *shepa* (*shes pa*), is actually a verb used in such expressions as "One knows," or "I know," so it indicates an activity. Thus one speaks of consciousness on the basis of the ability to know. In terms of the internal classifications of consciousness, we designate two categories of consciousness. The first of these is sensory consciousness, which has for its dominant contributing condition something physical. Second, there is mental consciousness, the dominant contributing condition of which is not physical.

Another classification distinguishes between the mind and mental factors. The mind apprehends the sheer presence, or nature, of its object, whereas mental factors apprehend specific attributes of the apprehended object.

The Vaibhāṣika school of Buddhist philosophy asserts that consciousness apprehends its object nakedly, or without mediation, implying the existence of "image-free" consciousness. In contrast, the Sautrāntika philosophical school and all of the higher philosophical systems (namely, the Yogācāra and Madhyamaka schools) assert that consciousness apprehends its object by way of images. Therefore, they state that consciousness arises with images.

Another classification is made in terms of conceptual and nonconceptual cognition. Conceptual cognitions apprehend their objects by way of generic ideas, whereas nonconceptual cognitions, such as perception, experientially apprehend their objects more directly, which is to say, not by way of generic ideas.

In terms of the ways in which consciousness apprehends an object, first of all there is false cognition, which simply misapprehends its object. It is totally mistaken. Second, there is doubt, or uncertainty, in which cognition waivers between two options. Then there is belief, which is simply an opinion, without any compelling rational or empirical basis. Next, there is inference based upon conclusive reasons or evidence. And, finally, there is perception, which apprehends its object experientially. So we have many types of cognition. It is extremely important to distinguish between mistaken cognition and valid cognition.

For the most part, those types of cognition that lead to suffering are mistaken cognitions, which do not accord with reality. Many states of consciousness that lead to suffering are out of accord with reality and are mistaken. The remedies for those states of consciousness are valid cognitions that do accord with reality. So it is very important to investigate the distinction between cognitions that are delusive and those that are accurate. How is this to be done? Both mistaken and valid cognitions are alike insofar as they both do exist, both arise and are experienced. Now our task is to investigate those that are and are not mistaken. This needs to be done with reference to reality, to those phenomena that are apprehended by the mind.

The question of the relationship between reality and appearances arises everywhere, for there can be a disparity between how things appear and how they exist. This must be examined closely. In light of the importance of investigating the nature of reality and not simply relying on appearances, within the context of the Buddha's own teachings, it is also crucial to investigate rationally whether or not a certain teaching is to be taken literally.

Such investigation is to be done with the mind, of course, and not simply with the instruments of technology. In order to counteract a completely mistaken cognition, one pursues logical consequences in order to bring about valid inference or one may use conclusive syllogisms. Syllogisms entail reasonings, sometimes used to affirm the existence of a given entity or the validity of a given proposition and sometimes to refute the existence of something or to show the fallacy of a certain proposition.

That is, at times one may infer the *existence* of a given entity, and sometimes the *nonexistence* of something may be inferred. Given that twofold distinction, the syllogisms are sometimes negative in the sense that they demonstrate the absence of something, and sometimes they are affirmative in the sense of affirming the existence of the given object. Therefore, analysis is central to logical reasoning. Because of the centrality of logical analysis and investigation within Buddhist philosophy, I think there is a great potential for dialogue and collaboration between Buddhist philosophy and Western philosophy.

I have had conversations with some philosophers who have told me that according to some schools of thought the very existence of universals is refuted, for the distinction between universals and specifics is rejected. I have also heard there are others who deny the Law of the Excluded Middle. In Buddhism we assert that if one apprehends the opposite of an affirmative entity this refutes the existence of that entity. In contrast, it seems, in some philosophical systems, the Law of the Excluded Middle is not accepted. This is definitely a topic for further discussion and collaborative investigation. If there is disagreement between Buddhist and Western philosophers on this point, we don't want simply to conclude that they are different. Rather, we need to investigate the reasons why philosophers take the positions that they do. So this calls for further investigation. If, upon careful investigation, it turns out there are compelling reasons for dispensing with the Law of the Excluded Middle, this would call into question many of the pivotal reasonings within the Buddhist philosophy. In that case, one would have to reassess many Buddhist beliefs.

From a Buddhist perspective the reason for engaging in such investigation is not simply to gain greater knowledge about the world. Rather, our goal is to bring about a transformation in the mind. This doesn't occur simply by prayer or by wishing that the mind will change. The mind isn't transformed by that alone but rather by ascertaining various facets of reality. For example, if you have a certain assumption about reality, and you subject this assumption to investigation and consequently find evidence that invalidates your prior assumption, then the more you focus on this evidence the more the previous assumption will decrease in power, and the power of your fresh insight will increase. Thus, most good qualities of the mind accord with reality, which is to say, they are reasonable. They are grounded upon sound evidence. The mind is transformed when one ascertains and thoroughly acquaints oneself with fresh insights into the nature of reality that invalidate one's previous misconceptions or false assumptions.

For example, within Buddhism we speak of faith or confidence. If one's faith is based simply upon authority—because the assertion one believes was stated by an authoritative person or scripture—such faith is not very stable or reliable. In contrast, there is another type of faith that arises in dependence upon careful, sustained investigation. Such faith is based upon knowledge. Qualities such as faith and compassion that are to be nurtured as one follows one's spiritual path are to be cultivated on the basis of reasoning and knowledge. They are actually supported by wisdom, even though they themselves are not wisdom. By means of such investigation one's mistaken cognitions are decreased and one's valid cognitions are increased. On the other hand, it is legitimate to approach the study of Buddhism purely academically in order to increase one's erudition.

Within Buddhist Tantra, or Vajrayāna, there are classifications of different degrees of subtlety of consciousness. For example, there is a threefold classification of waking consciousness, dreaming consciousness, and the consciousness of dreamless sleep. All these are investigated. The state of consciousness when one has fainted is subtler than any of those. Finally, the subtlest form of consciousness occurs during the dying process. I believe that it would be very fruitful to investigate the relationship between the mind and brain in relation to these various degrees of subtlety of consciousness.

It may be more appropriate to speak of these subtler mental states as types of *potential* consciousness. It seems that accounts of these subtler states of mind do not refer to consciousness having a clearly apprehended object or to which some object appears and is discerned. When the more coarse forms of consciousness—the five sensory consciousnesses and mental consciousness—manifest, these subtler states of mind remain latent. But when the appropriate conditions or catalysts arise, these subtler states of mind may become manifest and fully conscious.

In Vajrayāna Buddhism the subtlest state of consciousness is known as *clear light*. In terms of categories of consciousness, there is one type of consciousness that consists of a permanent stream or an unending continuity and there are other forms of consciousness whose continuum comes to an end. Both these levels of consciousness—one consisting of an endless continuum and the other of a finite continuum—have a momentary nature. That is to say, they arise from moment to moment, and they are constantly in a state of flux. So the permanence of the first kind is only in terms of its continuum. The subtlest consciousness consists of such an eternal continuum, while the streams of the grosser states of consciousness do end.

Within Buddhist philosophy there is another point about which there is considerable debate. On the one hand, if one looks at a stream of moments of consciousness, it is asserted that one moment of consciousness may apprehend another preceding moment of consciousness. But Buddhist philosophers raise the further question as to whether it is possible for a single moment of consciousness to apprehend itself. There is a lot of discussion around and investigation into this point.

That is a general overview of Buddhist theories concerning the nature of the mind. As there are issues that have remained unresolved concerning the nature of the mind, after more than two thousand years of Buddhist investigation into these matters, I suspect that some of these may still remain unresolved even after they have been subjected to the methods of modern research. But, finally, whether we really solve these problems or not, I think in this life we should have a more open mind or warm heart. That is, I think, more practical or useful.

TRANSFORMING THE MIND

I regard all the major world religions, especially Buddhism, as instruments, or methods, for training the mind, for overcoming problems, primarily of the mind, specifically negative forces in our emotions that create mental unrest, unhappiness, fear, and frustration. Such mental states result in various negative activities that bring more problems and suffering. *Dharma* means an approach for overcoming these long-term problems, so it has the connotation of protecting, or saving, one from unwanted things. Therefore, Buddhadharma is a system of transforming, or disciplining, the mind to bring about inner tranquillity.

Mental transformation is achieved by using the mind itself, for there is nothing else we can use to bring about such transformation. Most mental states that bring us unhappiness are out of accord with reality. All false views and afflictive emotions trace back to a misapprehension of reality, which is to say that they all stem from fundamental ignorance, namely, self-grasping or grasping at self-existence. The antidote for such ignorance is to develop a way of viewing reality as it is. Recognizing that the mental states that produce suffering are based on unreality, one must then apply a remedy for such ignorance. In short, false views are dispelled by correct views.

To do that, we first have to see which mental states are faulty and which are beneficial, then we need to avoid the former and cultivate the latter. Otherwise, we would not have the desire to turn away from harmful mental thoughts and emotions or apply antidotes to counteract them. The essential point is that the mind can be transformed with the use of remedies only if we *want* to do so; no one can *force* us.

It is very important to learn to distinguish between positive and negative mental states and their origins on the basis of our own experience. It is not enough just to read about this subject and form intellectual opinions on that basis. That is all too easy. This type of inquiry must be based on experience, observing one's own emotions. There is a great difference between those two approaches. With some exceptions, mere intellectual understanding alone is not so hard, whereas knowing something experientially is difficult. Such knowledge is gained only with sustained, diligent effort, which results in a kind of "felt" experience. When such experience arises, it emerges together with powerful emotions that may be either destructive or constructive.

If human beings had no emotions at all, there would be no basis for survival, so emotions are very much a part of being human. Moreover, emotions have a very powerful impact over the course of a lifetime. One class of emotions not only brings not only mental pain but is also detrimental to one's physical health. Another class of emotions immediately gives rise to an inner strength, or mental fortitude, and, due to that, one's physical health is also enhanced. Thus wholesome emotions are evidently beneficial within the context of this life, without taking into consideration any specifically religious perspectives on emotions. Likewise, if one can attenuate unwholesome, or destructive, emotions, that is pragmatically beneficial in this lifetime as well.

Therefore, the practice of mental transformation entails attenuating afflictive thoughts and emotions and familiarizing oneself with beneficial ones. Such transformation, or modification, of the mind is the meaning of *chö,* which is the Tibetan word for dharma. The practice of dharma involves improving the quality of one's mind, which means simply that one empowers those mental processes that are beneficial to oneself and disempowers those that are harmful to oneself.

On what grounds can we assert the possibility of such transformation in our minds? Two principles are basic to this possibility. The first is that change takes place due to causes and contributing conditions. Whenever change occurs, including positive transformation in the mind, this happens due to causes and conditions. If one carefully examines this process of transformation, one sees that it takes place moment by moment. There is nothing in the mind that remains static; everything is constantly in a state of flux. If one uses a powerful microscope to observe atoms, one may be able to see the momentary fluctuations of these minute particles. Conventionally, we may speak of a person who existed yesterday, as if that same person exists today, or of an event that occurred thousands of years or a place that existed a million years ago. However, we are all aware of the reality of impermanence on an experiential level in the sense of observing how the continua of things that exist over time are eventually destroyed. For example, we all know that our planet will eventually come to an end. However, if things and events did not change moment by moment we could not explain how their transformation could take place over time, or how they could eventually be destroyed. When we examine time precisely, we see that every instant things are changing, and it is this principle of impermanence that gives us the potential for change and progress. Thus the first principle is that the nature of things is change due to causes and conditions.

In the records of the Buddha's teachings, he asserts that the passing of phenomena moment by moment does not occur as a result of the production of fresh causes and conditions acting on them from the outside. Rather, it is in the very nature of things that they arise and pass from one instant to the next. When explaining subtle impermanence, it is not simply that the first instant of something does not carry over into the second instant. Rather, the first instant does not remain unchanged even during its own time. Thus the very cause that brings something into existence creates it in the nature of impermanence, so that as soon as an effect arises the process of its disintegration has already begun. This means that the first instant of a phenomenon is in the process of being destroyed even while it is present, as the result of its very nature and not some outside influence causing it to be destroyed.

If one considers the continuum of a thing or event, viewing impermanence at the gross level, there exists the following sequence. Something first arises and is brought into existence because of the convergence of conducive causes and conditions; then its continuum endures because of the perpetuation of conducive causes and conditions and the absence of unconducive influences; finally it ceases because of the cessation of supportive conditions and the occurrence of destructive conditions. But, from the perspective of subtle momentary impermanence, the very nature of creation is the nature of destruction, as contradictory as that may seem. This is incompatible with the assertion that things first arise, then exist, then disintegrate, and finally become nonexistent. On a subtle level, these processes occur simultaneously. This is one principle that allows for the transformation and development of the mind.

A second principle is that among both outer and inner phenomena there are some that are incompatible with others. When two phenomena are incompatible, then, depending on which one is stronger, changes occur. Take the outer examples of heat and cold and of darkness and light. When two such opposing forces meet, one or the other will change, either gradually diminishing or instantly disappearing. Likewise, among inner mental phenomena, including emotions, there are many events that are mutually incompatible. Therefore, when one type of thought or emotion is developed, it will naturally counteract those that are incompatible with it. So the fact that change occurs when one phenomenon meets with an incompatible phenomenon is the second principle that allows the mind to be transformed.

Thus, in dependence upon those two bases, the mind can be transformed. Another issue that pertains to such transformation is that of truth and falsehood. For any mental state, if there exists something in accordance with what is apprehended by that mental state, it is backed by a valid cognition. If nothing exists in accordance with what is apprehended by a specific mode of cognition, that cognition is deemed to be false. So those two modes of cognition are mutually incompatible, and, generally speaking, between the two a valid cognition based on reality is stronger than a cognition based on nonreality. However strong a false cognition may be in the short run, since it has no valid cognition to back it up, in the long run, it turns out to be weaker than a realistic way of apprehending reality. This is a very important point.

Consider two emotions that are mutually incompatible, one of which seems to be temporarily beneficial and satisfying, but that, essentially, over time, proves to be harmful. Such an emotion cannot be aided by the discerning intelligence by which one distinguishes between that which is helpful and that which is harmful over the short term and long term. The contrary emotion may temporarily feel a bit disturbing, but if it proves fundamentally helpful over time, then the more one develops the power of one's intelligence the more it will support that emotion. So between two emotions that may seem similar, one kind can be supported by intelligence, whereas the other cannot. Thus it is possible to enhance those emotions that can be reinforced with intelligence and wisdom, and, for this reason, such emotions may be said to be more powerful.

On the basis of those principles we can understand how emotions change. This suggests the importance of gaining insight into the nature of the mind. Specifically, it is important to recognize which mental processes, especially emotions, are incompatible with each other. Moreover, it is crucial to investigate with discerning intelligence which emotions are truly beneficial over the long run and which are harmful. In conjunction with that, one should study which emotions are in accord with reality and which are misleading. Given the importance of understanding this, it is apparent that one also needs to gain a precise understanding of the objects apprehended by the mind. This leads one to investigate whether an object that appears to the mind actually *exists* in accordance with the way it *appears*.

The things that influence us, for better or worse, include not only those phenomena that arise in our immediate experience. There are also unchanging, more abstract phenomena that appear to the mind and are not produced by causes and conditions. So we need to explore and understand all possible kinds of composite and noncomposite phenomena that appear to the mind. This is why in Buddhist treatises there are detailed explanations of all types of composite and noncomposite phenomena, including the eighteen constituents of reality and the twelve sense-bases. Among all those classes of phenomena, those that either help or harm us in terms of bringing us happiness and suffering are, for the most part, included among composite phenomena. This is why the five psychophysical aggregates are discussed in Buddhism. If the point of meditation were simply to cultivate deeper faith and belief alone, there would be no need for such presentations of these constituents, sense-bases, and aggregates. When we combine this pursuit of knowledge with the attempt to transform the mind, we must explore and analyze the nature of objective reality. This is why Buddhist treatises present specific classifications of all types of permanent and impermanent phenomena.

From this perspective, Buddhism presents itself as an exploration and resultant presentation of the nature of objective reality. In the course of such exploration, it is strongly emphasized that one must have an impartial, objective attitude. In the course of scientific exploration and research one must also be objective, not allowing one's work to be prejudiced by one's own beliefs and preferences. One's research must be guided by the empirical findings themselves. The same is true in Buddhism: one must be objective, identifying the extent of one's preconceptions and recognizing how they can get in the way. It is important to discover the actual nature of reality apart from one's preconceptions and conduct research with this goal in mind. In this way science and Buddhism are quite similar.

On the other hand, if one is considering a hypothesis that is simply conjecture, or a figment of one's imagination without any basis in reality, however much one may familiarize oneself with this view, one will not be able to hold to it indefinitely. It is in these ways that one engages in the task of transforming the mind.

After one has engaged in the Buddhist methods of sustained, careful research into a certain facet of reality, eventually one comes to a point of certain knowledge. A clear discovery is made. Within the Buddhist tradition one speaks of wisdom and skillful means, and, between these two, the faculties of the mind that are directly involved in the pursuit of knowledge are called "wisdom." Other mental faculties, such as mental fortitude and various mental processes that facilitate the pursuit of wisdom, are called "skillful means."

Notes

1. The first part of this essay, on understanding the mind, was presented on June 2, 1997, as the inaugural lecture for the Fourteenth Dalai Lama Endowment for Tibetan Buddhism and Cultural Studies at the University of California at Santa Barbara. It is printed here with the permission of Mr. Tenzin Geyche, representing the Private Office of His Holiness the Dalai Lama and Professor Richard Hecht, representing the Department of Religious Studies at the University of California, Santa Barbara. The second part of this essay, on transforming the mind, was presented on June 25, 2000, in H. H. the Dalai Lama's oral commentary on *Lamp for the Path* and *Lines of Experience*

in Los Angeles, California, and it is published here with the permission of Geshe Tsultim Gyeltsen, representing Thubten Dhargye Ling, and the Ven. Lhakdor, representing the Private Office of His Holiness the Dalai Lama. Both these lectures were originally translated orally by Geshe Thupten Jinpa, but they have been retranslated from the Tibetan for publication in this volume by B. Alan Wallace.

2. This is the Prāsaṅgika Madhyamaka school.—TRANS.

3. The five psychophysical aggregates are form, feelings, recognition, compositional factors, and consciousness. The twelve sense-bases are the faculties of the eyes, ears, nose, tongue, body, and mind, together with form, sound, smell, taste, tactile object, and mental object. The eighteen constituents of reality are the twelve sense-bases plus visual, auditory, olfactory, gustatory, tactile, and mental consciousness.—TRANS.

Understanding the nature of the self is a central theme of Buddhist theory and practice, and it is also a topic that has drawn much attention among Western philosophers and cognitive scientists. In this essay David Galin addresses views of the self from the perspectives of Buddhism and cognitive neuropsychology.

Galin briefly reviews the chaotic state of the multiple Western accounts of self, the problems they leave outstanding in neuropsychology and psychiatry, and some ways in which they are inadequate from the Buddhist point of view. He then summarizes a new body of work based on the cognitive structure underlying "natural" day-to-day speech, showing that abstract thought is built on metaphors drawn from elementary experiences of sensory perception and bodily movement. The many metaphoric systems are inconsistent, and this gives rise to paradoxes in thought, both nonconscious and conscious, nonverbal as well as verbal. Particularly in regard to the ideas of self, I, and person, this leads to some very strange concepts and thinking.

Then Galin proposes his own novel reframing of these issues, a new set of definitions, and his own account of self, all consistent with the neuropsychological and metaphor data. The Prāsaṅgika Madhyamaka school of Buddhist philosophy also places much emphasis on the "natural," or day-to-day, use of language.

Insofar as Galin sees person and self as much more changeable and less well bounded than in the conventional Western views, his account is closer to Buddhist tradition. However, he differs from Buddhist tradition when he argues that "person" includes more than "self," and "self" includes more than "I," and that therefore the terms person, self, *and* I *are not synonymous. In Sanskrit Buddhist literature and in colloquial Tibetan, the terms corresponding to* person *(pudgala, gang zag),* self *(ātman, bdag), and* I *(aham, nga) are commonly used interchangeably, as they are in common and professional English usage. However, Galin's proposed definitions of person, self, and I are meant, as he points out, for professionals, not for the layperson.*

Galin does not explicitly address one aspect of conceptualization that the Prāsaṅgika Madhyamaka school holds to be crucial regarding the identities of persons and other phenomena. This is the distinction between the basis of conceptual (or verbal) designation and the object that is designated upon that basis. For instance, one may designate the person John as tall or as intelligent on the basis of the height of his body or his degree of intellectual acumen. The bases of imputation here are the height of John's body and the quality of John's intelligence, but neither is a person or a self. Generally speaking, whenever one critically analyzes the designated object (including person, self, and I), one finds that it is imputed upon a basis that is not identical to that object. In short, there is no inherently existing, objective referent to the words person, self, I, *or any other phenomenon. According to the various Realist (vastusatpadārtha-vādin) schools of Buddhism, however, for the self to exist even conventionally it must be identifiable under analysis.*

In the Indo-Tibetan Buddhist view, misconceptions of the person, self, or I range from the crudest to the most subtle: one may regard oneself as 1. an independent, unchanging, unitary entity, 2. a semi-autonomous agent, changing from moment to moment, who possesses and controls the body and mind, while operating within a nexus of causal interrelationships, or as 3. a personal identity existing by its own nature independently of conceptual and verbal designation. The following essay addresses primarily the first of these misapprehensions of the self. Galin suggests that we regard a person as a dynamic system of quasi entities in an open network of relations rather than as a bounded, persisting entity; the self as the current organization of all the subsystems of the person; and the I as the perceptual and action repertoire of that self.

According to the Prāsaṅgika Madhyamaka view, ordinary people do not misconceive of themselves at all times. Rather, only when they reify themselves as existing independently of verbal and conceptual designation do they fall into delusion, which acts as the basis for all mental afflictions such as attachment and hostility. Galin, on the other hand, emphasizes that the Ordinary Man's view of person and self, reification and all, is complex and multifaceted, based on an essential evolutionary adaptation and therefore should not be maligned as simply ignorant or erroneous or deluded. Thus Galin presents interesting points of agreement as well as contention between his own concepts of the self and those posited in the Buddhist tradition.

David Galin

The Concepts "Self," "Person," and "I" in Western Psychology and in Buddhism

1. The goal of this collection of essays is to deepen the dialogue between Buddhism and Western science, two very different systems of thought, by focusing on areas where their core concerns intersect.[1] The concept of *self* is certainly at their core, pervading daily life and theoretical writings for millennia. Yet, for both systems, *self* remains problematic; there is much confusion over exactly what *self* means, for ordinary folk and for the academics and professionals who are supposed to be experts on it. Thus *self* is a promising meeting area to explore.

2. My purpose here is to map the relations and disjunctions between Western psychology and what I understand of the Buddhist concept of self, or more properly, of "no-self" (*anātman*). The idea of "no self" is counterintuitive to most Westerners. Buddhist tradition holds that the root cause of suffering is the Ordinary Man's erroneous view of self as an unchanging essence, and that the error is inevitable because it is based on inborn patterns, pretheoretic and unreasoned (e.g., Garfield 1995:88). It is this "erroneous" view of self that is the focus of this essay.

3. I introduce the idea that person, self, and I are not synonymous; person includes more than self, and self includes more than I. Some current confusions and controversies are presented to show the ways in which the old concepts are inadequate. Then I discuss two cognitive mechanisms under-

lying the natural view of person and self. The first concerns the role of metaphor in abstract thought, especially in our thinking about person, self, and I. Second, humans tend to seek and find, or project, a simplifying pattern to approximate every complex field in two nonconscious ways: by lumping (ignoring some distinctions as negligible) and by splitting (ignoring some relations as negligible). Both lumping and splitting create discrete entities useful for manipulating, predicting, and controlling at the sensorimotor level and at abstract levels too. Unfortunately, they may impose ad hoc boundaries on what are actually densely interconnected systems and then grant autonomous existence to the segments. As this occurs in experience of our own "inner life" and in concepts of the structure of "the person," we come to see the self as a bounded persisting entity rather than as a dynamic open network of relations. This view of self as entity or essence is maintained so strongly because it is rooted in these basic nonconscious cognitive approximations. However, the lumping part of our pattern seeking, which simplifies by finding more relatedness among things (unifying), can also be corrective to the creating of isolated entities. This second type of approximation may be the seed of the Buddhist "correct" view that all things are interdependent. Western perspectives of cognitive neuropsychology and adaptive evolution may add to Buddhist understanding of the inborn view of self, and why it is so difficult to transform, and of how the "correct" view is attained.

1. THE BUDDHIST CONCEPT OF NO-SELF ("ANĀTMAN")

4. I will sketch my understanding of the generic Buddhist view for those with little familiarity with Buddhism, drawing heavily on Collins (1982), Garfield (1995), Hopkins (1983, 1987), and Wallace (1989, 1998). In the Buddhist "correct view" the Self is seen not as an entity, or as a substance, or as an essence but as a dynamic process, a shifting web of relations among evanescent aspects of the person such as perceptions, ideas, and desires. The Self is only *misperceived* as a fixed entity because of the distortions of the human point of view. Ultimately, no separation is to be found between these dynamic processes and the universal frame of reference or ground of being; all is interdependent and changing. Thus, in this sense, there is no Self separable from a Nonself. This Buddhist declaration is misunderstood in the West because *anātman* meaning "self-is-not-an-essence-or-entity," is taken as "self-does-not-exist-at-all" by people who have not imagined any scheme of existence other than entities or essences.[2]

5. The Buddhist tradition holds that Ordinary Man's inborn erroneous view of self as an enduring entity is the cause of his suffering because he tries to hold on to that which is in constant flux and has no existence outside of shifting contexts. Therefore, a new corrective experience of self is needed. Buddhism takes great interest in how people experience their self, rather than just their abstract concept of it, because Buddhist practices are designed to lead to a new (correct) experience. It takes arduous training to modify or overcome the natural state of experiencing the self as persisting and unchanging. There is a great literature on the theory and practice of the three main paths leading to a changed experience of self. One path is via meditation trainings (changing mind processes or mind controls, e.g., attention, awareness, arousal). Another is via theoretical argument (changing structure of concepts, the contents of mind). The third path is social-behavioral, the life of active service (Deikman 1996, 1997, 2000). The three paths of meditation, scholarly study, and service in the monkhood or wider community are usually intertwined in practice.

6. This doctrine of no-self has major theoretical implications for two other central components of Buddhist doctrine: karma and rebirth. With not even a temporary self, how can we understand the apparent continuity and coherence of personality in the present life? Without a permanent self, just what is reborn in another life? If there is no self, to whom or to what does karmic ethical responsibility belong, and to whom or to what is it transferred in a later life? These issues are beyond the scope of this essay (see Garfield 1995, Collins 1982). But *anātman* raises other issues, more pragmatic than theoretical or doctrinal: how can the seeker go about developing "right views" of self and person, particularly since the erroneous view is held to be inborn and pretheoretic, particularly resistant to rational discourse or scholarly philosophical argument? Buddhist schools differ in their beliefs as to the effectiveness of simply quieting the mind and introspecting, versus developing intense and continuous attention, versus nonrational dialogue and interaction, versus directing introspection with rational analysis and conceptual framing (e.g., Japanese Soto and Rinzai Zen, Indo-Tibetan Vaibhāṣika and Prāsaṅgika).

2. IN WESTERN PSYCHOLOGY: MULTIPLE CONCEPTIONS OF SELF

7. To the Ordinary Man in Western cultures, as in Buddhist cultures, the question "What is a self?" may seem trivial; it is casually believed that every

person has one, or is one, and that it is the self that acts or experiences. "Normal" folks have a vivid sense of themselves as distinct from not-self, from objects, or other selves, and, most important, as single, unitary. Two elegant statements of this sense of unity in the ordinary view are quoted (in order to refute them) by Joseph Bogen, the neurosurgeon-scholar most knowledgeable about separating the two hemispheres of the human brain, in his papers on the unrecognized *disunity* in normal people (Bogen 1986, 1990):

> 8. Sherrington (1947): "This self is a unity. . . . It regards itself as one, others treat it as one. It is addressed as one, by a name to which it answers. The Law and the State schedule it as one. It and they identify it with a body which is considered by it and them to belong to it integrally. In short, unchallenged and unargued conviction assumes it to be one. The logic of grammar endorses this by a pronoun in the singular. All its diversity is merged in oneness."

> 9. Descartes (cited in Bogen 1986): "There is a great difference between the mind and the body, in that the body is, by its nature, always divisible, and the mind wholly indivisible. For in fact, when I contemplate it—that is, when I contemplate my own self—and consider myself as a thing that thinks, I cannot discover in myself any parts, but I clearly know that I am a thing absolutely one and complete."

10. Consider the possessives *me* and *mine*. "Me" seems to refer to self, and "mine" seems to refer to objects: my car, my hair, my hand, my thoughts, my intentions, my mind. But the boundary is not clear. While "my car," "my hair," and "my hand" are all treated equivalently in syntax as "mine," most people feel that their hair is a more substantive part of "me" than their car, and their hand or mind more so than their hair. Furthermore, they believe that if they were to lose their hand their self would remain, that their *essential nature as an entity* would not be diminished. William James referred to this as "the self of all selves" (James 1950 [1890]:297).

11. Thus the question arises, "Are there degrees of self?" When we speak of self-development do we mean that there was a little self before and now there is more self? Or that there was qualitative change? If self varies in amount or quality, how do we measure these dimensions? Indeed, what sort of losses do we have to sustain to experience a diminishment of self, or for others to recognize it? And what do we mean when we say, "I just don't feel myself today"? If there is a difference between how you usually feel and how

you feel today, does that mean that there is a qualitative difference in selves? Is the difference substantive or just a trivial difference in appendages to the self, like a coat or a hairdo? When a psychiatric patient says, "I do not feel like it is me," or hears his own thoughts but perceives them as external voices, is that qualitatively different? Apparently, the concept is not clear. The idea of self is elusive; lay people are surprised that they cannot easily articulate it. But their conception of self as an entity, and as unchanging for life or even beyond, is not really shaken.

12. Professionals in psychology and its neighbor disciplines (cognitive science, philosophy of mind, psychiatry, behavioral neurology) are no more coherent about self than lay people. When I have asked psychologist friends for their definition, many of them narrowed their eyes as if suspecting a trick question. Sometimes the professionals use the term *self* synonymously with self-concept, with self-awareness, with consciousness itself, and with volition. Sometimes it is used in the sense of personality or social roles. The literature is voluminous. Different disciplines have focused on different aspects of self, and within disciplines different factions have models with little in common—e.g., in psychiatry: Janet, Freud, and Jung; in social and personality psychology: G. H. Mead, K. Lewin, and H. Markus. In philosophy, of course, there is even less consensus. One philosopher who had written a keynote journal paper summed up *four* special issues of discussion and rebuttal: "The result was a festival of misunderstanding. . . . Large differences in methodological and terminological habits gave rise to many occasions on which commentators thought they had disagreed with me although they had in fact changed the subject" (Strawson 1997, 1999).

13. In contrast to a plethora in some disciplines, the technical literatures of neurology, cognitive psychology, or neuropsychology show only sporadic concern with self or the wholeness of a person. A few exceptions do stand out: Wm. James (1950 [1890]), K. Goldstein (1939), E. Hilgard (1977), O. Sacks (1984), J. Kihlstrom (1993, 1997), and too few others. Perhaps the most dramatic and most scientifically significant are the studies by Sperry, Bogen, Zaidel, and their colleagues (1969, 1979) of the "split-brain" patients whose cerebral hemispheres have been surgically disconnected for the treatment of epilepsy. After the surgery each hemisphere is separately conscious and can perceive, learn, and remember, *without knowing what the other hemisphere is experiencing*. Nevertheless, both the patient and the patient's family report that they seem to be as much "themselves" as ever (Sperry 1968; Sperry, Gazzaniga, and Bogen 1969; Sperry, Zaidel, and Zaidel

1979). The "split-brain" studies demonstrate how interconnections at the neurological level contribute to the wholeness at the psychological level. To understand how these observations were made, it must be remembered that each hemisphere controls feeling and movement only on the opposite side of the body, and sees only the opposite half of visual space. Only the left can talk. One dramatic film records a patient trying to match a colored design with a set of painted blocks. The film shows the left hand quickly carrying out the task; the left hand is controlled by the right hemisphere, which is good at spatial relations. Then the experimenter disarranges the blocks and the right hand (left hemisphere, poorer at spatial relations) is given the same task. Slowly and with great apparent indecision it arranges the pieces. In trying to match a corner of the design, the right hand corrects one of the blocks, and then shifts it again, apparently not realizing that it was correct: the viewer sees the left hand dart out, grab the block, and restore it to the correct position . . . and then the arm of the experimenter is seen, reaching over to pull the intruding left hand off camera. The left hand repeatedly tries to intrude, and the experimenter finally makes the patient sit on the left hand while the right hand continues trying to arrange the blocks.

14. In another experiment a picture is shown to one hemisphere and the patient is asked to point to a matching object in a row of objects before him. Both hemispheres can see the objects; only one was shown the picture. In one case, when the picture was shown to the right hemisphere, as expected, the left hand pointed to the correct object, but the patient said, "I know it wasn't me that did that!" presumably by way of the left (speech) hemisphere.

15. In these incidents, just who are the "persons" involved? What has become of the apparently unified self that existed before the surgery? Is it now two, or was it always two, but now the duality has been made obvious? Until now, the language of Western psychology has been too fuzzy even to formulate these questions clearly, and there has been no consensus on a model that will describe the "I" who knows that "me" did not do it and who *did* do it. Furthermore, we need to account for the testimonies of the patients that their experience of self has not changed. We also need to account for such phenomena as they occur in "normal" people (Bogen 1986, 1990; Galin 1974; Galin et al. 1979). Our present theories of self do not address such phenomena. Now let us turn to another window on the panorama of confusion and apparent paradox that pervades Western thinking about self and person.

3. COGNITIVE LINGUISTICS THEORY OF METAPHOR, AND THE FOLK MODEL OF PERSON, SELF, AND I

16. Over the last twenty years cognitive psychologists and linguists have carried out paradigm-busting work with powerful implications for all of psychology. They show that metaphor is fundamental to nearly all abstract thinking. Citations of representative contributors can be found in two influential books, *Fire, Women, and Dangerous Things,* (Lakoff 1987) and *Philosophy in the Flesh* (Lakoff and Johnson 1999), which summarize much of this research and theory. What concerns us here are observations on how metaphor determines ordinary people's concepts of *person, self,* and *I.* Understanding how metaphor normally works in thought explains much that seems incoherent in common and professional talk and thinking about person and self. Lakoff and Johnson's analysis of common speech uncovers a nonconscious complex system of a dozen metaphors, many incompatible with the others, and quite different from the consciously reported notion of self as unitary, unchanging, and essential, like a soul. This section is mostly paraphrased or quoted from *Philosophy in the Flesh.* Limited space allows only a too condensed overview.

3A. HOW METAPHOR WORKS

17. We are not aware of how much our thinking is based on metaphor, even in science (e.g., in concepts of number, time, force, and category). Metaphor is related to reasoning by analogy. It gives our thinking enormous power, because we can extend our knowledge of the complex relations in a concrete domain to make inferences in an abstract domain (e.g., thinking of a love relationship as like a journey, thinking of numbers as like positions on a line). For example, consider our commonly repeated experiences moving along a path. We learn "logical entailments" that are true of all paths, such as that going from start to end entails passing through all the other points on the path. Thus if love is a journey, we will have to pass through a series of stages on the way to the destination. But analogies fit only partially; love is not only like a journey but also like a rose and like an invincible conqueror; numbers are not only like positions on a line but also like collections of objects and like containers that hold collections. Therefore, our metaphors break down if stretched too far from the original context and must be replaced (usually unconsciously) by new ones more apt

for the new context. Because we are not aware of how much our thinking is based on metaphor, we are peculiarly prone to take our metaphors literally. Taking metaphors literally can have very serious effects, as when we take teaching stories and fables as historical.

3B. *PERSON, SELF,* AND *I* ACCORDING TO LAKOFF AND JOHNSON

18. Our "inner life" includes experiences that we often want to refer to verbally such as

- conflicts between our conscious values and our behavior
- inner dialog and monitoring
- disparities between what we know about ourselves and what other people believe about us
- controlling our bodies and ways of "getting out of control"
- taking an external viewpoint or imitating someone

19. One might expect the whole domain of "inner life" to be handled by a unified conceptual structure. It is really shocking to find that what we have is a system of *inconsistent* metaphorical conceptions drawn from very different elemental experiential domains such as space, possession, force, and social relationships. Furthermore, several terms for ourselves as persons with "inner lives," which we thought were synonymous (*self, I, me, myself*) are not synonymous at all. Nevertheless, people seem to have no difficulty intuitively understanding these metaphor systems and switching among them, although they are unaware of what it is they are doing.

3C. THE GENERAL "DIVIDED PERSON" METAPHOR

20. Lakoff and Johnson present convincing evidence that every metaphor for our inner life is a special case of a single general schema of ourselves as split (269).[3] According to this unconscious schema, a person is divided into an *I* and one or more *Selves*. The *I* is the aspect of a person that is the experiencing consciousness (as the *subject*) and, by its nature, *it exists only in the present*. It is always conceptualized as a humanlike being and is usually but not always the locus of reason and values, it is the locus from which will is exercised (as the *agent*), although the acts must be carried out by one of the selves.

21. A *Self* includes those parts of a person not picked out by the *I*, such as the

body, social roles, past and future states, and actions in the world. There can be more than one Self. Unlike the *I*, each Self can be conceptualized metaphorically as an object, or a location, as well as a humanlike being.

22. The general schema, then, contains a humanlike being (the conscious I), one or more entities (one or more Selves), and a specification of who is in control and who judges whom. There are many specific varieties of the general metaphor, grounded in types of everyday experience such as 1. manipulating objects, 2. being located in space, 3. entering into social relations, 4. taking other points of view, and 5. as an instance of the Folk Theory of Essence: Each person is seen as having an "essence" that gives the I its qualities. A person may have more than one Self, including their Inner Self and Outer Self, but only one of those selves is compatible with their essence. This is called the "real" or "true" Self. These variations give rise to the extraordinary richness of our metaphoric conceptions of our "inner life." I will give illustrations of only a sample of metaphoric types. Please note that in spite of their apparent ambiguity, these sentences are all immediately understood in common speech.

23. *I, me, and myself are not always synonyms*   Notice the difference in meaning of these two sentences:

    a) If I were you I'd hate me.
    b) If I were you I'd hate myself.

In a), me refers to the subject (*I*) of the speaker; in b), myself refers to a self of "you," the person addressed.

24) Sometimes the locus of judgment is the *I* and sometimes it is the *self*:

    a) I was disappointed in myself.
    b) I disappointed myself.

25. *Self-control expressed as moving and possessing objects*   Control as forced movement:

    I held myself back
    I dropped my voice
    You're pushing yourself too hard
    I've got to get myself moving on this project

Control as possession:

I got a grip on myself
I didn't let myself wiggle out of that
I let myself go, and lost myself in dancing
I was seized by anxiety
I was carried away by fear
I was possessed by Demon Rum

26. Self-Control and Location   When *self* is seen as a location in space, self-control can be expressed as the *I* and the *self* being together at the usual place or contained space, such as the body, home, or on earth:

- I did not *get high*; I kept *centered*; I feel *well-grounded*
- I was *beside* myself; *ecstasy*; *out of my head*; *out to lunch*; *off in space*
- I was *scattered*; I must *get myself together*; He is *all over* the place

3D. THE ESSENTIAL SELF

27. The Essential Self is another important set of metaphors based on the Folk Theory of Essence rather than on perceptual-motor experiences. According to this nearly universal belief system, every object has an ssence that makes it the kind of thing it is and that is the causal source of its natural behavior. It is the essence that makes one behave like himself and not like somebody else. (It was this same folk theory that was formalized by Plato and others). A person's essence is part of her subject (the I)—ideally it should determine her behavior, but often it is incompatible with what she actually does. This incompatibility is conceptualized as having two selves. One self (the "real" or "true" self) is compatible with one's essence and is always conceptualized as a person. The second self ("false" self) is incompatible with one's essence and is conceptualized as either a person or a container in which the real self is hidden (282).

28. *The inner self*

He won't reveal himself to strangers.
She rarely shows her real self.
Whenever anyone challenges him, he retreats into his shell to protect himself.
Her sophistication is a facade.
The iron hand in the velvet glove.
His petty self came out.

29. *The real me*

I'm not myself today.

That wasn't the real me yesterday.

That wasn't my real self talking.

30. Very little research has been done on metaphor's role in structuring the concept of self in other languages and cultures. Much work remains to be done to support generalizing beyond the English or Western observations. However, there are Japanese examples that look like and are understood in the same way as the English (see Lakoff and Johnson 1999:284–287). The Japanese conception appears remarkably like the American one. It is only the socially proper relationship between self and other that is radically different in Japanese culture.

31. Does this analysis of covert metaphoric concepts of self revealed in everyday language tell us anything about how our inner lives are "really" structured? I emphasize that it does not mean that we are literally divided up at a neurological or microcognitive level into an I and one or more selves or into essences. But these metaphors do seem to capture much of the qualitative feel of inner life. When we say "I'm struggling with myself over whom to marry" or "I lost myself in dancing" or "I wasn't myself yesterday," the metaphors seem apt because they conform in a significant way to the *phenomenological* structure of our inner lives and capture its logic and how we reason about it.

32. However, there is no correspondence between the Divided Person metaphor and any of the psychiatric, neurologic, sociologic, or philosophic models of self considered in the previous section. Thus, now we have further questions to ask: Why have we evolved with this strange conceptual system? How can it be so apparently effective in daily situations? Why is the Ordinary Man unaware of its workings and its paradoxes?

4. SOME PROPOSED CLARIFICATIONS: PERSON, SELF, I, SELF-MONITOR, SELF-AWARENESS, SELF-CONCEPT

33. I have presented this review of confusion and controversy in Western notions of the self to justify my call for radical changes in the way professionals talk and think about these matters. It would be foolish to expect lay people to change their common speech, but for professionals this is not just scholastic wordplay; terms are tools. A relatively easy first step is to sharpen our terms, taking care to preserve the insights into structures and functions

on which these metaphors and intuitions are built. My purpose here is to arrive at a set of terms consonant with the broad "scientific" frame of reference, the cognitive sciences, and contemporary philosophy of mind. The need makes the gamble worthwhile. I ask the reader to inhibit the "Yes, but . . ." reflex, and perhaps she will find her objections handled a bit further on.

4A. PERSON AND SELF: DEFINITIONS

34. The concept *person* can be distinguished from the concept *self*, and self can be separated from hyphenated derivatives with which it is often conflated or confused: *self-monitoring*, *self-concept*, and *self-awareness*. The concept I can be distinguished from self and from person.

> A person is a complex system, made up of component subsystems. Person always includes the entire self-organizing,[4] multilevel, causal thicket,[5] including bodily, mental, and social aspects, and representations of past and future organizations (selves). This list is meant to be open-ended, and other dimensions can be added as needed. Person, of course, is embedded in a larger complex environment (the universe).

My new definition contrasts with common usage, in which just how much is included by *person* and *personal* always depends on the speakers' conventions and on their purposes. Thus, under the common definitions, sometimes *person* refers to the body ("touched his person"), sometimes to intimate feelings ("religion is too personal"), and sometimes to social relations ("separate personal and professional life"). Since this is not made explicit in ordinary discourse, it contributes to the confusion.

35. Our common speech sometimes distinguishes between person and self, but with multiple, context-dependent senses, as illustrated by Lakoff and Johnson's observations presented above. I propose that we explicitly designate that

> person is extended over time, and *self is the current organization* of the person. Self is the way all subsystems of person are related to each other at this moment.

I emphasize that self is a characteristic of the person as a whole, rather than just another subsystem or constituent as some psychological models would have it. In discussing the complex we call *person* it is very useful to be able to refer to its organization (self) in contrast to its instantiation (embodiment). Organization is simply the set of all the relations among the constituents of the system. This can include such relations as membership, connection, control, and many others.

36. It is readily apparent that a person, like any complex system, might be capable of several different patterns of organization (selves); e.g., with subsystems tightly integrated or with semiautonomous subsystems. Defining the self as the pattern of organization emphasizes that it changes over time—over a period of years, as in the maturation of adolescence, or over minutes, as in multiple personality disorders. By defining self as the *organization of subsystems* rather than as just another subsystem, we get a clear referent for common phrases such as "more integrated self," "development of the self," or "loss of self." The concept of self *as organization*, varying dynamically in degree and quality, works at the neurological level of description as well as at the psychological level and across levels.

4B. HYPHENATED DERIVATIVES OF SELF

37. Self is often confused with several of its own hyphenated derivatives: self-monitor, self-concept, and self-awareness. I have previously detailed the differences (Galin 1992, 1994); here there is only space for a brief mention.

38. *Self-monitor*: Self-regulating systems that adapt to their environment may have subsystems that keep track of their own state and the adaptations made. We infer our self-monitors because we know a lot about our present "mode" of organization: drowsy or alert or drunk, imagining versus remembering, passive versus active. When the doctor taps our knee and elicits a reflex knee jerk, we can say, "I didn't do that." Deficits of self-monitoring follow certain brain injuries (Galin 1992), and it is important in understanding hypnosis, hallucinations, and how the left and right cerebral hemispheres relate to each other. However, it is different from self-awareness, self-concept, or self.

39. *Self-awareness*: Self is often treated as synonymous with self-awareness. Logically, self-awareness simply means that information about the self from self-monitoring or self-concept has been brought into awareness, but it is only a small part that ever enters awareness. Numerous experiments have

demonstrated that even very complex information processing goes on without our being aware (Kihlstrom 1987), and only their final product appears in consciousness.

40. *Self-concept*: The self-concept can be thought of as a body of knowledge, beliefs, attitudes, etc., about "who or what one is" as an entity in the world. The self-concept refers to the self but is just a subsystem. It is similar in form to our concepts of any other objects, usually nonconscious, often incomplete or incorrect. Some of it must come from the self-monitor, but much of it comes from other sources, such as the opinions of our relatives. We can have many, even contradictory self-concepts. Like any other concept, it can be brought into consciousness from time to time, but it is clearly distinguishable from self, from self-awareness in general, and from self-monitoring of current states as defined above.

4C. THE I: ENTITY, KNOWLEGE, AND POINT OF VIEW

41. The first-person point of view is a surprisingly subtle concept (Galin 1999). The novel issue here is, "What *kind* of a thing is the I"? First I will define three common but uncommonly difficult terms: *entity, knowledge,* and *point of view.* Then I will propose that I is a kind of point of view, or perspective.

42. *Entity*: This is a key concept, almost always mistakenly assumed to be intuitively obvious. In general, an entity (a unit, a wholeness) does not have a sharp boundary. An implicit or virtual border is created by the pattern of relationships among elements "belonging to each other" in some sense. According to the analyses of Simon (1969:209ff.), and of Wimsatt (1974 and 1976a:242, 261), we call a set of parts an entity if there is *sufficient inter-relatedness* among them.[6] "Sufficient" is decided by some criterion *chosen for our purpose.* Functional relations, spatial and temporal relations, social relations are examples of aspects by which we commonly decide that some distribution is an entity or not. For example, a group of people is an entity we call "family" if the people have sufficiently close relations. It depends on our purposes whether we set the criterion to include second cousins, adoptees, steps, and pets, or only include parents and their natural children. Thus, *entiticity is a matter of convention as well as a matter of degree.* I believe this greatly softens, but does not quite extinguish, the usual hard self/object boundary that has been such a contentious point in Western and Buddhist metaphysics and psychology.

43. I stress the rather surprising conditionality of entities because properties such as point of view, or knowledge, or consciousness belong to a specific entity, and to understand the property we must be clear about just what we think constitutes the entity that hosts it. Problems arise when we forget that what we are treating as an entity is only more or less an entity, of limited duration, and only by agreement for the present purposes (e.g., a marriage, a corporation, the Republic of Congo). Many confusions about "person" and "self" arise here. Humans have a great passion for entifying; we frequently turn verbs and adjectives into nouns, and turn processes and relations into things.

44. *Knowledge, to Know*: Attempts to define this concept (or conceptual cluster) have a very long history in philosophy. It is currently used in many senses: to perceive directly,[7] to be capable of, to be fixed in memory, to be acquainted or familiar with, to be able to distinguish (*Webster's Unabridged Universal Dictionary*, 2d ed.). Note that none of these usages specifies consciousness; they all make sense for instances of nonconscious as well as for conscious knowings. The sense that best captures the phenomena that concern us here is one of the most general; *to know is to be able to distinguish*.

45. *Know* implies a knowing entity, and that which is known. For an entity *to know* something XYZ (to know how to XYZ or to know that XYZ), is for it to distinguish XYZ from at least some not-XYZ, to act discriminatively toward XYZ. No consciousness is implied. The minimum discrimination is detection; something happened or didn't; something is or isn't. Knowledge is the capacity to discriminate, *a potential to use information*.[8] It is a functional property, and it does not imply any particular underlying mechanisms. such as time-coded and source-labeled memories or discrete representations.

46. *Point of view*: The terms *point of view* and *perspective* are metaphors taken from the domain of visual-spatial perception and applied to the more general domain of knowing. These metaphors express the intuition that people operate within a frame of reference or coordinate system (analogous with a spatial coordinate system) made up from their repertoire of concepts. Much of the conceptual repertoire is quite abstract and not visual-spatial or sensory-perceptual at all. For example, we can speak of having a particular political, ethical, or pragmatic point of view. Thus *point of view* applies to domains such as values as well as to domains of spatial perception and action. An event or object is good or beautiful or moral *from my point of view*, just as an object is above or below, or on the right or the left,

from my point of view. Note that, as with the previous concepts, awareness is not required for a point of view (Galin 1999).

47. Like knowledge, a point of view belongs to a specific entity. A point of view is *the total set of possible discriminations that an entity can make* in its present state or over some period of time specified for our purpose. Point of view determines the "set of discriminations that are available"; e.g., if you are in a windowless room, you can see the walls but not the street outside, or if you have your eyes shut you can hear but not see. It can be thought of as the current working frame of reference, or as the set of variables, or the set of dimensions. If an entity changes, some knowledge may temporarily be unavailable. For example, in an enzyme molecule the critical receptor region may become folded inside, unable to interact. Similarly, different capacities for discrimination may be available to you in the context of a street mob than in the context of your private study. Thus the point of view will vary as the properties of the entity are affected by the time, place, and other contexts.

4D. DEFINING THE I, THE SUBJECT, AND THE AGENT

48. With these definitions in hand, I propose that

the *I is a kind of point of view*. It is the point of view of the entity *person*, given by the person's present organization (its *self*). Thus, *I* is not equivalent to self. *I* includes both the *subject* and *agent*. *Subject* is the input point of view, that is, the set of currently available discriminations from which perceptions are selected. *Agent* is the output point of view, the set of currently available discriminations from which actions are selected.

The psychological construct we call the present is a reference point in "real time" or "clock time" (time that changes in one direction, at a constant rate, for all entities). The self's I must be in or at the present because the self's perception and action machinery must use the same synchronizing reference point as the entities they perceive or act upon. For example, speech is severely disrupted if you hear your own voice through earphones with even a slight delay. I hypothesize that the process of adopting the real-time point of view is the basis for what we call subject and agent. Subject and agent can be lumped as a single entity named I because they both share the real-time point of view.

49. *Selves* other than the current one are reference models (representations

of possible states of self). They are either not at the present (like the past or future states of *I*), or not in real time at all, like ideal self (what you should be like but are not), shadow self (what you should not be like but are), longed-for self (desired), inner self (hidden from *I* or others), and the real or true selves (many meanings, contextual). These are best thought of as multiple *self-concepts,* named so we can refer to the richness and contradictions of our physical, mental, and social life that cannot be denoted by the I anchored in clock time and action.

50. This multiplicity would confuse Aristotelian rationalists, but ordinary people do not notice any difficulty in daily life. As Lakoff and Johnson have shown, in common speech the multiplicity is managed nonconsciously through the Divided Person metaphor, discussed above in section 3c. Note that my definitions, arrived at by analysis and for professionals' analytic purposes, are different from the nonconscious heuristics that Lakoff and Johnson found empirically in the natural speech of daily life. What the Divided Person metaphor names as "I" includes much more than my subject-and-agent point of view, and what it names as "other selves" is what I have identified as self-concepts (compare sections 3c and 4b above). In what follows I will stick to my terminology, developed for internal consistency.

51. Note that the I point of view includes only those discriminations made by the entity as such, not by one of its parts acting for the time autonomously and without relation to the entity as a whole. Consider, for example, the knee-jerk reflex response that the doctor elicits by putting the knee joint in a relaxed posture and tapping on the kneecap tendon. A person generally does not have a sense of agency about the leg movement, does not feel that "I did that," and claims that the "leg" did it. In this posture the leg is *less related* to the rest of the person (not participating in support) and thus can be interpreted as an entity acting on its own. In contrast, when the person is standing, the leg is integrated into the rest of the superordinate entity and cannot be allowed to act autonomously; the reflex is much more difficult to elicit in this position. Thus we see that the feeling of agency claimed by *I* is associated with the superordinate entity *person.*

52. *I-Awareness*: The definitions developed thus far reveal a previously hidden conflation. I itself is not in awareness; it is a point of view, not a content. The point of view I is the total set of discriminations that *could* be brought to bear in the person's interactions, not only the ones actually in operation at the moment, and certainly not only the few the person may be aware of at the moment. Because of the limited channel capacity of aware-

ness, rather little of the person's total point of view could be in awareness at any moment. The part that does get into awareness can be called *experience of I*, or *I-awareness*. Note that the term *self-awareness* must include more than just I-awareness; it should cover any of the experiences related to the self, such as feeling alert or sleepy, lonely or in love, healthy or sick, that do not refer to the subject and agent points of view per se. Self, defined as the organization of all the subsystems of the person, includes much more than the person's point of view.

53. The I (the point of view of the person) may be changed a little or a lot as the total organization (self) changes in response to the environment (e.g., if you see a police car in your rearview mirror). Therefore, the I may change from moment to moment or may stay relatively "the same" for some time. Perhaps this is an aspect of "stabilization of mind" that is sought in the Buddhist meditative practice of *śamatha* (Wallace 1998, 1999).

4E. MISIDENTIFICATION OF THE RELEVANT ENTITIES

54. Now that we have clarified the concepts of I, subject, and agent as based on the point of view of a particular entity, we can untangle confusions that arise from misidentification of the relevant entity, such as in talk about "to whom" a perception or action belongs. Complex entities may include relatively autonomous subentities with points of view discrepant from each other or from the whole. The superordinate entity must reconcile such conflicts; its point of view is not simply the sum or the union or the intersect of the points of view of its parts. To illustrate the problem of assigning a "to whom," consider that even our literally spatial point of view is not as simple as people assume. What we take to be *directly in front of us* is sometimes determined with respect to our body midline, sometimes with respect to our direction of gaze (i.e., how the eyes are turned in the head), and sometimes with respect to the direction in which our head and neck are turned (tonus in the neck muscles) or by our vestibular apparatus (i.e., whether we are standing vertical or lying horizontal) (see Vallar et al. 1993 for extensive references). The eyes, head, and torso midline can each have a different *front*! That is, we have multiple frames of reference for 3-D space, which can give conflicting points of view. When we cannot reconcile them, as in motion sickness or the hemineglect following a parietal lobe injury, we have trouble (Bisiach 1991; Galin 1992). Similarly, we have multiple frames

of reference for values and we have trouble if we cannot reconcile them (e.g., different ethical standards in business and in the family).

55. The superordinate entity (e.g., a person) may usefully combine the multiple points of view of its parts (subentities, e.g., the two eyes). If the eyes are on the sides of the head as in a rabbit, their two points of view can be added to give a wider field of view. If, as in humans, the eyes both face front and their two points of view overlap, then the discrepancy between them can be used for information about depth. But when there is too much conflict, all but one may be suppressed (e.g., amblyopia, the loss of vision in one eye in a severely "cross-eyed" person). Another strategy for conflict resolution is to alternate among points of view (e.g., in multiple personality disorder or in contextual ethics).

56. Ordinarily we are good at keeping track of which of our multiple points of view are in control at the moment. It must be very important to our species, since children can do it by age eighteen months; they can take another point of view in imaginative play without losing track that the new one (the "pretend") is nested inside of the usual one (the "real") (Leslie 1987). However, even adults can lose track of a nested hierarchy of points of view, as when one gets "carried away" at the theater or movies and reacts as if it were real or in hypnosis, where the I accepts the hypnotist's point of view as overriding the self-monitors.

57. A similar analysis untangles the confusion and paradox concerning subject and agent in "split-brain" patients (see paragraphs 14–15). Remember that entiticity is a matter of degree. After the surgery there is still a single superordinate entity that we call "the patient," even though its parts are less related than they were. The two hemispheres still share the same brainstem connections, body, mouth, relatives, history, and many other unifying factors (Sperry 1968). What is changed is the degree to which each hemisphere (and mind) is related to the other and to the superordinate entity (person). When the superordinate entity *patient* says, "It wasn't *me* that did that," it is the point of view of the left (speech) hemisphere that is being adopted and articulated by the superordinate *person* through the shared midline vocal organs. *Who* selected the correct object? We can attribute it to the superordinate entity *person* making the discrimination from the right hemisphere's point of view. My terminology makes it possible to refer unambiguously to each component in this drama and thus to resolve the debates between Sperry and philosophers like Puccetti (1981).

5. SPECULATIONS ON THE BASIS FOR THE ORDINARY MAN'S VIEW OF PERSON, SELF, AND I

58. Now, after this long detour to develop the needed technical vocabulary, we take up the inborn, pretheoretic notions of self. This is a highly speculative account of just why the Ordinary Man's view of person and "inner life" has the surprising hidden structure revealed by Lakoff and Johnson's linguistic research, and why the hidden structure is different from what is consciously reported. I will also suggest how the basis of the Buddhist's "correct" view can be found in a part of the structure of ordinary awareness usually neglected by cognitive psychology.[9]

59. The central idea is that in order to deal with the enormous complexity of the world a mind/brain routinely and unconsciously makes use of ad hoc approximations. Whatever complex field we encounter, we simplify it into a small number of discrete entities to make perception, action, and abstract thought easier. Even the contents of consciousness are structured by the need to limit information (Galin 1994; Mangan 1991; James 1950 [1890]) and to facilitate action (Deikman 1971). This results in the usual appearance of the world to us as if it were made up of distinct independent objects. We do not generally tidy up the approximations and correct the round off errors; if they work *well enough* we do not look back and try to reconcile. We only try to rationalize discrepancies when pressed.

5A. THE BASIS FOR THE OBJECT WORLD

60. Our experience of the world as full of separate objects does not correspond in a simple way to "just what is out there." It is the result of complementary operations, some of which seek information and some that seek to filter, reduce, and compact information. For example, the retina is built to report in terms of dots of light of a few discrete colors in discrete positions. Even though the stimulus field may be continuous, it is encountered in discrete terms. The array of bits is then segmented into regions, in each of which the bits are considered to constitute a single entity (an object, form, or pattern). In human vision information is again segmented, this time into shapes, locations, colors, and movements, with each property in a separate region of the cerebral cortex, and then once again compacted into the unified perception of a single particularly shaped multihued object at a particular location moving in a particular direction.

61. Just how we segment the manifold depends on many factors. For example, we are particularly sensitive to borders and outlines, perhaps because we interact with an entity at its edge or surface; that is where we can grasp or bite it. Also, our movement systems are adapted to work with entities of a certain scale, not too big or too small, so we tend to segment the manifold into regions (entities) of manageable size. Large, medium, and small might be thought of in terms of what you can walk around, what you can grasp and move, and what you can bite and swallow.

62. Recall that what we include in an entity depends on how stringent a criterion we choose for "sufficient relatedness" (e.g., for "family": sister versus cousin). Shifting the criterion is the basis of two cognitive strategies for segmenting a field into appropriately sized entities. The first is called analysis or splitting or decomposition. It creates smaller entities by assuming that some relations explicitly shown between elements in the field are "minor" and therefore can be treated as negligible, at least for the present purpose. The second strategy is called synthesis or lumping or constructing. It creates larger units by assuming some relations are more significant than explicitly shown. Thus these strategies differ only in that they add or delete relations from the complex manifold presented to our point of view. These two abilities can be complementary, and can also keep us in paradox whenever we forget that they are heuristics and that the entities they create are approximations, not absolutes. Application of either lumping or splitting in excess can lead to serious errors.

63. It should be emphasized that Buddhist tradition also holds that entification in itself is useful or indispensible in the "conventional world"; it is only to the extent that we *reify* the entities (treat them as "real") that attachment, selfish craving, and aggression arise, causing all suffering. To reify something is to attribute to it an essence or independent existence (see footnote 1). Hence, reification violates the basic Buddhist principle that all things are interdependent. Similarly, reification is contrary to my analysis above, which holds that all entities are conditional approximations, only heuristic segmentations of a manifold, and therefore they cannot have the unconditional nature of an essence. Unfortunately, I believe that reification (a.k.a. belief in essences) derives from still another of our foundational simplifications: we assume constancy wherever possible. It is convenient (and usually correct) to act as if most of our environment is not changing much from minute to minute. In practice, we do not actively make any such assumption; rather, we just do not bother to compute new values for anything in

the field except our topic of interest, ignoring the background field or scanning it only for big changes. When pressed to reflect, we may give a pseudo-explanation in terms of essence: "If left alone, these things stay the same." This is just repeating that the old values still worked well. The notion of essence is deeply connected to the notions of cause and of explanation. See Rosch (1994) on explanation and cause, Wimsatt (1976a, 1976b) on reality and reductive explanation, and Garfield (1995) on conventional versus ultimate explanations by Nāgārjuna and other Buddhists.

64. It is natural that we would use the same basic strategies of entifying by lumping and splitting to deal with the complexity of our inner life. Thus person, present self, I, and all of the self-concepts and the not-me and others' points of view (other people's Is) appear as independent entities, created with boundaries that are computationally convenient for our day-to-day purposes. If on subsequent occasions we create a different segmenting for a conceptual field, we ignore the differences if possible. We do not generally tidy up the ad hoc approximations; if they work well enough, we do not look back and try to reconcile all the quasi-arbitrary boundaries into a single, consistent frame of reference. This describes how Science actually works, just as it describes how Ordinary Man deals with daily life. We only try to rationalize the differences when forced, by two kinds of circumstances. First, when our nonconscious approximations fail seriously, we may examine what other choices of boundaries were successful in the past. Second, when someone asks for an account of what we are doing, we try to come up with a socially acceptable and linguistically comprehensible formulation. In both cases we switch out of the present-centered action orientation and into what has been called an abstract attitude, or reflective mode, operating on internal data. The abstract attitude is not constrained by "clock time" (as any philosopher's spouse can testify), and thus we can access nonpresent-centered points of view such as the "me" of past circumstances, or the "me" of hypothetical circumstances. Nevertheless, our attempts to reconcile our past ad hoc approximations often fail. Gallagher and Marcel (1999), drawing on evidence from brain damage rehabilitation studies, argue convincingly that psychological laboratory experiments on the experience of self limit the subject to abstract attitudes and a detached stance, just as philosophical reflection does, and that unless subjects are embedded in pragmatically and socially contextualized situations their reports about self must be incomplete, distorted, or mistaken.

65. Although decomposition of a complex field into isolated entities is pow-

erfully useful, it sometimes leads to ignoring relations or differences that are functionally important. This *over*simplification is often associated with single-cause reasoning and its familiar disasters and, some would say, leads to the atomization and alienation of modern life. Only occasionally do we experience the world (and ourselves) primarily as a densely interconnected manifold. This bias against seeing the world as a pattern of relations (code-pendently arising) may be due in part to the constraints of our action orientation on the evolution of language and of awareness itself.

5B. THE STRUCTURE OF AWARENESS

66. Contrary to customary view, awareness is not monolithic; it includes several parts. The form of its contents is determined by the same old need to approximate. In his classic *Principles of Psychology* William James described awareness as based on two qualitatively different kinds of information (James 1950 [1890]:chapter 9). One hundred years later Mangan revived and reframed James's model in cognitive terms (Mangan 1991), recognizing that what James had described were two complementary types of abbreviation made necessary by the limited channel capacity of consciousness. My thinking along these lines was initiated by Mangan's thesis (1991). I have discussed my critiques, revisions, and extensions of James's and Mangan's views on the structure of awareness at length in previous papers (1994, 2000); here I can give only the briefest sketch to introduce terms and working hypotheses.

67. The first type of abbreviation in awareness is to display only those features of an object or event that are most relevant to our current goal, the ones that maximally discriminate among items or choices. I call it *feature awareness* (James called it "the Nucleus"). For example, on one shopping errand color may be the feature that most determines choice, and other features may not enter into awareness at that moment at all, although they are perceived and processed nonconsciously. Note that this is not just the selectivity of "attention" in general, which applies to both conscious and nonconscious processes. Feature awareness is *the form* in which this kind of content appears in awareness.

68. The second type of abbreviation is a large group of awarenesses that explicate the bare features. They carry the context and relational information that has been stripped away from feature awareness. I call them *explicating awarenesses* (James called them "the Fringe," and I previously called them

nonfeature awarenesses). One type presents feelings of the *relations among the features* (e.g., greater than, pair of, rhymes with). Another type presents *value judgments*. A third type was first described explicitly by Mangan (1991, 1993); it presents feelings that *implicitly condense or summarize nonconscious knowledge* related to the selected bare features but too extensive to fit explicitly in the limited capacity (e.g., as the *feeling of the meaning* of a word condenses or summarizes the extended semantic network related to the word's bare phonemes). These "explicating" awarenesses are felt contents of consciousness, specific, often vivid, worthy to be called qualia as much as any sensory content. They may or may not be at the center of attention. This group includes many experiences; e.g., the feeling of "if" and "but," of familiarity, of being on the right track, of knowing, of intending a specific movement, the feeling of specific semantic meaning of waiting for something unknown, and (I add) the felt component of emotions. I stress that both feature awarenesses and explicating awarenesses are usually present at the same time, may be brief or persistent, and either one may be the primary object of attention. For a full account see my taxonomy of awarenesses (Galin 1994).

69. Casual introspectors seldom report their explicating awarenesses explicitly. Language lends itself more to communicating features (via nouns, verbs, properties) and, unless we are verbally gifted, we make less use of grammatical mode, manner, and aspect to communicate explicating awarenesses. Further, we are drawn to the feature awarenesses because we are preoccupied with "doing," and the features are selected to be discriminative for our intended action. I believe the spiritual disciplines of contemplation, devotion, and service all promote and develop the explicating awarenesses of interrelatedness (Deikman 1997, 2000). Although explicating awarenesses are continually and necessarily part of Ordinary Man's subjective experience, deep and extended awarenesses of relatedness are infrequent, transient, and incomplete. Nevertheless, this may be the seed for Buddhism's "correct" view, which stresses the conditionality and interdependence of all things, person and self included. Meditation training takes one out of the stream of physical activity that demands the use of feature awarenesses. A saying attributed to the Zen tradition is that an enlightened man is an ordinary man who has nothing more to do. Explicating awarenesses can also be made prominent (sometimes inappropriately) by psychotropic drugs,[10] rituals and ceremonies, and some brain injuries (e.g., déjà vu) and hyperreligiosity in temporal epilepsy (Bear and Fedio 1977).

6. RELATING THE PRESENT THEORY TO BUDDHIST PRACTICES AND DOCTRINES

70. Summary thus far:

At the beginning of this essay I proposed that the concept of self is at the core of both Buddhism and Western psychology, and is problematic for both in many different ways. The Buddhist tradition holds that the suffering of unenlightened people is caused by an inborn erroneous view of the self as a persisting entity or essence. I suggested that we could begin to map the relations and disjunctions of the Buddhist and Western psychological accounts of self by focusing on this alleged inborn error. As background, section 2 reviewed some of the disarray in contemporary scientific thought about self. Section 3 introduced the surprising observations and theory from cognitive linguistics on how people actually use a rich and inconsistent system based on the nonconscious Divided Person metaphor in order to think about their selves and persons as perceiving subjects making willful acts.

71. In section 4 I clarified the foundation concepts of *entity, knowledge,* and *point of view,* and argued that *person, self,* and *I* are quite different kinds of things. I called for professionals to go beyond sectarian formalizations in how they talk and think about these matters and adopt a set of terms (preferably mine) consonant with a multidisciplinary frame of reference, commensurable with the cognitive sciences and contemporary philosophy of mind, still consistent with the intent of the most common usages. I accounted for the Divided Person metaphor's fundamental separation of "I" from one or more "selves" (self-concepts). Examples of how my terms and schemata can be applied to resolve apparent paradox were drawn from neuropsychology, and in particular from observation of split-brain patients.

72. In section 5 I returned to examining the "inborn" pretheoretic notions of self. I proposed that the central theme of ordinary human life is its *action orientation* and the consequent focus on entities. Our evolutionary adaptations for acting have shaped our nonconscious mental life and account for the structure of our consciousness itself as a combination of feature awareness and explicating awareness, which in turn accounts for the limited kinds of communication that language is good for. To reduce information overload we abbreviate immense complexity into manageable entities and do not tidy up the approximations if they work well enough. We apply the same heuristics to the complexities of our inner life, and the result is the view of self as an object in a world of objects.

73. Here I consider how this account, developed within the framework of Western cognitive science, relates to Buddhist views and practices in three areas: the necessary multiplicity of Ordinary Man's views and their pragmatic usefulness rather than their error, the issue of absolutism in the Buddhists' "correct view," and the centrality of the action orientation to human life generally and thus to Buddhist practice.

74. First, Ordinary Man frames the self and person in many ways because his limited information processing capacity cannot handle a single view comprehensive enough to deal with the complexity of our physical, mental, and social lives. Thus, it is necessary to use partial and approximate views that change as the context requires. This is a wonderful pragmatic solution, although it gives rise to a zoo of partially overlapping, partially contradictory self-concepts: the subject and the agent, the conscious and the unconscious, the real me, the outer and the inner, the ideal, the anticipated future me, the possibles, the many selves of the past, the anima and the animus and the shadow, etc.

75. Although it is recognized that Ordinary Man's views are inborn and useful, in some varieties of Buddhist discourse there still seems to be some moral opprobrium clinging to these "errors," although not as severe as the opprobrium of Augustinian Christianity for the similarly unavoidable Original Sin. This tone persists, even though Buddhist traditions like Madhyamaka hold that heuristics of entification and even conceptual philosophizing are indispensable in the "conventional world." Perhaps more emphasis could be given to the multiple inborn views as useful approximations rather than as errors. When suffering does arise it is from misuse of these heuristics. Buddhist doctrine moves in this direction when it declares that the danger is in the "reification" of entities rather than in entification per se. It is only to the extent that we *reify*, and in particular reify the self, that attachment, selfish craving, and aggression arise, causing all suffering. Unfortunately, I believe reification itself derives from another of our inborn, foundational simplifications, that of *assuming constancy* as much as possible (see paragraph 65, above). Reification therefore is not just due to ignorance or perversity. Nevertheless, according to some Buddhist schools, people who achieve Buddhahood can actually experience the world without the illusions caused by simplification, entification, conceptual designation, and reification. Those of lesser attainment can recognize their experience as illusion, but they will still experience the illusion, just as man who is familiar with the desert may see a mirage but know enough not to chase after it.

76. The most common misuse of approximations is the overemphasizing of

features and entities and the neglect of explicating contexts and relations. It is certainly the most common in Western "scientific" culture. But suffering can arise from misuse of any of these heuristics. Although explicating awarenesses may be developed into the Buddhist's "correct" perception of ourselves and the world as a matrix of dependently related events, yet even this can be overemphasized. Thus Buddhist tradition warns against neglecting the entities by getting stuck in "forest quietism" or "blissing-out" in a sense of oceanic connectedness.

77. Second, another pernicious error, dominant among Western scientists and philosophers alike, is failing to see that *all* our views are necessarily heuristics, not absolutes. It is for this failure that cognitive science can refute rationalism and naive realism (e.g., see Lakoff and Johnson 1999:74ff., 94ff.). So pervasive is this misuse that even some Buddhist schools and texts seem to take codependent arising as an ontological absolute. There is a ubiquitous distinction in the texts between "conventional reality" and "ultimate reality." It seems that *ultimate* is sometimes used in the sense of "better" as well as "final"; the ultimate is "more real" and has more authority and value than the conventional. Garfield notes the debates over this interpretation in ancient and modern commentaries (299). Collins attributes these debates to two different uses of the doctrine of *No-self*, "as the description of reality vs. as an instrument of salvation," i.e., he suggests it is understood as a heuristic by sophisticates concerned with preserving and clarifying the conceptual content of Buddhist theory, and in contrast is used as an absolute by novice monks earnestly engaged in meditative reflection to achieve *nirvāṇa*. He also notes the subtle difference between "the 'right view' of no-self, which opposes other 'wrong views' . . . and Nāgārjuna's 'no-view' approach, a . . . moral and epistemological attitude towards the activity of conceptualization per se," i.e., that right and wrong are themselves conditional categories (see also Garfield 1995:299, 304–305). Collins suggests that the same finessing of the issue may be served by the Buddha's refusal to reply to the famous "unanswered questions" in the scriptures (Collins 1982:131–138). The refusal "does not mean that such an analysis is not possible, nor that Buddhism as a whole rejects all speculation. . . . What is rejected is harmful speculation based on mistaken premises. . . . This is not a universal recommendation to an 'empiricism' or practical antimetaphysical agnosticism, but a piece of advice given to an enthusiastic but misled admirer." This is another debate beyond my limited scholarship, but the beginning student should be aware that it awaits her.

78. Third, I want to highlight the importance of humans' *action orientation,*

and how and why it is addressed in Buddhist practice. Academics, in particular, seem to have little appreciation for how deeply action is embedded in our human nature, and what a pivotal role action plays in determining our mental life. The reader may find it instructive to consider how different are the life challenges for a person than for a tree, simply because we are highly mobile. It seems reasonable to presume that evolutionary development *necessarily* proceeds from simple physical mobility to relocating adaptively within the environment, to finely differentiated ability to manipulate the environment, which in turn provides an opportunity to evolve the ability to model future states based on our possible acts and to choose among them. Each succeeding development incorporates rather than replaces the prior ones. I have discussed how we segment the manifold into those entities that will afford the best possibilities for action, and how even our abstract conceptual thinking is based on action metaphors, which perhaps grow out of the underlying action neurophysiology. The paradoxes and inconsistencies of I and self-concepts that derive from the central Divided Person metaphor were accounted for in part by the I/agent's need to be in present time for the action-perception-action loop. We are very much occupied with *doing,* pressed into action by the imperatives of the changing world.

79. Thus, since Buddhism seeks to deconstruct the Ordinary Man's view of the world and himself in it, and provide corrective experience, it is reasonable to find that many Buddhist practices serve to modify the seeker's action orientation. In many ways, *nonaction* is promoted. This is a subtle concept, found in many contemplative disciplines. It is not simply immobility; Buddha preached specifically against the Forest Quietists (see also Suzuki 1949 on Hui Neng's doctrine of No-mind and Smith 1992 on Taoist problems with slipping from *wu wei,* "no-action," into quietism). Nonaction has more to do with will, intention, and agency than simply with movement. The intertwining of action with will, voluntariness, agency, intention, motivation, and desire is a thorny thicket, not yet resolved by Western philosophy or science. Currently there is a growing and productive concern with will and agency in neurology and cognitive neuroscience (Frith and Done 1989; Goldberg 1985; Jeannerod 1994; Lhermitte 1983; Libet 1985). I note in passing that the complex ideas of karma, so deep in Vedic as well as Buddhist thought, concern these same aspects of action, not simply the physical *doing.* Physicist John Wheeler (1991) also finds surprising links in a quantum theory framework between future events and present choosing.

80. Sitting meditation, which is central to Buddhist practice, particularly

eliminates physical doing and progressively limits mental doing. The action-observation-action loop is interrupted, which allows the *explicating awarenesses* to come to more prominence than *feature awareness*. In the absence of an action program, all that the explicating awareness can report on is the state of the system itself, and such reparsing of the self itself creates the opportunity for new, more integrated structure to form. This may be the core of the "corrective" experience of the self.

81. Getting out of the action loop without slipping into discursive thought or sleep is not a trivial achievement; this is one aspect of the stabilization, focussing, and intensifying of mind that is sought in *śamatha* attention training. But quiesence, or *śamatha*, only permits but does not produce the "corrective" experience, and so other practices may be added. Meditation traditions differ in the degree to which they stress passive observation. In some traditions, after mental stabilization is achieved, didactic exercises may include an active exploration and testing of states as described in the texts, just as the Jesuits might encourage active philosophical exploration of doctrine. Whereas in ordinary philosophical reflection one introspects within the usual framework of categories, in meditation one is often asked to abandon categories or forced to abandon them by exercises such as the Zen koan. The Madhyamaka school may make more use of exploring specific categories to direct the explicating awareness to various parts of the system.

82. Other traditional practices also serve to modify the *action orientation*. The imperatives of the changing world are somewhat decreased by joining a monastic community or temporarily going into retreat. Submitting to a community or a master to some extent shifts the burden of agency to them (as seen most dramatically in hypnotic suggestion). Performing compassionate service to others promotes a shift in the I to incorporate the other's point of view (Deikman 1996, 2000). Repetitive group chanting also shifts control of action to the group and to the chant itself, and the repetitiveness loosens the connection to the present in clock time. See Deikman's personal report of the transfiguration of his experience as "his self vanished" while chanting during a prolonged Zen retreat (1997). All of these practices seem to weaken the action-oriented experience of self, thereby allowing a new experience and a new integrated structure to emerge. However, in general, the goal is not to remain in a meditative state dominated by the explicating awarenesses but to achieve a new point of view and return to action in the world.

CODA

83. The central theme emerging is that humans are participant quasi entities in the dynamics of the world manifold, and this circumstance mandates our action orientation. We adapt to constant change and information overload by heuristic approximations, by entifying and simultaneously mitigating the oversimplifications with contextual knowledge presented in explicating awarenesses. Perhaps the human species should be renamed as *Homo approximatus, subspecies pragmaticus.* The same approximating, of course, is applied to the complexities of our "inner life," and leads to the confusion, internal contradictions, and paradoxes in our thought and speech about person, self, and I. A second theme in this essay is my defense of the maligned Ordinary Man's inborn view of person and self. It is better seen as an essential evolutionary adaptation rather than as erroneous, as more complex and multifaceted than simply ignorant. Although it is open to misuse, the remedy of the misuse is to rebalance rather than replace.

84. The reconception of *person, self,* and *I* that I propose is compatible with the main lines of Buddhist concepts and indicates several points of contact between Buddhism and contemporary psychology. I have suggested definitions that may be conceptually challenging to some, but are much more explicit than the usual:

> A person is a dynamically changing, self-organizing, multilevel, quasi entity without sharp boundaries, and embedded in a causal thicket; self is the current organization of the person; and I is the self's point of view, its set of currently possible discriminations.

My account of self-as-dynamic-organization resembles Buddhism's whirling aggregate of *skandhas* but may have more structure; I focus more on the various degrees of stability rather than the stark dichotomy of ephemeral versus eternal, transience versus timelessness. My definition of entity as always conditional, dependent on our purposes, constrained by context, is completely in accord with the Buddhist view of the phenomenal world. However, I also note the degree to which invariance or stability *does* occur. A state that persists within some range for an hour is different than one that persists for a second or a decade, and the difference may be worth our while to talk about.

85. Both my schema and Buddhist tradition situate persons in the rest of the natural world, rather than considering them to be in some way unique (as

for example, by possessing a soul or a special essence). In the key Buddhist concept of *anātman,* self is not an essence, not a substance, not unchanging, not independently arising. As expressed by Nāgārjuna (Garfield 1995), there is no independent being, only relationship and its dynamics. The key idea in my schema of self is organization, which is also nothing but dynamic relationship. As such, both systems imply the possibility of transformation, of reorganization.

86. Buddhism emphasizes meditation as the way to provide a corrective experience of self and of the manifold in which self is embedded. I have introduced the idea that meditation breaks up the action orientation and trains the ability to focus on explicating awarenesses rather than on feature awareness. In all humility this essay develops such bridging concepts *merely to begin* translating into Western psychological terms those experiences and underlying cognitive changes that occur in the course of training toward enlightenment. A preliminary framework such as this makes it possible to hypothesize about how transformational practices like *śamatha,* service and compassion, all-night group dancing, and drugs might work in terms of cognitive psychology and brain subsystems. I have limited myself, as much as possible, to discussing the processes and experience of seeking transformation and said nothing about the experiences of transformation itself, whether by sudden insight, slow reorganization, or other means. I hope this framework will be useful to those who have personal knowledge of transformation. It is up to them to distinguish the experiences related to transformation per se from the secondary emotional and physical reactions to it and, further, to characterize the experiences of living in ordinary contexts after that reorganization.

87. The doctrine of no-self, *anātman,* is often misunderstood as yet another example of Buddhism's alleged pessimism and nihilism. On the contrary, the Buddhist solution to the modern suffering of alienation and anomie is to completely contextualize self, not to simply erase it. This seems to be remarkably consonant with trends toward holism in Western thought that go beyond psychology. The West's new appreciation of context is shown by mounting interest in ecology, in sustainable practices in relation to the natural environment, and even to society conceived as an environment. There is a broadening emphasis on understanding the relation of the part to the whole that seems to me all of a piece with efforts more conventionally identified as psychological or spiritual work. The exploration of person and self in this essay is intended to contribute to it.

Notes

For e-mail: dgalin@itsa.ucsf.edu. For snail mail: 5 Mount Hood Court, San Rafael, CA 94903–1018.

1. Please note that neither "Buddhism" nor "Western Psychology" is homogeneous, and these sweeping collective terms hide important and even contradictory internal differences. Nevertheless, some generalizations apply, and that is the main concern of this essay. Furthermore, I acknowledge the limitations of conceptual efforts to treat what many believe to be beyond concepts, beyond self/object and self/other; this essay is for those who want to explore whatever may be possible within these limits.

2. For a helpful account of the notion of *essence* in Western folk psychology and philosophy, see the index entries in Lakoff and Johnson 1999. The essence of something is the postulated timeless totally intrinsic aspect of the thing which gives it its properties. When essence is used as causal explanation it is just tautology, adding nothing; e.g., "Opium makes one sleepy because it has a *dormative essence.*" The idea of essence is at the core of Greek thought and its legacy down to the present. Conversely, it is negatively central to Buddhism, as the idea which is specifically denied and rejected from the Vedic and Dravidian milieu in which Buddhism arose, and as that which is to be overcome by each practitioner. See Garfield 1995:89, and his other index entries. For a deeper analysis of what makes an explanation satisfying, see Rosch 1994.

3. The experience of self as deeply divided is well known in many branches of psychology, religion, and the arts. This cognitive psychology analysis which focusses on how this split manifests itself in language is just the latest.

4. Self-organizing systems can become much more complex than the elements they started with. See Holland 1998 for principles governing the development of emergent properties.

5. A thicket is a kind of organization with so many feedback loops that it is not orderable into clear levels. The term was introduced by Wimsatt to contrast with the more familiar notion of a hierarchy. Wimsatt 1976a:254; see also pp. 251–256, 237 ff.

6. For a penetrating analysis of the differences between entities and aggregates see Wimsatt 1974; for an account of parts and wholes, levels of analysis and their components and contexts, see Wimsatt 1976:237ff.; and Simon 1969.

7. I have never understood what is entailed or excluded by "direct" in this usage. It seems to mean *im-mediate*, without mediation. As discussed in Galin 1999, I have a problem with disembodied forms.

8. A form contains unique information in so far as it differs from all other distributions in that frame of reference. How the information is embodied, represented, or encoded in particular cases does not concern us here. See Wheeler 1991.

9. I want to acknowledge Arthur J. Deikman's original contributions to much of the present account. My thinking in this area evolved in the course of our thirty years of conversations. Deikman pioneered empirical psychological studies of meditation, distinguished two "modes of consciousness" that served complementary purposes, and called my attention to the central role of action in human organization and the significance of service as a spiritual practice. See Deikman 1963, 1971, 1996, 1997, 2000.

10. For example, this report is typical of casual marijuana users: "When I smoke it, I see the poetry in what otherwise would appear to be the random placement of objects in space" (Tart 1972).

References

Bear, D. M. and P. Fedio. 1977. "Quantitative Analysis of Interictal Behavior in Temporal Lobe Epilepsy." *Archives of Neurology* 34:454–467.

Bisiach, E. 1991. "Understanding Consciousness: Clues from Unilateral Neglect and Related Disorders." In A. D. Milner and M. D. Rugg, eds., *The Neuropsychology of Consciousness*, pp. 113–137. London: Academic.

Bogen, J. E. 1986. "Mental Duality in the Intact Brain." *Bulletin of Clinical Neurosciences* 51:3–29.

——— 1990. "Partial Hemispheric Independence with the Neocommisures Intact." In C. Trevarthen, ed., *Brain Circuits and Functions of the Mind: Essays in Honor of Roger W. Sperry.* Cambridge: Cambridge University Press.

Collins, S. 1982. *Selfless Persons.* Cambridge: Cambridge University Press.

Deikman, A. J. 1963. "Experimental Meditation." *Journal of Nervous and Mental Disease* 136(4): 329–343.

——— 1971. "Bimodal Consciousness." *Archives of General Psychiatry* 45:481–489.

——— 1996. "Intention, Self, and Spiritual Experience." In S. Hameroff, A. Kaszniak, and A. Scott, eds., *Toward a Science of Consciousness: The First Tucson Discussions and Debates*, pp. 30–35. Cambridge: MIT Press.

——— 1997. "The Spiritual Heart of Service." *Noetic Sciences Review* (Winter), pp. 30–35.

——— 2000. "Service As a Way of Knowing." In E. Hart, ed., *Noetic Sciences Review.* Albany: State University of New York Press.

Frith, C. D. and D. J. Done. 1989. "Experiences of Alien Control in Schizophrenia Reflect a Disorder in the Central Monitoring of Action." *Psychological Medicine* 19(2): 359–363.

Galin, D. 1974. "Implications for Psychiatry of Left and Right Cerebral Specialization: A Neurophysiological Context for Unconscious Processes." *Archives of General Psychiatry* 31(4): 572–583.

——— 1992. "Theoretical Reflections on Awareness, Monitoring, and Self in Relation to Anosognosia." *Consciousness and Cognition* 1(2): 152–162.

———— 1994. "The Structure of Awareness: Contemporary Applications of William James's Forgotten Concept of 'the Fringe.' " *Journal of Mind and Behavior* 15(4): 375–400.

———— 1999. "Separating First-Personness from the Other Problems of Consciousness." *Journal of Consciousness Studies* 6:222–229.

———— 2000. "Comments on Epstein's Neurocognitive Interpretation of William James's Model of Consciousness." *Consciousness and Cognition* 9:576–583.

Galin, D., J. Johnstone, L. Nakell, and J. Herron. 1979. "Development of the Capacity for Tactile Information Transfer Between Hemispheres in Normal Children." *Science* 204(4399): 1330–1332.

Gallagher, S., and A. J. Marcel. 1999. "The Self in Contextualized Action." In S. Gallagher and J. Shear, eds., *Models of the Self*, pp. 273–299. Thorverton: Imprint Academic.

Garfield, J. L. 1995. *The Fundamental Wisdom of the Middle Way: Nāgārjuna's Mūlamadhyamakakārikā.* New York: Oxford University Press.

Goldberg, G. 1985. "Supplementary Motor Area Structure and Function: Review and Hypotheses." *Behavioral and Brain Sciences* 8:567–616.

Goldstein, K. 1939. *The Organism, a Holistic Approach to Biology Derived from Pathological Data in Man.* New York: American Book.

Hilgard, E. R. 1977. *Divided Consciousness: Multiple Controls in Human Thought and Action.* New York: Wiley.

Holland, J. H. 1998. *Emergence.* Reading, Mass.: Addison-Wesley.

Hopkins, J. 1983. *Meditation on Emptiness.* London: Wisdom.

———— 1987. *Emptiness Yoga: The Tibetan Middle Way.* Ithaca: Snow Lion.

James, W. 1950 [1890]). *The Principles of Psychology.* New York: Dover.

Jeannerod, M. 1994. "The Representing Brain: Neural Correlates of Motor Intention and Imagery." *Behavioral and Brain Sciences* 17(2): 187–245.

Kihlstrom, J. F. 1987. "The Cognitive Unconscious." *Science* 237:1445–1452.

———— 1993. "The Psychological Unconscious and the Self. *Ciba Foundation Symposium* 174:147–156.

Lakoff, G. 1987. *Women, Fire, and Dangerous Things: What Categories Reveal About the Mind.* Chicago: University of Chicago Press.

Lakoff, G. and M. Johnson. 1999. *Philosophy in the Flesh: The Embodied Mind and Its Challenge to Western Thought.* New York: Basic.

Leslie, A. M. 1987. "Pretense and Representation: The Origins of 'Theory of Mind.' " *Psychology Review* 94:412–426.

Lhermitte, F. 1983. " 'Utilization Behaviour' and Its Relation to Lesions of the Frontal Lobes." *Brain* (part 2) 106(2): 237–255.

Libet, B. 1985. "Unconscious Cerebral Initiative and the Role of Conscious Will in Voluntary Action." *Behavioral and Brain Sciences* 8:529–526.

Mangan, B. 1991. *Meaning and the Structure of Consciousness: Essay in Psycho-aesthetics.* Ph.D. diss., University of California, Berkeley. University Microfilms no. 92033636.

————— 1993. "Taking Phenomenology Seriously: The "Fringe" and Its Implications for Cognitive Research." *Consciousness and Cognition* 2(2): 142–154.

Puccetti, R. 1981. "The Case for Mental Duality: Evidence from Split-Brain Data." *Behavior and Brain Sciences* 4:93–123.

Rosch, E. 1978. "Principles of Categorization." In E. Rosch and B. B. Lloyd, eds., *Cognition and Categorization*. Hillsdale, N.J.: Lawrence Erlbaum.

————— 1994. "Is Causality Circular? Event Structure in Folk Psychology, Cognitive Science, and Buddhist Logic." *Journal of Consciousness Studies* 1:50–65.

Sacks, O. W. 1984. *A Leg to Stand On*. New York: Summit.

Sherrington, C. S. 1947. *The Integrative Action of the Nervous System*. Cambridge: Cambridge University Press.

Simon, H. 1969. "The Architecture of Complexity." *The Sciences of the Artificial*, pp. 193–229. Cambridge: MIT Press.

Smith, H. 1992. *Tao Now: Essays in World Religion*, pp. 71–92. New York: Paragon House.

Sperry, R. W. 1968. "Hemisphere Deconnection and Unity in Conscious Awareness." *American Psychologist* 23(10): 723–733.

Sperry, R. W., M. S. Gazzaniga, and J. E. Bogen. 1969. "Interhemispheric Relationships: The Neocortical Commissures, Syndromes of Disconnection." In J. J. Vinken and G. W. Bruyn, eds., *Handbook of Clinical Neurology* 4:273–290. Amsterdam: Elsevier.

Sperry, R. W., E. Zaidel, and D. Zaidel. 1979. "Self Recognition and Social Awareness in the Deconnected Minor Hemisphere." *Neuropsychologia* 17(2): 153–166.

Strawson, G. 1997. "The Self." *Journal of Consciousness Studies* 4:405–428.

————— 1999. "The Self and the SESMET." *Journal of Consciousness Studies* 6:99–135.

Suzuki, D. T. 1949. *Zen Doctrine of No-Mind*. London: Ryder.

Tart, C. T. 1972. "States of Consciousness and State-Specific Science." *Science* 176:1203–1210.

Vallar, G., G. Bottini, M. L. Rusconi, and R. Sterzi. 1993. "Exploring Somatosensory Hemineglect by Vestibular Stimulation." *Brain* 116:71–86.

Wallace, B. A. 1989. *Choosing Reality: A Buddhist View of Physics and the Mind*. Ithaca: Snow Lion.

————— 1998. *The Bridge of Quiescence: Experiencing Tibetan Buddhist Meditation*. Chicago: Open Court.

————— 1999. "The Buddhist Tradition of Samatha: Methods for Refining and Examining Consciousness." *Journal of Consciousness Studies* 6:8–16.

Wheeler, J. A. 1991. "Information, Physics, Quantum: The Search for Links." In W. H. Zurek, ed., *Complexity, Entropy, and the Physics of Information*, pp. 1–28. Redwood City, Cal.: Addison Wesley.

Wimsatt, W. C. 1974. "Complexity and Organization." In K. F. Schaffner and R. S. Cohen, eds., *Boston Studies in the Philosophy of Science* 20:67–86. Dordrecht: Reidel.

———— 1976a. "Reductionism, Levels of Organization, and the Mind-Body Problem." In G. Globus, G. Maxwell, and I. Savodnik, eds., *Consciousness and the Brain*, pp. 205–266. New York: Plenum.

———— 1976b. "Reductive Explanation: A Functional Account." In C. A. Hooker, G. Pearse, A. C. Michealos, and J. W. v. Evra, eds., *Proceedings of the Meetings of the Philosophy of Science Association, 1974*. Dordrecht: Reidel.

According to Buddhism, confusion about the nature of the self is deeply related to the origins of evil and human-inflicted suffering, and in the following essay William S. Waldron presents a cogent comparative study of these themes from the perspectives of Buddhism and the cognitive, biological, and social sciences. The strategy he adopts is to move the discussion of human identity from philosophy to an investigation of the profound impact upon human lives of our tendency to reify ourselves as substantial, independent, unchanging entities. It is the Buddhist view that our misguided attempts to sustain such constructed "identities" in our changing and interdependent reality lead to the preponderance of suffering caused by human actions; they lead, in short, to "evil." Like David Galin, Waldron argues that, from the perspectives of both Buddhism and important parts of modern science, we may understand ourselves and our world more deeply and fully if we conceive of all phenomena, including ourselves, as interconnected patterns of relationships rather than as reified entities somehow existing independently of their own developmental history, their internally differentiated processes, or their surrounding, supporting conditions.

In his examination of the interface between Buddhism and evolutionary biology, Waldron points out the disturbing discrepancy between behaviors that were successful in our ancestral past and the needs and conditions of our current situation in the modern world. Like Galin, Waldron argues that our sense of an independent I or self may not be simply a disastrous mistake, as asserted in Buddhism, but an eminently practical construct that, conventionally speaking, fulfills important purposes and hence must have enjoyed evolutionary advantage. That is to say, the actions (karma) instigated by this sense of an independent I led to reproductively successful results that, while evoking the mental afflictions, reinforced and strengthened that very sense of self. And it is only by experientially realizing the constructed, illusory nature of all phenomena, including oneself, that one is liberated from all mental afflictions and their resulting evil and suffering. Here, certainly, is a crucial point for dialogue and collaborative research between Buddhist scholars and evolutionary biologists and developmental psychologists.

Waldron presents a prolegomenon to an interdisciplinary social theory of evil based upon the three basic Buddhist principles: 1. the dependent nature of all phenomena, 2. the constructed nature of "self-identity," and 3. the apprehension that human evil and ill are caused by attempts to secure our "selves" at the expense of "others." Despite important differences between Buddhist and scientific ideas concerning the nature of human identity and the origins of evil, this essay points out numerous areas of convergence, where each of these fields of inquiry may be deepened by engaging with the other.

The world in its variety arises from action ("Karmajaṃ lokavaicitrayam").
—*Vasubandhu*

A path is made by walking it ("Tao hsing chih er cheng").
—*Chuang-tzu*

All sentient beings are deranged ("Sabbe sattā ummattakā").
—*Gotama Buddha*

William S. Waldron

Common Ground, Common Cause: Buddhism and Science on the Afflictions of Identity

It is no small task to understand the vast, variegated world we humans have carved out for ourselves on this small planet. How does one know where to begin, what to interrogate, and to what end? Events, however, have a way of imposing themselves. As the cold war melts down and bitter ethnic and religious conflicts heat up the world over,[1] as endless images of death and violence flash daily across the globe, the multiple faces of human evil and suffering stare steadfastly into our own, intimating, we fear, an inescapably inhumane reality. Our task then, our moral imperative, is as urgent today as it was when Albert Camus (1971:11) expressed it nearly fifty years ago, just as many millions of murders ago: "One might think that a period which, within fifty years, uproots, enslaves, or kills seventy million human beings, should only, and forthwith, be condemned. But its guilt must also be understood." This essay is an attempt to take this challenge seriously, an attempt to understand the awful dynamics of human-inflicted suffering, of "man's inhumanity to man" in traditional parlance, of—in a word—evil. Human beings make war and kill each other in a way that no other species does, that no other species could, that no other species would. Somehow, we must make sense of it all. We must be able to discern some pattern, some common dynamic, behind behaviors that are repeated so terribly often, in so many times, in so many places. As Camus suggests, such an understanding—however repugnant its details, however unpleasant its conclusions—is required to even begin preventing them.

Understanding, however, is not only what we require, it is also what we must interrogate. For, we shall see, it is understanding itself, imperfect, wrong-headed understanding of our human condition, that lies deeply and malignantly behind these unholy dynamics of human evil. It is this mistaken understanding of ourselves—as individuals, as members of social groups, and as a contingent, historical species—that we must address. We must understand not only the passions that drive men to evil but the confusion over our condition that makes such evil possible.[2] The tenacity and pervasiveness of these tragic strains in the human condition—our "fallen state" as it were—have been recognized and addressed by nearly all religious traditions. In seeking to understand these darker sides of human life, however, we shall draw upon the conceptual resources of only one such tradition, Indian Buddhism, in dialogue with comparable areas of inquiry from the biological and social sciences.[3] As with any dialogue, we appeal to no external or superordinate authority; it is the cogency of the arguments that count, their compelling and persuasive power, whatever their provenance.

This dialogue is only possible because recent developments in Western thought and science have begun to find common ground with traditional Buddhist perspectives on the human condition, including the underlying conditions of human evil. There is a growing consensus that we may understand ourselves and our world more deeply and fully if we conceive of things in terms of interconnected patterns of relationships rather than as reified entities existing somehow independently of their own developmental history, their internally differentiated processes, or their enabling conditions. There exists, that is, an increasing recognition that thinking in terms of unchanging essences, entities, and identities deeply misconstrues the human condition—a misunderstanding that inadvertently leads to, rather than alleviates, human evil and suffering.

Although expressed differently in various fields, the relationship between our misunderstanding of the human condition and its causal influences upon evil and suffering have been articulated exceptionally clearly, directly, and comprehensively in the principles of classical Buddhist thought, which provide the conceptual framework for this essay: 1. that all "conditioned phenomena" (*saṃskṛta-dharma*) are radically dependent (*pratītya-samutpāda*) and hence lack any fixed or unchanging "essence" (*svabhāva*); 2. that what we are, rather, are assemblages of dynamic yet wholly conditioned "constructs" (*saṃskāra*) that have been painstakingly carved out (*upādāna*) of these contingent dependent relationships; 3. that we tend to construe these assembled constructs as substantial "selves" or fixed identities (*ātman*); 4. that in our efforts to fashion and secure such an "identity" we actively ignore and attempt to counteract its contingent, constructed nature; and, finally, 5. that these efforts effectively channel human activities (karma) into the repetitive behavioral patterns that actually bring about more evil and suffering. These activities, in short, represent misguided and futile efforts to deny our dependence, to counteract our impermanence, and to attain lasting security for this putative, substantial "self"—attempts, as the Buddhists would say, to "turn reality on its head."[4]

While the basic ideas of essencelessness, contingency, and construction of identity are straightforward enough, it requires considerable thought— and sufficient specifics—to appreciate the profound implications these have for our understanding of human life. We shall therefore draw upon various Western sciences for many of the details to support and flesh out this perspective, attaining in the process, we hope, a more compelling understanding of the dynamics of human evil than either the Buddhists or the sciences have yet to articulate on their own.

THE AFFLICTION OF IDENTITY: THE BASIC BUDDHIST PERSPECTIVE

In the classical Buddhist perspective the sufferings endemic to human life are ultimately brought about by the construction of and a deep-seated attachment to our sense of a permanent identity, what we mistakenly take to be a unitary, autonomous entity, independent of and isolated from the dynamically changing and contingent world around us. This common but misguided apprehension of such a self is succinctly defined in one of discourses attributed to the Buddha: "It is this self of mine that speaks and feels, which experiences here and there the result of good and bad actions; but this self of mine is permanent, everlasting, eternal, not subject to change, and it will endure as long as eternity" (Ñāmoli 1995:92, M I 8).[5]

From the Buddhist perspective what we all are, rather, are ever-changing conglomerates of processes (*skandha*) formed in self-organizing patterns that are ever open,[6] like all organic processes, to change, growth, and decay based upon the natural functions of assimilation, interpenetration, and dissolution.[7] What we often think of as an essential or fixed "identity" or "nature" is, in this view, actually a complex construct generated by misunder-standing, forged by emotional attachments, and secured by endless egocentric activities. Identity is constructed, that is, by what the Buddhists call the "three poisons," the primary afflictions of ignorance, attachment, and aggression.[8] But since any identity constructed and construed within a dynamic and contingent environment is necessarily unstable and insecure, it requires constant reinforcement and protection. From the Buddhist point of view, it is our ultimately futile efforts to permanently secure such fragile, fabricated identities through *activities* driven by ignorance, attachment, and aggression that ironically create our insecurities, dissatisfactions, and frustrations—consequences to which we *tend* to respond with yet further efforts to shore up our shifting selves through further attachment, aggression, etc.[9] We are caught, in short, in an unending and unhealthy feedback cycle of repetitive, compulsive behavioral patterns, what the Buddhists call the vicious cycle of *saṃsāra*.

This bears repeating: in the Buddhist view it is our misguided attempts to sustain such constructed identities in our changing and contingent reality that lead to the preponderance of suffering caused by human actions, that lead, in short, to "evil." Evil and suffering, that is, are the unintended yet inevitable consequences of our tendencies to reify relationships and processes into unchanging things, to abstract characteristics and qualities in terms of fixed essences or natures, and, most egregiously, to identify ourselves as singular, substantive selves in contrast to and standing apart from our surrounding, sustaining environments. Identity is thus not only a construct based upon an ignorant and untenable dichotomy between self and not-self but also, almost inevitably, leads to attachment to "us" and "ours" and aggression toward "them" and "theirs," the very processes that lead, in the extreme, to the horrible inhumanity we are attempting to understand.

We shall explicate these dynamics of human evil by first describing in Buddhist terms the multidimensional influences that the three poisons—ignorance, attachment, and aggression—impart on the evolution of life in general and of humans beings in particular. Identity and its incumbent afflictions are, we shall argue, inseparable from and most explicable in terms of the long-term conditioning processes of evolutionary development. We will then extrapolate the influences of these afflictions into the terms and concerns of several of the biological and social sciences, in what amounts to a prolegomenon to a "natural history of the affliction of identity."[10]

COMMON GROUND: BUDDHISM AND BIOLOGY ON EVOLUTION, EMBODIMENT, AND ENACTION

The world in its variety arises from action. ("Karmajaṃ lokavaicitrayam")
—*Abhidharmakośabhāṣya IV 1*

The three poisons, and their specific expressions, the afflictions (*kleśa*),[11] are arguably the core concepts in the Buddhist explanation of the origins, development and functioning of our psychophysiological processes—what corresponds (roughly) to phylogeny, ontogeny, and psychology, respectively. Similar to theories of evolutionary causation, the Buddhists envision a deep interdependence between the long-term processes that have brought about the human species, the behavioral patterns specific to our present human embodiment, and their particular enactment in our ongoing activities that are enabled and influenced by the first two. Though for the sake of analysis and exposition we will discuss these three—evolution, embodiment, and enaction—separately, we must bear in mind their profound interdependence. For, ultimately, in both the Buddhist and biological perspectives, it is the actions of living beings inseparable from their sustaining environments that, over the long term, give rise to evolutionary change. We shall first trace the influences of the three poisons in the Buddhist worldview and then examine analogous models of causality within evolutionary biology.

COSMOGONIC DIMENSIONS: PROPELLING THE VICIOUS CYCLE (*SAṂSĀRA*)

Buddhists contend that actions informed by the three poisons and the other afflictions play a prominent role in bringing about the structure and conditions of the sentient world we inhabit.[12] At first blush, it may be difficult to imagine how such seemingly "psychological" processes could bring about the forms of life on this planet. But not only is this notion central to classical Buddhist cosmogony, it is also, after a fashion, deeply congruent with the perspectives of evolutionary biology.

Though not a cosmogonic myth proper,[13] one early discourse of the Buddha, "On Knowledge of Beginnings," describes how heavenly beings living in an interim period between world-cycles gradually devolved into hu-

man beings living in the world we know (Walshe 1987, *Aggañña Sutta*, D iii
81f.; I have paraphrased the relevant passages).

> At first, the heavenly beings were "mind-made, feeding on delight, self-lumi-
> nous, moving through the air, glorious." At that time, everything was an un-
> differentiated mass of water and darkness, with neither sun nor moon nor
> stars, neither day nor night, nor months, years or seasons. The heavenly be-
> ings had no distinguishing sexual characteristics.
>
> A sweet earth spread out over the waters and one of the beings, who was
> greedy, broke off a piece and ate it. Liking it, craving arose for it, and others
> soon followed suit. Their self-luminance gradually diminished, the sun and
> the moon appeared, day and night, the months, the years and the seasons
> arose. The world gradually re-evolved.
>
> As they fed upon such food, their bodies became coarser and physical dif-
> ferences appeared among them. The good-looking disparaged the ugly, and,
> as they became arrogant and conceited about their looks, the sweet earth
> slowly disappeared. There followed a succession of coarser foods, leading to
> yet more physical differences among them, and more arrogance and conceit
> in turn.
>
> There gradually arose a kind of huskless rice, which grew all by itself and
> could be repeatedly harvested for every meal. As the beings ate this coarser
> food, they became sexually differentiated and altogether preoccupied with
> each other. "Passion was aroused and their bodies burnt with lust." Those who
> indulged themselves accordingly were expelled from their communities and
> forced to live apart from others, although everyone eventually developed
> these same habits.
>
> Now one of them, through laziness, decided to gather up enough rice for
> two meals instead of one; others followed suit. Eventually, husks began to de-
> velop around the rice grains and it would no longer replenish itself naturally.
> Labor was now required. And as the rice no longer grew just anywhere, rice
> fields were eventually established and parceled out to those who worked
> them. With ownership thus instituted, envy and then stealing appeared, and
> punishment, lying,and false accusation ensued. Eventually, social rules, the
> mechanisms for their enforcement, and the distinctions underlying them de-
> veloped, resulting in the complex and stratified social world we now inhabit.

We need not consider this Rousseauian "fall from grace" as an early
Buddhist analogue of Genesis in order to appreciate its central theme: that
we and the world we inhabit came about because of arrogance, greed, pas-

sion, and envy.[14] In Buddhist terms, our particular world developed as a result of the particular actions (karma) of sentient beings that were instigated by specific afflictions (*kleśa*). These same causal principles were succinctly expressed by the fifth-century Indian Buddhist, Vasubandhu (AKBh V 1a; Shastri 759; Poussin 106):

> The world in its variety arises from action (karma). Actions accumulate by the power of the latent afflictions (*anuśaya*); because without the latent afflictions [they] are not capable of giving rise to a new existence. Consequently, the latent afflictions are the root of existence.[15]

In other words, in the Buddhist view our entire sentient world, including the structures and capacities of our embodied existence, are the cumulative result of the prior activities (karma) of living beings instigated by the afflictions and their latent counterparts, i.e., ignorance, desire for sensual pleasure, thirst for existence, grasping onto identity, etc. In brief, the Buddha declared, our bodies together with their faculties should be regarded as the results of "former action (karma) that have been constructed (*abhisaṅkhataṃ*) and intended and are now to be experienced" (S II 65).[16] In order to draw out the implications of these statements, we shall examine one already widely accepted account of how such behavioral patterns have been built up, and are built into, our mental and physical structures—this is the view of evolutionary biology.

THE INTERDEPENDENCE OF EVOLUTION AND EXPERIENCE

Self-protection begins at the beginnings of life, manifesting in the processes of attraction and aversion implicitly based upon the distinction between self and nonself. At the most basic level of life, single-cellular organisms distinguish between what is threatening and what is beneficial to them in their environment, aggressively repulsing the one and engulfing and absorbing the other. This discrimination by semiporous membranes is a primary prerequisite of life. Without it, single-cellular life forms would never have survived and gradually developed into more complex and multicellular organisms such as our present species, homo sapiens.

We are all descended, through the extended processes of evolution, from those creatures whose successive transformations produced successful biological organisms. This occurred through the processes of differential reproductive success, in which those organisms that reproduce more prolifically over successive generations pass on more of their heritable[17] characteristics than those who reproduce less. The theory of evolution thus depicts a positive feedback loop in which those specific behavioral patterns that lead to greater reproductive success are steadily reinforced over extended periods of time. As biological creatures, we all therefore embody the cumulative results of whichever behaviors facilitated more reproductively successful interactions between our forebears and their natural and social environments. That is to say, the characteristics we embody today reflect, for the most part, behaviors that have successfully furthered their own reproduction in the past.

Chief among these behavioral patterns are the physical and mental capabilities that allow us to acquire food and shelter and the cognitive and emotional wherewithal necessary for reproducing and raising offspring. In other words, the will to preserve personal existence, a desire for those activities that lead to reproduction, and sufficient attachment to the people and things necessary to achieve these objectives are all essential for producing, preserving, and reproducing human life. That these drives, this thirst for life, are constitutive of the very form of existence we embody right here and now follows from the simple yet profound postulate at the heart of evolutionary theory: what has been more (re)productive in the past is more plentiful in the present. These include as well our acute social sensitivities, our abilities to think, feel, and empathize, to wonder and to worry, to love and to hate, to compete and to cooperate;[18] none of these are, in theory, wholly outside the broad scope of the extended, interdependent, and self-reinforcing processes known as evolution.

It is easy to overlook the implications of this relationship between the past actions and experiences of our ancestors and the particulars of our present species, since they so radically implicate our unique human capacities, our special modes of knowing, feeling and thinking, within the constructive processes of the past. As evolutionary biologist David Barash explains:

> If evolution by natural selection is the source of our mind's a priori structures, then in a sense these structures also derive from experience—not the immediate, short-term experience of any single developing organism, but rather the long-term experience of an evolving population. . . . Evolution, then, is the result of innumerable experiences, accumulated through an almost unimaginable length of time. The a priori human mind, seemingly pre-

programmed and at least somewhat independent of personal experience, is actually nothing more than the embodiment of experience itself. (1979:203)

The Buddhists and biologists thus largely concur that the very forms and structures of human life result from the accumulative actions of innumerable beings over countless generations.[19] Like all species, we too have been formed and conditioned by an immensely long and complex series of transformations. In this respect, we are contingent and historical creatures through and through, lacking any unchanging "species-essence" or fixed "human nature."[20] What we are, rather, are assemblages of dynamic yet wholly conditioned structures (*saṃskāra*) forged through the crucible of past actions and experience. While Faulkner's famous dictum—"The past is never dead, it is not even past"—may be more poetic, the biological view is startlingly similar, since "the structure of the organism is a record of previous structural changes and thus of previous interactions. Living structure is always a record of previous development" (Capra 1997:220). To more fully appreciate the continuing influences these previous interactions impart on the present, however, we must examine the historically conditioned structures that constitute human embodiment, the very structures "that have been constructed and intended and are now to be experienced" (S II 64).

EMBODYING, ENACTING, AND ENHANCING *SAṂSĀRA*

A path is made by walking it ("Tao hsing chih er cheng").

—*Chuang-tzu, chapter 2*

We have seen some common ground between evolutionary biology and Indian Buddhism on the general causal dynamics whereby we have come about as a species. We must now examine some of the capacities we are actually endowed with, paying particular attention to the problematic processes of identity formation occurring in the personal, social, and cultural arenas. We shall see that here, too, the themes of insight into the interdependent nexus of identity formation and the deleterious results of ignoring it arise again and again.

One dimension of this interdependence is expressed in the feedback cycle between our propensities, our actions, and their long-term consequences. Embodying the postulate that what has been more (re)productive in the past is more plentiful in the present, the three poisons and other afflictions find expression in our proclivities or dispositions to certain behaviors whose supporting physiological structures are either present at birth or mature within critical developmental periods. These behavioral structures thereafter facilitate various activities in life, whose results—in the long-term and in the aggregate—are indispensable elements in the evolutionary processes whereby reproductively successful capabilities are strengthened and developed. Actions thus constitute an indispensable link in a positive feedback cycle: our inherited capacities, which largely result from previous actions, enable and influence the range of our current activities, which thereafter condition future evolutionary developments.[21] We will examine this crucial feedback cycle in Buddhist terms before transposing it into the discourses of the biological and, eventually, the social sciences.

In the Indian Buddhist view human beings are endowed at birth with, among other things, the underlying tendencies or dispositions toward actions motivated by lust, greed, aggression, and ignorance, the very kinds of afflictive activities that were so instrumental in bringing about our bodily structures in the first place. In one famous discourse the Buddha explains that even though an innocent baby boy lying on the grass lacks a developed view of self-identity, a notion of sensual pleasure, or aggressiveness toward others, nevertheless, the child still has the dispositions "toward a view of self-identity," "to desire sensual pleasure," and "to aggressiveness to others," etc. All of these dispositions lie latent within him, awaiting their full development as he grows and matures.[22] Once they have matured, these latent dispositions continuously inform one's moment-to-moment cognitive and affective activities, adversely influencing one's actions whenever they manifest. In a passage that could well have been drawn from a modern psychology textbook, one discourse of the Buddha depicts how the latent dispositions to the three poisons are instigated by everyday perceptual experience, using vision as the prototypical example:

> Dependent on the eye and forms, eye-consciousness arises; the meeting of the three is contact; with contact as condition there arises [a feeling] felt as pleasant or painful or neither-painful-nor-pleasant. When one is touched by a pleasant feeling, if one delights in it, welcomes it, and remains holding to it, then the underlying tendency to lust lies with one. When one is touched by a painful feeling, if one sorrows, grieves and laments, weeps beating one's breast and becomes distraught, then the underlying tendency to aversion lies with one. When one is touched by a neither-painful-nor-pleasant feeling, if one

does not understand as it actually is the origination, the disappearance, the gratification, the danger and the escape in regard to that feeling, the underlying tendency to ignorance lies within one. (Ñāṇamoli 1995:1134, M III 285)[23]

As long as they persist, these dispositions to the three poisons of attachment, aggression, and ignorance (as well as, according to the same text, the view of self-existence)[24] are evoked or "activated" by nearly all our ordinary cognitive and emotional experiences. These processes of activation constitute the crucial connection between the sheer potential for our inherited dispositions to respond to certain things in certain ways and our actually responding to them in that way. Since they are dispositions, they are not absolute determinants ("if one delights in it . . . if one sorrows . . . if one does not understand"),[25] nor do they by themselves entail consequences unless or until they instigate intentional actions (karma).[26] But they do exert, the Buddhists say, powerful influences upon nearly all our activities, disposing us to act in certain ways instead of others. Especially fateful among these, as we shall see, is the strong connection between ignorance and our disposition toward self-identity, an underlying sense of "I am" (asmīti anusayo): "Touched by the sensation born of contact with ignorance, there comes to the untrained ordinary man the view 'I am,' there come the views 'I am this,' 'I shall be,' 'I shall not be,' 'I shall have a body'" (Johansson 1979:167, S III 46).

In the Buddhist view, then, human beings are innately endowed with the dispositions to the three poisons and the sense of "I am," which persist in a latent state ever ready to be activated by everyday experiences and to instigate fresh actions, which in turn give rise to further causal consequences, etc.[27] As the core of the feedback cycle that perpetuates sentient existence, the Buddhists consider the afflictive dispositions to be "the root of [cyclic] existence"[28]—an idea that in the light of evolutionary biology has more cogency than one might first imagine.

OVERCOMING THE AFFLICTION OF IDENTITY: BUDDHISM

We are, it seems, strongly conditioned by the persistent, ongoing influences of the three poisons and the latent afflictions. They are central to the innate capacities, sensitivities, and proclivities that instigate and enable the basic activities (karma), whose results, in the aggregate and over the long-term, have given rise to the kind of beings that we are. Being so deeply involved in our ongoing activities, their pernicious influences are almost impossible to evade. But as long as they persist we can never be truly free from misguided actions. It is "impossible," the Buddha asserts,

> that one shall here and now make an end of suffering without abandoning the underlying tendency (*anusaya*) to lust for pleasant feeling, without abolishing the underlying tendency to aversion towards painful feeling, without extirpating the underlying tendency to ignorance in regard to neither-painful-nor-pleasant feeling, without abandoning ignorance and arousing true knowledge. (Ñāṇamoli 1995:1134, M III 285f.).

But by completely eradicating these latent tendencies (*anusaya*) toward ignorance, attachment, aggression, self-identity, etc.,[29] the Buddha insists, it is entirely possible to end the vicious cycle of compulsive behavioral patterns (*saṃsāra*).[30] None of these tendencies, however, is more difficult to eradicate, even for the advanced Buddhist practitioner, than the fundamental sense of self-identity. As one text (S III 131) declares: "Though a spiritually advanced being (*Āryan*) has eliminated the five lower fetters, he still has not eradicated the subtle remnant of the conceit 'I am,' of the desire 'I am,' of the disposition of 'I am' ("Asmīti māno asmīti chando asmīti anusayo) toward the five aggregates of attachment."[31] Despite the difficulty of altogether eradicating this self-identification of "I am," the early Buddhist traditions nevertheless unequivocally affirm the possibility, the necessity, and the desirability of doing exactly that.[32]

OVERCOMING THE AFFLICTION OF IDENTITY:
EVOLUTIONARY PERSPECTIVES

We find common ground between Buddhist thought and evolutionary biology regarding the important influences that behavioral patterns such as the three poisons and latent afflictions impart over vastly extended periods of time, in their continued presence as heritable emotional and cognitive capacities, and as predispositions active in our moment-to-moment psychological processes—that is, their role in evolution, embodiment, and enaction, respectively. And although Buddhist traditions energetically testify to the challenging possibility of eradicating these afflictions, much of modern life and history tragically testifies to their utter tenacity—their activity is so instinctive, their origins so obscure, and one's self-identity so seemingly self-evident that we can barely even acknowledge their constructive character, let alone eradicate their pernicious powers. Even with an intellectual understanding of the evolutionary or developmental history of their structure and dynamic activity, our latent dispositions and sense of self-identity remain deeply entrenched, underlying and influencing nearly all our cognitive and emotional processes.

Consonant with this Buddhist view, we also find sound evolutionary reasons for an "innate ignorance" of our contingent, constructed nature and the dispositions that underlie so much of our behavior. Evolutionary processes have largely delimited the range of both our interests and our awareness, strengthening our obvious obsessions with sex, success, and survival while simultaneously obscuring their underlying aims. As anthropologist Barkow states:

> Since you are a product of biological evolution, your conscious and unconscious goals presumably are linked to the kinds of activities that would have tended to enhance the fitness [i.e. reproductive success] of our ancestors. This linkage—which may be exceedingly indirect—should be there regardless of the effects of your (sub)goals/plans on your current fitness. (1989:112)

This disturbing discrepancy between behaviors that were successful in our ancestral past and the needs and conditions of our current situation are correlated with specific physiological structures of the human brain. These structures slowly evolved over successive stages of animal and human evolution and, although functionally inseparable, correlate with relatively distinct modes of behavior. Self-preservation and reproduction are largely governed in the older structures, sometimes called the reptilian and old mammalian brains, while the distinctively human capacities of language, reasoning, long-term planning, etc. are processed primarily in the neocortex, the most recently evolved component of the brain's architecture. Although these newer capabilities favor deliberate and dispassionate action, the deeper drives associated with self-preservation and reproduction often override our more rational calculations. "We experience these overrides, subjectively, as emotions," which represent, Barkow asserts, echoing age-old allusions to our animal nature: "Limbic system overrides of the neocortex, of the old mammalian brain overriding the new . . . All our strong emotions . . . —rages, panics, lusts—represent such overrides. In a certain sense, they are the levers by which . . . our evolutionary past controls our present" (121f.).

These emotional "overrides," these erupting afflictions, are, however, hardly the only way that structures from the past tend to supersede the needs of the present. The conditions in which human intelligence evolved have duly circumscribed the range and content of our understanding of the world as well. This should hardly be surprising. "The conventional view," Trivers chides (1976:vi) "that natural selection favors nervous systems which produce ever more accurate images of the world must be a very naive view of mental evolution." Rather, our highly selective view of the world not only hinders our understanding of the underlying aims and motives of our own behavior, but it hinders our capacities for self-understanding as well.

It is widely and persuasively argued that the capacity to fashion an internal representation of one's "self"—of a continuous, predictable locus of experience by reference to which one could "map and order the physical and social universe and our own place in it" (Barkow 1989:110)—must have greatly assisted early human beings to more successfully negotiate their physical and social environments, conducing to greater reproductive success, and thereafter continuously developing through the positive feedback cycles of evolutionary causation. But even this "representation of self will not be some kind of miniature image," accurately representing the world, because, Barkow warns (103), our *self-awareness extends only to aspects of the self that in our evolutionary past have strongly and directly affected inclusive fitness* [i.e., reproductive success]" (95).[33] That is, we tend to be acutely concerned with aims we have neither consciously chosen nor whose motivating forces we fully comprehend. Evolutionary biologists thus conclude that self-awareness typically entails a degree of built-in blindness, an innate ignorance about who we are and what we do, especially concerning the illusion that "the 'self' is a miniature, controlling 'person,' a homunculus" (94).

The conclusions we must draw here are obvious, if not agreeable. What is it that has been reproductively successful in the past, which affects our current behavior, sometimes adversely, and about whose underlying influences we remain blithely oblivious? The Buddhists have suggested that our sense of self, of the "I" as an enduring, subjective locus of experience in contrast to and independent of our encompassing environment, is both a tragically inaccurate view and the most deeply entrenched of the afflictive dispositions. With limited awareness and blind emotion, we grasp onto such an illusory will-o'-the-wisp in order to "map and order the physical and social universe and our own place in it." These arguments from evolutionary biology therefore suggest strong evolutionary rationales for the following Buddhist-like critiques of self-identity:

1. Our sense of an independent "I" or "self" is an eminently practical construct that performs important biological functions[34] and hence must have enjoyed evolutionary advantage.

2. This sense, however, is illusory,[35] insofar as it typically presumes more functional unity, permanence, and independence than logical or scientific analysis bears out.[36]

3. And being illusory, it is inherently fragile and insecure, requiring constant reconstitution through psychological machinations, social reinforcements, and cultural conventions.[37]

4. Moreover, we are so largely "blind-sighted"[38] to these conditions that the constructed and interdependent processes underlying self-identity are themselves largely obscured.

A fragile, constructed, yet functional illusion whose originating conditions remain obscure: this is the stuff of which madness is made.[39] Avoiding these plain, unwelcome facts is the madness of which history is made.[40] But before we examine these same dynamics within the discourses of history, society, and culture, let us reassess the afflictions and their relationship to the "nature" of human nature in the light of Buddhist and evolutionary biology.

THE AFFLICTIONS AND THE CONDITIONED "NATURE" OF HUMAN NATURE

Evolutionary biology and Buddhist thought concur that our animate world is largely created by the constructive energies of past activities and we inherit powerful dispositions at birth that predispose us to act in certain, often harmful, ways. These afflicting dispositions, however, although "innate" in the literal sense that we are "born with" them, are neither "essential" nor "inherent" to us as a species or as individuals. They are interdependently produced phenomena that result from the aggregated effects of past actions, are activated under specific (albeit relatively common) conditions, and can be controlled or eradicated, to varying degrees and with varying difficulty, through concerted efforts thereto. As with the species as a whole, the dispositions are contingent phenomenon requiring supporting conditions to arise and rearise.

Comparatively speaking, the perspectives surveyed here suggest a common middle ground between the extremes of those determinists who maintain, on the one hand, an inherent, ineradicably evil side of human "nature" and those behaviorists who deny, on the other hand, that there are any innate dispositions whatsoever, who maintain that human beings are primarily products of their immediate environment, veritable blank slates upon which "society" can do all its dirty work. In our perspective, however, this "nature versus nurture" debate is based upon a false dichotomy:[41] "nature," in the sense of a fixed species "essence," is nothing but conditioned phenomena (*saṃskāra*), however remote some of those conditioning influences may be from their present results; while "nurture," i.e., the social conditioning incumbent upon one's upbringing and environment, could not even occur without our innate abilities to grow and to learn, which are themselves highly developed capacities constructed through our evolutionary past. Pure nurture, then, is as incoherent as unconditioned nature is unexampled.

On the other hand, although the perspectives outlined here agree that our inherited behavioral capacities—such as ignorance, attachment to self and aggression toward others, craving for self-existence, etc.—are powerfully productive influences in the interdependent, evolutionary processes that give rise to all life forms, it should be stressed that for the Buddhists (like other religions traditions) these are the very malevolent factors to be eliminated or, rather, radically transformed. This perspective, therefore, while fully recognizing the natural, i.e., biological, supports of these afflicting tendencies, neither endorses nor condones them on the simplistic assumption that whatever is "natural" is good (the Naturalist Fallacy). Rather, the Buddha's explanation of suffering and its causes (the first two Noble Truths) insists that we unflinchingly examine the entrenched nature of these afflictions without losing sight of the possibility and desirability of transforming and constructively channeling their considerable energies toward freedom from suffering (the second two Noble Truths)—an aim fully consonant with the ameliorative thrust of much of the social sciences to which we now turn.

COMMON CAUSE: THE CONSTRUCTION OF IDENTITY AND ITS DISCONTENTS

In this last section of the essay, we will attempt to flesh out the relationship between the construction of identity and the generation of evil by an inevitably sketchy excursus through several of the social sciences, for it is only at the collective level that the uniquely human scale of evil mentioned at the beginning of this paper manifests. We will again briefly touch upon topics in evolutionary biology and developmental psychology before moving into cultural, social, and political spheres. We will also find here overwhelming consensus that the construction, maintenance, and protection of a secure identity—at the personal, social, and political levels—are carried out with increasing complexity and vulnerability, requiring ever more strenuous and artificial supports, which in turn lead to yet more complexity and vulnerability, etc. We are caught, in short, in a vicious cycle of increasing and frightening intensity. This represents an extrapolation to the collective level of the Buddhist ideas sketched at the beginning of this paper: that our misguided efforts to secure an ordered, unchanging, and singular identity within our chaotic, changing, and pluralistic world not only expresses our futile attempts to "turn reality on its head" but also, tragically, leads to the preponderance of human evil, of "man's inhumanity to man."

What follows, therefore, is a prolegomenon to an interdisciplinary social theory of evil based upon the three basic Buddhist principles outlined above: 1. the dependent nature of all phenomena, 2. the constructed nature of "self-identity," and 3. the apprehension that human evil and ill are caused by attempts to secure such "selves," often at the expense of "others." Although similar ideas are found in a diverse array of disciplines, this Buddhist framework provides the overarching conceptual rubric for the discussion that follows. Again, for the sake of analysis, we will discuss the construction of identity in separate and sequential stages, from its evolutionary origins, its genesis within the family unit and dependence upon the processes of socialization and acculturation, and finally to its tragic political expression in modern times. We should not allow this mode of exposition to lead us, however, to overlook the crucial and often unstated "coevolutionary" (i.e., interdependent) relationships between our innate capacities for constructing identity, the evolution of the human brain, and the development of human sociality and culture. For ultimately, these were, and are, inseparable, mutually reinforcing processes.[42]

IDENTITY AS INTERDEPENDENT CONSTRUCT

All sentient beings are deranged ("Sabbe sattā ummattakā").

— *Gotama Buddha*

We mentioned earlier that the processes of aversion and attachment based upon an implicit distinction between self and not-self were essential to all biological life, informing and influencing the more evolved cognitive and affective structures found in higher life-forms. Thus, although an explicit sense of "self" only reaches its apogee in human beings, it nevertheless represents an evolution of elementary cognitive capacities all organisms enjoy. An acute self-awareness of "oneself in relation to others" was already highly evolved in our primate cousins, for example, and was surely a selective factor in both primate and early hominid evolution. However it may be conceived, a self-conscious sense of "self" is an indispensable part of the complex web of agency, organization, and order that constitutes human identity.[43] But it is the vagaries of this sense of identity—its evolved origins, its dependent development, and its precarious persistence—that are so problematic, so fraught with frailties, tensions, and conflict.

What is unique about human beings is the extent to which our sense of order and identity depends upon the common experienced world that arises out of the regularities of our interaction with others and the shared, symbolic means we have of expressing, communicating, and transmitting that world—that is, culture.[44] This world of social interaction and communication, however, is never simply "cultural" as opposed to "natural," for, as we mentioned above, what is cultural or social has irrevocably configured our physiology: by most accounts our uniquely human brain structures evolved roughly simultaneously with the development of culture, which itself could only have developed based upon the social and cognitive capacities this evolving brain facilitated.[45] Culture and human biology, therefore, are inseparable, interdependent, coevolutionary phenomena. Culture, and the social order that engenders it, is therefore not something added on or extraneous to human life; it is constitutive of human existence itself.[46] And it is these cultural and social worlds that, always inseparable from our biological endowments, provide the context and content of our constructed "realities."[47] The sociologist Peter Berger:

> To be human means to live in a world—that is, to live in a reality that is ordered and that gives sense to the business of living. . . . This life-world is social both in its origins and in its ongoing maintenance: the meaningful order it provides for human lives has been established collectively and is kept going by collective consent. (Berger, Berger, and Kellner 1973:63)

But humans of course did not evolve in generic social groups. As children we have always first lived in particular groups embodying specific cultures, through whose specific social and cultural patterns we grew into our unstated understandings of reality, our implicit, and usually unexamined, worldview. It is only within this larger context of a meaningfully ordered reality provided by social life that we develop a sense of identity, one that is irredeemably individual and social at the same time.

This sense of identity is first and foremost forged during our prolonged childhood dependency, through constant interaction within the limited social nexus of the family unit. According to developmental psychologists, our sense of "I-ness" develops in fairly specific stages during infancy. Piaget, for example, thinks that infants develop a conception of the ongoing existence of external objects that may be temporarily out of sight ("object permanence") more or less at the same time they develop a conception of a separate "self" experiencing these objects. Starting from a presumed original symbiosis, the complex conceptions of self and object undergo a figurative mitosis and thereafter continue to determine each other in parallel developmental stages.[48] Both the distinction between self and nonself and their interdependence are therefore not only logical but ontological as well, for they are intrinsic to the notion of "self" identity from its very inception.[49]

And since the family can never truly be separated from the larger social and cultural contexts in which it too operates, all senses of self are not only individually but also socially or culturally conditioned. For such inescapably social creatures as ourselves, self-identity is never simply given; it is forged in the crucible of interaction with others. Identity is thus a product not only of the evolutionary development of species-specific behavior but of the developmental processes of growth, maturation, and socialization of each individual as well. It is thus ontogenetic as well as phylogenetic. As with our species itself, individual identity is a contingent and conditioned construct, a *saṃskāra* in Buddhist terms. Thus Bertalanffy (1968:211f.), the founder of general systems theory, concludes:

> "I" and "the world," "mind" and "matter," or Descartes's "res cogitans" and "res extensa" are not a simple datum and primordial antithesis. They are the final outcome of a long process in biological evolution, mental development of the child, and cultural and linguistic history, wherein the perceiver is not simply a receptor of stimuli but in a very real sense creates his world. . . . "Things" and "self" emerge by a slow build-up of innumerable factors of gestalt dynamics, of learning processes, and of social, cultural, and linguistic determi-

nants; the full distinction between "public objects" and "private self" is certainly not achieved without naming and language, that is, processes at the symbolic level.

This sense of "I" as distinct from "other" therefore depends not only upon an evolved innate capacity for self-identity but also upon conceptions of identity that are culturally and socially acquired,[50] conceptions that have also undergone their own history of development, articulation, and often conflicted expression and that, in turn, introduce still further problems and conflicts.

These are the paradoxes of identity: as self-making, culture-creating, symbol-processing organisms, we require meaning and order at multiple levels—personal, social, cultural, etc. It is within and around these overlapping dimensions of identity that meaning and order coalesce and cohere. Identity thus serves important, perhaps indispensable purposes: it provides that continuous, predictable locus of experience, that sense of agency and organization, which allows us "to map and order the physical and social universe and our own place in it" (Barkow 1989:110). But it is this very dependence upon social interaction and cultural construction that gives the lie to the assumptions of independence, unity, and stability upon which our deepest sense of identity depends. For identity is inherently unstable, its instability grounded in the social and cultural nature of its origins, and any cultural symbol system is similarly and necessarily fragile and vulnerable. We are always changing our minds, our feelings, our modes of expression, our established patterns of interaction, and our complex symbolizations of reality. Identities, meanings, and shared symbols proliferate and disperse with distressing regularity, ever prone to differentiation, dissolution, and decay.[51] And it is precisely this tension between the sheer necessity for such overlapping levels of identity and the inherent fragility of all such constructions that drives the underlying compulsions behind humanity's massive, engineered inhumanity. "Identities," the Buddhists remind us, are constructs designed to counteract the impermanent, restless, and identity-less nature of things, to, in short, turn reality on its head.

SECURING IDENTITY BY CONSTRUCTING EVIL

How do human beings respond to this instability, to the inescapably provisional nature of our constructed identities? How do we shape and sustain these distinct personal, social, and cultural dimensions of order and identity, constructed on such shifting sands? How do we "fix" the basic groundlessness of identity in order to sustain stable, established modes of being?

Identities at all levels are constructed by establishing order and security within a radically impermanent and interdependent world. These processes involve dynamics similar to those we observed at the individual level—discomfort with insecurity, desire for perpetuation, and disavowal of dependence. To create order, we exaggerate differences between peoples, create dichotomies between "us" and "them," and reify these into fixed and independent entities set off from one another by imputedly intrinsic and insurmountable differences. Fostering intense and distorting emotional attachments, these fixations provide the basis for strong identification with our group and antagonism toward the other.[52]

Our identification with the social and cultural realities in which we are raised is so powerful and so ingrained that social scientists, following Durkheim, consider it the fundamental religious reality. A culture's definitions of reality provide both the regularities required for meaning and order to cohere in our chaotic and confusing world and the symbols that represent that compelling and enduring sense of reality, that aura of objective and eternal truth, that sacralized reality[53] within which we find our place in the cosmos, our ultimate "identities." Cultural symbols thus express the "sacred" meaning, order, and permanence through whose mediation we, as mere mortals, may symbolically participate in something transcendent or immortal.[54] As Becker (1975:64) so eloquently states:

> All cultural forms are *in essence sacred* because they seek the perpetuation and redemption of the individual life. . . . Culture *means* that which is *supernatural;* all culture has the basic mandate to transcend the physical, to permanently transcend it. All human ideologies, then, are affairs that deal directly with *the sacredness of the individual or the group life,* whether it seems that way or not, whether they admit it or not, whether the person knows it himself or not.

But the sacralized "realities" and identities that culture provides are arguably compelling and effective only to the extent that they are not acknowledged to be mere constructs, mere human fabrications. "The institutional order," argues Berger (1967:33), must "be so interpreted as to hide, as much as possible, its constructed character." Our culturally constructed realities are "sacred" not in spite of but *because* of their obscured nature. They require mystification.[55]

These are by no means wholly evil processes, but neither are they incidental. They are essential and constitutive of identity formation. They involve, in fact, the same characteristics found in individual "self-making": they are eminently functional, yet constructed, conflicted, and concealing. But while the direct consequences of individual identity construction are relatively simple, limited, and short-lived, the consequences of collectively forging social, cultural, and political identities are complex, massively disruptive, and disastrously enduring.

In the modern world our definitive "sacred reality" has increasing come to be represented in the nation or nation-state. The nation has been sacralized by the same processes through which individuals, societies, and cultures are reified into selves or entities[56]: by creating boundaries dichotomizing the world into us and them, coercing homogeneity within and excluding foreignness without, and imbuing all this with an emotionally charged aura of eternal truth and goodness that simultaneously sanctifies and obscures its contingent, constructed nature. We thereby populate our complex, interdependent human environment with such imagined entities as cultures, subcultures, and nation,[57] even though, as Wolf (1982:3) points out,

> Inquiries that disassemble this totality into bits and then fail to reassemble it falsify reality. Concepts like "nation," "society," and "culture," name bits and threaten to turn names into things. Only by understanding these names as bundles of relationships, and by placing them back into the field from which they were abstracted, can we hope to avoid misleading inferences and increase our share of understanding. (cited in Carrithers 1992:26)[58]

And it is this falsification, this reification of aggregated individuals into independent entities separated from their encompassing human contexts that provides the locus of identification required for sacralizing order and security. The modern nation is thus a socially constructed "ultimate reality" imbued with the same implicit sacrality that all societies share: a cosmic order that provides ultimate meaning, purpose, and identity—an order, that is, one could give one's life to. Thus, by creating order and purpose at the social and political levels, while simultaneously providing a locus of sacralized identity and belonging at the personal level, the nation is the modern sacred order par excellence.[59] As Becker (1975:113) so eloquently observes:

We couldn't understand the obsessive development of nationalism in our time,—the fantastic bitterness between nations, the unquestioned loyalty to one's own, the consuming wars fought in the name of the fatherland or the motherland—unless we saw it in this light. "Our nation" and its "allies" represent those who qualify for eternal survival; we are the "chosen people." . . . All those who join together under one banner are alike and so qualify for the privilege of immortality; all those who are different and outside that banner are excluded from the blessings of eternity.

We can now see how the animosities evoked by ethnic, cultural, or national conflict draw upon the deepest dynamics of identity formation: "others" play an indispensable role in defining "us"; they provide both the contrasting boundary by which we can distinguish who we are and the common threat that unites us in our sacred cosmos.[60] These attempts to establish and protect definite, substantial ethnic or national identities do not merely define evil, they require it.[61]

Identity is thus a tragically double-edged sword. It is the juxtaposition of the sheer fragility of any symbolic order with the magnitude of our need for it, the juxtaposition of our deep dependence upon a larger consensual reality with its constructed nature, and the constant threats to its integrity and validity that helps explain the endless violence over sacred symbol systems. Sacred symbol systems proliferate, mix, and mutate in the vast marketplace of competing and incompatible worldviews. The radical pluralism that is so part and parcel of this untidy world of ours is deeply disturbing and destabilizing. Since our sense of order and meaning is so bound up with the sacralized symbol system of our specific cultures, worldviews, ethnic identities, religious beliefs, or national belongings, threats to our sacred symbol systems threaten our very existence. As Berger (1967:39) warns, "When the socially defined reality has come to be identified with the ultimate reality of the universe, then its denial takes on the quality of evil as well as madness." The implications of this are obvious and ominous: when each particular "socially defined reality" is sacred, then pluralism produces endless evil and madness. We have no need to merely imagine what this entails.

These processes have generated a vicious cycle of mind-boggling proportions and heart-numbing consequences. As indispensable institutions for sacralizing identity within an eternal sacred order that provides meaning, value, and belonging, modern nation-states have filled the vacuum created by the diminished influence of institutionalized religion in the early modern era.[62] As a reflection of the desire to participate in immortality, such cultural or national identities have come to supersede even one's own personal survival. Men fight and die so that their group, and by extension themselves, may have immortal life.[63] "Men seek to preserve their immortality rather than their lives," Otto Rank observed (Becker 1975:65). It is this intense, "sacred" identification with groups larger than ourselves that sanctifies the massive, incalculable blood sacrifice at the altar of the nation-state, our modern secular, ersatz religion. Millions have died defending abstract symbols such as a flag[64] or ideologies such as fascism, socialism, or even democracy; millions more have killed in the name of the "fatherland," "racial purity," or even, most ironically and tragically, in the name of a loving God. As Duncan (1962:131) declares, in this century of ideological warfare, "All wars are conducted as holy wars."

We have finally reached the bloody irony of our modern era. Our attempts to turn reality on its head results in "the paradox . . . that *evil comes from man's urge to heroic victory over evil*" (Becker 1975:136), from our ill-chosen means of constructing sacred identities whose very existence requires that we continuously create and vanquish opposing "evil" entities in the world. Human beings make war and kill each other in a way that no other animal species does because no other species is as dependent upon sacralizing symbols of consensual reality in order to make sense of their lives.[65] No other species has the capacity, or the need, to externalize identity out into the wide-open world where its fate, our fate, blows so helplessly in the wind.

INTERDEPENDENCE, IDENTITY, AND UNDERSTANDING

In the aggregate these observations from the biological and social sciences not only resonate with but resound classical Buddhist notions of the construction of identity as the locus of self-grasping and ignorance in the face of the radical impermanence and interrelatedness of all phenomena. These ideas have provided a comprehensive framework from which we may make some sense of the massive perpetration of evil and suffering we inflict on each other each and every day.[66] We can see how the interdependent nature of phenomena, the fabrication of identity, and attachment to our selves at the expense of others all function equally effectively, and nefariously I might add, at the biological and individual levels as well as at the sociocultural levels organized around sacred symbol systems. Essential to these processes are the three "poisons" and other afflictions, major motivating forces underlying many human behaviors, as well as our deeply entrenched sense of self-identity. They are based upon our emotional and cognitive modes of behavior, in particular on both a self-centered sense of identity (which is at the same time inseparable from the complex and symbolic interactions that constitute human society and culture) and on powerful limitations to our awareness of their originating conditions. At the individual level these capacities, particularly our self-centeredness and ignorance, are universally recognized to underlie many of the interpersonal problems in life, and much of traditional religious or moral culture is geared toward mollifying their expression or ameliorating their excesses. These processes, however, are but the ground level, the bare prerequisites, of our human capacities toward evil.

To understand how these are transmuted into the scale of violence and hatred unique to our species, we have considered the "sacralization of identity" (*satkāyadṛṣṭi*) and its attendant poisons: the inordinate emotional attachment and irrational belief adhering to our social, cultural, and political constructions of reality and the disastrous aggressions resulting from in-group/out-group discrimination and its exclusionary loyalties. It is the very interdependent nature of human identity, however, that enables it to be molded into vehicles of self- and group-aggrandizement, with its concomitant projection of enmity, that conduces to the unprecedented scale of violence characterizing this last blood-soaked century. For built into the self/other dichotomy is the tragic blindness to our ontological interdependence, our reciprocally conditioned and contingent nature. It is this ignorance that facilitates our blind belief in independent, autonomous entities, whether individuals or groups, clans or ethnicities, or, as in our modern era, nation-states, the apogee of belonging and belief in whose name so much senseless suffering has been instigated, so many deaths decreed, such unimaginable evil administered.

Having gained at least some purchase on these unholy dynamics, we cannot avoid asking what a cross-cultural dialogue between Buddhism and science might do to help ameliorate these ills, to circumvent these vicious circles. Since, as both anthropologist Gregory Bateson (1979:66) and the Buddhists may agree, "when causal systems become circular, a change in any part of the circle can be regarded as cause for change at a later time in any variable anywhere in the circle," there are several points at which these

vicious behavioral patterns can be broken. As was suggested in the introduction, these can be divided into understanding the human condition and
overcoming the baneful influences of our afflictive self-centered activities.
From this perspective the abstract and theoretical understanding pursued
herein possesses no mere academic import, but points to a potentially powerful analytic tool for overcoming our ingrained ignorance concerning the
constructed and conflicted nature of human identity. Without such an understanding we can hardly approach these issues in a comprehensive, constructive fashion. That is, we cannot fully appreciate the indispensable
meaning-making functions the construction of identity clearly does serve
without at the same time unreservedly interrogating the destructive dynamics into which the sacralization of identity all too often degenerates.[67]
This Janus-face quality of human identity must be an explicit component
of any serious attempt to understand our human condition. "Ignorance,
thirst for illusion, and fear," Becker (1975:143) avers, must all be "part of the
scientific problem of human liberation." Such a science, he continues (162),
"would share a place with historical religions: they are all critiques of false
perceptions, of ignoble hero systems. A science of society, in other words
. . . will be a critique of idolatry." Such a science, put in Buddhist terms,
would be a critique of our concerted but futile and ultimately frustrating efforts to "turn reality on its head": to misconstrue the impermanent as permanent, the unsatisfactory as satisfactory, and what is not-self as self. The
collective recognition of both our interdependence and of the alienation
created by all false identity constructions is thus an essential component of
a new, and yet very ancient, mode of understanding ourselves and our place
in the world.[68]

Such understanding, however, needs to issue in action. This is no easy
task, nor is it to suggest that the Buddhists or anyone else have a single
panacea for all that ails our world. Buddhists traditionally say that the Buddha taught eighty-four thousand practices directed toward alleviating
eighty-four thousand kinds of afflictions. This traditional stock figure expresses the necessity, one could say, of understanding all the particulars of
our complex world in order to address its multifarious ills. If, as many have
emphasized, we are continuously constructing our "worlds," then we also
have collective responsibility for the kind of world we construct. We have
little choice but to exercise the weighty responsibility of our "knowledge of
good and evil" with intelligence and compassion, fully appreciative of our
creative possibilities, fully cognizant of their demonstrated dangers. We

have recently reached some consensus on both the grounds and causes of some of the most egregious of these dangers. If this hard-won understanding remains ensconced in the academy or laboratory we may not survive to develop its more promising possibilities.

Notes

This paper is dedicated to my late father, without whose inspiration and example I surely would never have striven so persistently, so systematically, to understand such inescapable, unpleasant aspects of life.

1. The sheer scale of organized violence is evident in these appalling statistics: "In 1973, there were a handful of conflicts spread around the globe. By 1980 there were thirty or so, and today there are more than forty. Most statistics agree that an average of one thousand soldiers are killed every day throughout the world. If anything, this is a conservative figure, kept down by the impossibility of collecting statistics from many of the ongoing wars. A thousand casualties a day is approximately the same as the average number of French soldiers killed daily during World War I. That conflict lasted only five years. Our current levels of violence have been with us for more than a decade. Some five thousand civilians also die every day as a direct or indirect result of war. Three and a half million dead soldiers over the last decade and twenty million dead civilians." John Ralston Saul (1992:180f), cited from John Gellner, editor of the *Canadian Defense Quarterly, Toronto Globe and Mail*, December 31, 1980.

   William Eckhardt of the Lentz Peace Research Institute estimates 13.3 million civilian and 6.8 million military deaths between 1945 and 1989, while Nicole Ball of the National Security Archives (*Toronto Globe and Mail*, September 30, 1991) figures 40 million deaths since 1945 in 125 wars or conflicts. See Saul 1992:599. These figures predate the breakup of the Soviet Union, the Yugoslav wars, or the Central African genocides, and they show little sign of abating.

2. These correspond closely to the two obstacles to liberation in classical Buddhism, the obstacles consisting of the afflicting passions (*kleśa-avāraṇa*) and the obstacles to correct knowledge or understanding (*jñeya-avāraṇa*).

3. I am not implying that there is a single "Indian Buddhist" view. There are vast differences among Buddhist traditions regarding many matters. I am limiting the scope of this essay, however, to aspects that have both been widely accepted in Indian Buddhism as well as show some promise of productive dialogue with areas of the modern sciences.

4. This last phrase is close to the literal meaning of the four "inversions" (*viparyāsa*): regarding the impermanent as permanent, the dissatisfactory

as satisfactory, the impure as pure, and what is not self as self. Since in the Buddhist view only *nirvāṇa* promises lasting satisfaction, the four inversions describe our attempts to find—in a world lacking permanence, satisfaction, and any abiding self (the three marks of existence, *trilakṣaṇa*)—what is not possible to find, to secure what is not possible to secure, and to attain what is not possible to attain. This essay could be considered an extended meditation upon the implications of these inversions, especially in terms of their traditional classification into the inverted perceptions, thoughts, and ideas (*saññā-, citta-, diṭṭhi-vipallāsa*) that impute permanence, satisfaction, and self where they cannot be found. Nyanatiloka 1977:196. In addition to the page numbers of the respective English translations, we also refer to the various collections of Pāli texts using their standard abbreviations: A = *Aṅguttara Nikāya;* M = *Majjhima Nikāya;* S = *Saṃyutta Nikāya.* These are followed by volume and page number of the PTS edition. This formula is found in many texts, i.e., A IV 49 ("Anicce niccan ti, dukkhe sukhan ti, anattani attā ti, asubhe subhan ti"); *Abhidharmakośabhāṣya* (henceforth AKBh) V ad 9; Poussin 21; Shastri 888 ("Anitye nityam iti, duḥkhe sukham iti, aśucau śuci iti, anātmani ātma iti"). The centrality of these inversions is succinctly expressed in the following passage: "As long as their minds (*citta*) are turned upside-down by the four inverted views, beings will never transcend this unreal cycle of birth and death" (saṃsāra) ("Caturbhir viparyāsair viparyasta cittāḥ sattvā imam abhūtaṃ saṃsāraṃ nātikrāmanti. *Prasannapadā,* xxiii, cited in Conze 1973:40, 276, n. 31; see pp. 34–46 for an extended discussion of the three "marks" and the inverted views.

5. *Sabbāsava-sutta,* Ñāṇamoli 1995:92f. The psychologist Ernest Wolf (1991:169) describes the same universal "conviction that I am the person who was born in a certain place at a certain time as the son of the parents whom I knew and that I am the person who has had a history in which I can identify the 'I' of yesteryear as the 'I' of yesterday and, hopefully, of tomorrow." As cited in Mitchell 1993:109.

6. There is a close affinity between the Buddhist theory of dependent arising and self-organization theory and its close cousins, general systems and chaos theory. For a good synthesis of recent developments in self-organization theory, especially as it applies to evolution, see Capra 1997. General systems theory and Buddhist thought are compared in Macy 1991. Varela, Thompson, and Rosch 1991 astutely use Buddhist concepts to bridge cognitive science and Western phenomenology.

7. Buddhists do not hold a materialist worldview, that consciousness is merely an epiphenomenon of neural states. Most Buddhist traditions hold that some processes of an individual's mind stream persist from one life to the next (usually *vijñāna*), and whose "descent into" and "exit" from the body defines the span of a single lifetime. This does not vitiate, I believe, a comparison between evolution and Buddhist thought as causal theories regard-

ing the dependent origination of species through the aggregate activities of sentient beings, the focus of our comparison here. There are, of course, major divergences between Buddhism and biology concerning how the effects of these activities may be transmitted from one generation to the next. On this and other issues, Buddhist perspectives appear largely incommensurate with current scientific assumptions.

8. The "three poisons" ("greed, hatred, delusion" are variant terms) effectively epitomize a much larger range of human emotion and behavior, itemized more specifically as follows:

> "Greed: liking, wishing, longing, fondness, affection, attachment, lust, cupidity, craving, passion, self-indulgence, possessiveness, avarice; desire for the five sense objects; desire for wealth, offspring, fame, etc.
> Hatred: dislike, disgust, revulsion, resentment, grudge, ill-humour, vexation, irritability, antagonism, aversion, anger, wrath, vengefulness.
> Delusion: stupidity, dullness, confusion, ignorance of essentials (e.g. the Four Noble Truths), prejudice, ideological dogmatism, fanaticism, wrong views, conceit" (Nyanaponika 1986:99).

9. As the Buddha observed:

> He regards feeling as self . . . apperception as self . . . volitional formations as self . . . consciousness as self, or self as possessing consciousness, or consciousness as in self, or self as in consciousness. That consciousness of his changes and alters. With the change and alteration of consciousness, his mind becomes preoccupied with the change of consciousness. Agitation and a constellation of mental states born of preoccupation with the change of consciousness remain obsessing his mind. Because his mind is obsessed, he is frightened, distressed, and anxious, and through clinging becomes agitated. (Bodhi 2000, S III 16f)

10. The disheartening dynamics discussed in this paper touch only upon the first two of the Four Noble Truths of Buddhism: the universality of human suffering/dissatisfaction and its basic underlying cause, our attachment to a constructed yet ultimately untenable sense of permanent self-identity. The last two of the Four Noble Truths declare the radical possibility of complete freedom from such suffering and the path toward that freedom through eliminating the heartfelt belief in an independent, autonomous, fixed self and the actions that belief instigates. Using Buddhist ideas to examine this particular set of issues no more vindicates persisting stereotypes of Buddhism as "pessimistic" than using the findings of epidemiologists to understand the conditions for the spread of disease demonstrates their incorrigibly morbid mentalities. Findings are relative to the questions being posed, and the epidemiologists' findings are valued for their ameliorative effects. Buddhists traditionally understand the Four Noble Truths in similarly medical terms. An equally compelling and nearly exact mirror image of this pa-

per could well have been written, especially from the Mahāyāna Buddhist point of view, on the beneficent influence of compassion and cooperation on evolution, history, and social, cultural, and political life. But that is not the question we are pursuing here.

11.   The three "poisons" of attachment, aggression, and ignorance are depicted in the very center of traditional representations of the Wheel of Life, the cycle of death and rebirth, where they are represented by the cock, snake, and pig, respectively. Since the afflictions (*kleśa*) are basically elaborations or specifications of the three poisons, I am treating them interchangeably for the purposes of this essay.

The afflictions per se were not fully enumerated in the earliest Buddhist literature attributed to the Buddha himself but were elaborated in the later Abhidharma traditions. The ten afflictions in the Theravādin Abhidhamma (*Dhamma-sangani, Visuddhimagga*, etc.) are 1. greed (*lobha*), 2. hatred (*dosa*), 3. delusion (*moha*) (1–3 are the three poisons), 4. conceit (*māna*), 5. speculative views (*diṭṭhi*), 6. skeptical doubt (*vicikicchā*), 7. mental torpor (*thīna*), 8. restlessness (*uddhacca*), 9. shamelessness (*ahirika*), 10. lack of moral dread, or unconscientiousness (*anottappa*). See Nyanatiloka 1977:77.

The *Abhidharmakośa* (V 1c-d) also enumerates six, parallel to the Theravādin list (attachment, aggression, ignorance, pride, false view, and doubt), as well as an expanded list of ten (V 3) wherein "false view" (*dṛṣṭi*) is divided into five—roughly: 1. view of self-existence (*satkāyadṛṣṭi*), 2. extreme views (*antagrāhadṛṣṭi*, i.e., eternalism and nihilism), 3. false views based on wrong ideas (*dṛṣṭiparāmarśa*) 4. false views about the efficacy of rules and rituals (*śilavrataparāmarśa*), 5. false views about causality (*mithyādṛṣṭi*). See, among others, Guenther and Kawamura 1975:64–81. The "view of self-existence" (*satkāyadṛṣṭi*), upon which much of the discussion that follows focuses, plays a more fundamental role in Buddhist thought than its position here might suggest. Although its etymology is disputed, it uniformly refers to the range of misguided views regarding, and the attributions of a permanent "self" onto, the five groups (*skandha*) of physical and mental processes that constitute human existence.

There are challenges and controversies surrounding the most suitable way to translate *kleśa*. Snellgrove (1987:109) succinctly outlines these vexing issues:

> Difficulty in fixing suitable terms in English . . . is caused by the word *kleśa*, which means literally "anguish" or "distress," but which in Buddhist usage comes to mean whatever is morally distressful and thus in effect "sinful emotions." Although quite sure that it comes close enough to the actual meaning of this difficult term, I have tried to avoid the translation "sin" out of deference to a new generation of westernized Buddhists, who react very quickly against the introduction into Buddhist texts of terms with a specialized Christian application. "Affliction" may be regarded as a tolerable translation in that it can refer to

anything that upsets the equanimity of the mind, although it misses the moral aspect of the disturbance, which must be understood as also included.

12. Though many South Asian cosmogonies attribute the formation of the material universe to karma as well, we are limiting discussion to conceptions of the development of sentient life, which provide the most grounds for productive comparison on the influences of the afflictions.

13. In all likelihood, this was not even originally a Buddhist account but rather an ironic parody of traditional Vedic cosmogonies. See Collins 1993; and Carrithers 1992:117–145.

14. "The first person who, having fenced off a plot of ground, took it into his head to say this is *mine* and found people simple enough to believe him, was the true founder of civil society." J-J. Rousseau, *The First and Second Discourses* (New York: St. Martin's 1964),p. 141, as cited in Becker 1975:40.

15. AKBh ad V 1a. Poussin 1; Shastri 759. ("'Karmajaṃ lokavaicitrayam' ityuktam / tāni ca karmāṇi anuśyavaśād upacayaṃ gacchanti, antareṇa cānuśayān bhavābhinirvartane na samarthāni bhavanti / ato veditavyāḥ mūlam bhavasyānuśayāḥ"); Tib.: ("las las 'jig rten sna tshogs skyes zhes bshad pa / las te dag kyang phra rgyas gyi gbang gyis bsags par 'gyur gyi / phra rgyas med par ni srid pa mngon par 'grub par mi nus pa de'i phyir / srid pa'i rtsa ba phra rgyas yin par rig bar bya ste").

16. "'Nāyam . . . kāyo tumhākaṃ na 'pi aññsaṃ. purāṇam idaṃ bhikkhave kammaṃ abhisaṅkhataṃ abhisañcetayitaṃ vedaniyaṃ daṭṭhabbaṃ." Unless indicated otherwise, Pāli texts are cited from the translations of the Pali Text Society; technical terms altered for the sake of consistency.

17. As we shall see below, the genetic component of heritable characteristics is just one factor within a complex of factors that give rise to life in general and to behavioral patterns in particular. The persistence and causal influences of natural and social environments, importantly including behavior itself, are also indispensable conditions in the circular feedback processes of evolution.

18. The narrow focus of this paper precludes any *comprehensive* account of human psychology, which would need to give all these behaviors and capabilities their due consideration. Current biological thinking, in any case, is at quite the opposite pole from nineteenth-century Social Darwinism: "Life is much less a competitive struggle for survival than a triumph of cooperation and creativity. Indeed, since the creation of the first nucleated cells, evolution has proceeded through ever more intricate arrangements of cooperation and coevolution" (Capra 1997:243).

19. We are not suggesting that these are comparable in all respects, but only that we may find some common ground in their conceptions of long-term causal or conditioning influences. Evolutionary biology typically considers these questions in terms of the development of particular gene pools and populations while Buddhist thinking largely speaks in terms of the trajectories of individual mind streams coursing through multiple lifetimes.

20. "Darwinism . . . banished essentialism—the idea that species members in-
     stantiated immutable types" (Richards 1987:4).

     The ability to make the switch from essentialist thinking to population think-
     ing is what made the theory of evolution through natural selection possible.
     . . . Organisms . . . have a mechanism for the storage of historically acquired
     information. . . . The genotype (genetic program) is the product of a history
     that goes back to the origin of life, and thus it incorporates the 'experiences' of
     all ancestors. . . . It is this which makes organisms historical phenomena.
     (Mayr 1988:15f.)

21. The study of behaviour has now emerged as one of the most central issues in
     modern evolutionary analysis. With hindsight, it is easy to see why this should
     be so. After all, natural selection and genetic change depend, as we now inter-
     pret Darwin, upon the way in which an animal behaves since its behaviour, in
     particular everything leading up to the act of reproduction and later the pro-
     tection of offspring, determines the direction of evolution as a result of differ-
     ential breeding rates. (Nichols 1974:264)

22. Ñāṇamoli 1995:537f., *Malunkya-sutta*, M I 433.

     For a young tender infant lying prone does not even have the notion "personal-
     ity" [*sakkāya*, or "self-identity"], so how could personality view arise in him?
     Yet the underlying tendency (*anusaya*) to personality view lies within him. A
     young tender infant lying prone does not even have the notion "teachings," so
     how could doubt about teachings arise in him? Yet the underlying tendency to
     doubt lies within him. A young tender infant lying prone does not even have
     the notion "rule," so how could adherence to rules and observances arise in
     him? Yet the underlying tendency to adhere to rules and observances lies with-
     in him. A young tender infant lying prone does not even have the notion "sen-
     sual pleasures," so how could sensual desire arise in him? Yet the underlying
     tendency to sensual lust lies within him. A young tender infant lying prone does
     not even have the notion "being," so how could ill will towards beings arise in
     him? Yet the underlying tendency to ill will lies within him.

     These tendencies comprise the five lower fetters, enumerated in note 31 be-
     low.

23. Later Buddhist analysis of the arising of the afflictions is more nuanced
     than this. Abhidharma traditions, for example, analyze the relationship be-
     tween a latent disposition and the particular objects by which it is triggered
     in ways that closely resemble psychoanalytic conceptions of "invested" or
     "cathected" (*besetzen*) objects. 1. First, each type of affliction is "bound up"
     and attached to certain objects and reacts to them in certain conditioned
     ways. 2. Then, whenever an appropriate object appears in its respective
     sense field, it evokes an "outburst" of that affliction. So, for example, sensu-
     al desire arises whenever an object (*dharma*) that "provokes an outburst of
     sensual desire" (*kāmarāgaparyavasthānīya-dharma*) appears in the sense

fields and one has not abandoned or correctly understood the latent disposition toward it (*rāgānuśaya*). 3. This latter condition explains why ignorance is said to be the root of them all. (1) AKBh ad V 22; Shastri 801;
Poussin 48. "Yasya pudgalasya yo'nuśayo yasmin ālambane 'nuśete sa tena
tasmin samprayuktaḥ." AKBh ad V 18c-d; Shastri 793; Poussin 39. *yena yaḥ
samprayuktas tu sa tasmin samprayogataḥ // . . . te cānuśayāḥ* samprayoga-
to 'nuśayirannālambanataḥ; (2) AKBh ad V 34; Shastri 829; Poussin 72f.;
"Tat yathā rāgānuśayo 'prahīno bhavati aparijñātaḥ kāmarāgaparyavas-
thānīyaś ca dharmā ābhāsagatā bhavanti. tatra ca ayoniśo manaskāra evaṃ
kāmarāga utpadyate"; 3) AKBh ad V 36c-d; Shastri 831; Poussin 74; sarveṣāṃ
teṣāṃ *mūlam avidyā*.)

24. We tend to identify with all our bodily feelings, sense objects, psychological
processes, etc.: "Now, monks, this is the way leading to the origination of
personality [*sakkāya*, or 'self-identity']. One regards the eye thus: 'This is
mine, this I am, this is my self.' One regards [bodily] form and so on [i.e., all
the sensory and mental processes comprising human life] thus: 'This is
mine, this I am, this is my self.'" Ñāṇamoli 1995:1133, M III 285.

25. The distinction between merely conducing to certain behaviors and wholly
determining them is crucial. There are relatively indirect causal influences
between what brings about the dispositions and the way that they in turn
conduce to specific behaviors, i.e., the dispositions are important factors
within a larger, more multifaceted network of interdependent links no one
of which is capable of unilaterally determining behavior. Barkow (1989:74)
emphasizes this important distinction: "It will surprise no one that we are
capable of selfishness, deceit, and other such behaviours. It should not be
surprising that the capacity to act in such ways is a product of natural selection. Whether or how such behavior can be moderated or even eliminated
depends on the nature of the mechanisms that produce it—our psychology—and not on the selection pressures that produced the psychology. This
distinction, between our evolved behavioural mechanisms—our social psychology, our human nature—and the selection pressures that have generated them, is crucial. Human psychology is a product of natural selection, but
human behavior hardly reduces to a calculus of selection pressures."

26. If this were not the case, the theory of karma would amount to a narrow determinism and negate the very possibility of liberation. The Buddhist theory of causality, however, depicts neither the absolute inescapability of consequences nor a strict determinism. Rather, karmic activities set into
motion patterns of energies that conduce to effects consonant with the motivations that instigated them, which in turn tend to instigate further actions. Otherwise, the Buddha warned, there would be no way out of the vicious cycle and hence no point in religious practice: "If anyone should say:
'just as this man performs an action, just so will he experience the consequence'—if this were correct, there would be no pure life and no opportunity would be known for the stopping of suffering" (A I 249; "Yo . . . evaṃ

vadeyya—yathā yathāyaṃ puriso kammaṃ karoti tathā tathā taṃ paṭisaṃvediyatī ti—evaṃ santaṃ . . . brahmacariyavāso na hoti okāso na paññāyati sammā dukkhassa antakiriyāya") (Johansson 1979:146). Such a disheartening interpretation, the Buddha warned, would lead to an ill-advised passivity, a fatalistic and defeatist attitude that was the antithesis of the Buddha's exhortation to work toward one's own liberation: "For those who fall back on the former deed as the essential reason (*sārato paccāgacchataṃ*) [for their present actions], there is neither desire to do, nor effort to do, nor necessity to do this deed or abstain from that deed. So then, the necessity for action or inaction not being found to exist in truth and verity [for you], you live in a state of bewilderment with faculties unwarded." A I 174 (PTS translation).

27. One Pāli sutta (S II 65) states: "If one does not will, O Monks, does not intend, yet [a disposition] lies latent, this becomes an object for the persistence of consciousness. There being an object, there comes to be a support of consciousness. Consciousness being supported and growing, there comes to be the descent of name-and-form; conditioned by name-and-form, the six sense-spheres arises, etc. Such is the arising of this entire mass of suffering."

28. Vasubandhu describes this classic account of cyclic causality in terms of one's "mind stream": "the mind stream (*santāna*) increases gradually by the mental afflictions (*kleśa*) and by actions (karma), and goes again to the next world. In this way the circle of existence is without beginning." (AKBh III 19a-d; Poussin 57–59; Shastri 433–434. "Yathākṣepaṃ kramād vṛddhaḥ santānaḥ kleśakarmabhiḥ / paralokaṃ punaryāti . . . ityanādibhavacakrakam.")

29. These become a sine qua non of Buddhist liberation. Ñāṇamoli 1995:133, M I 47:

> When a noble disciple has thus understood the unwholesome and the root of the unwholesome, he entirely abandons the underlying tendency to lust, he abolishes the underlying tendency to aversion, he extirpates the underlying tendency to the view and the conceit "I am," and by abandoning ignorance and arousing true knowledge he here and now makes an end of suffering. In that way too a noble disciple is one of right view, whose view is straight, who has perfect confidence in the Dhamma, and has arrived at this true Dhamma.

30. "Abandon what is unwholesome [i.e. the three poisons], O monks! One can abandon the unwholesome, O monks! If that were not possible, I would not ask you to do so." Nyanaponika 1986:127, A II 19.

31. The five lower fetters that tie beings to the sensuous world were mentioned above in the *Maluṇkya-sutta* (Ñāṇamoli 1995:537f., M I 433). They are 1. a belief in self-identity or self-existence (*sakkāyadiṭṭhi*), 2. skeptical doubt (*vicikicchā*), 3. attachment to rules and observances (*sīlabbataparāmāso*), 4.

sensuous craving (*kāma-rāga*), and 5. ill will (*vyāpāda*). See Nyanatiloka 1977. These expand into the ten afflictions (*kleśa*) in later Abhidharma literature.

32. "A well-taught noble disciple . . . does not abide with a mind obsessed and enslaved by personality view [or "view of self-existence," *sakkāya-diṭṭhi*]; he understands as it actually is the escape from the arisen personality view, and personality view together with the underlying tendency to it is abandoned in him." Ñāṇamoli 1995:538, M I 434.

33. This point follows naturally from the principles of evolutionary biology and applies to all the sensory faculties: "The visual system is not built to represent an exact copy of the actual world; it is built to work by cues that maximize its function" (Gazzaniga 1998:87).

34. The psychologist and Buddhist scholar Rune Johansson (1979:173) concurs that, however ultimately illusory it may be, such a sense of self nevertheless serves important practical functions: "The ego-illusion is the glue or, rather, the structural tension that keeps the person together in a certain form. It gives a feeling of unity. A person without this feeling of identity or a wish to keep his identity or assert it will easily get a sense of unreality and of falling apart."

35. The psychologist Henry Stack Sullivan, in Mitchell's (1993:106) words, "also stressed repeatedly the illusory nature of the self we ordinarily take ourselves to be—singular, unique, in control of our self-revelations and self-concealments—which he felt was at enormous odds with what we actually do with other people. . . . Sullivan came to regard the experience people have of possessing a unique personal individuality as essentially a narcissistic illusion—'the very mother of illusions'—in the service of allaying anxiety and distracting attention from ways in which people actually operate with others." Mitchell considers the utility of this: "What may have begun as an illusion often becomes an actual guide to living by virtue of our necessary belief in it" (111).

36. These are the main Buddhist criticisms of a fixed "self." Though differing substantially on the ultimate nature of the body-mind relation, many scientific works on brain and consciousness also argue that our sense of a "unified, freely acting agent" is illusory because: 1. consciousness is merely a witness, not an agent, accompanying the mostly unconscious processes in the brain; 2. the notion of a "self" in control of these processes is therefore illusory; but that, interesting enough, 3. this illusion evolved because it served important survival needs.

    For example, brain scientist Richard Restak (1994: xvi) argues: "Modular theory . . . holds that our experience is not a matter of combining at one master site within the brain all of separate components into one central perception. . . . There is no master site, no center of convergence. . . . This means that no 'pontifical' cell or area holds sway over all others, nor do all areas of the brain 'report' to an overall supervisory center. Thus . . . the General Manager is a fictional character." Restak concludes (111–121): "Brain research on consciousness carried out over the past two decades casts important doubts on our traditional ideas about the unity and indissolubility of our mental lives" (121), particularly "the concept of ourself as a unified, freely acting agent directing our behavior."

The neurophysiologist Michael Gazzaniga concurs, arguing for conclusions remarkably similar to the Buddhist idea of no fixed "self" (*anātman*) and the implications we are drawing from it:

> Split-brain research . . . revealed that the left hemisphere contains the interpreter, whose job is to interpret our behavior and our responses, whether cognitive or emotional, to environmental challenges. The interpreter constantly establishes a running narrative or our actions, emotions, thoughts, and dreams. It is the glue that unifies our story and creates our sense of being a whole, rational agent. It brings to our bag of individual instincts *the illusion that we are something other than what we are.* It builds our theories about our own life, and these narratives of our past behavior pervade our awareness. . . . The insertion of an interpreter . . . that asks how infinite numbers of things relate to each other and gleans productive answers to that question can't help but give birth to the concept of self. Surely one of the questions the device would ask is "Who is solving these problems?" Call that "me," and away the problem goes! . . . The interpretation of things past . . . produces the wonderful sensation that our self is in charge of our destiny. . . . The interpreter . . . *creates the illusion that we are in control of all our actions and reasoning.* . . . Is it truly a human instinct, an adaptation that supplies a competitive edge in enhancing reproductive success? I think it is and my guess is that the very device which helped us conquer the vicissitudes of the environment enabled us to become psychologically interesting to ourselves as a species." (1998:174f., 151; emphasis added.)

37. Sociologists also find the notion that "self" is a function of narrative continuity useful for understanding identity in the modern world (Giddens 1991:53): "Self-identity, in other words, is not something that is just given . . . but is something that has to be routinely created and sustained in the reflexive activities of the individual. . . . Self-identity is . . . *the self as reflexively understood by the person in terms of her or his biography.* . . . A person's identity is not to be found in behaviour . . . but in the capacity *to keep a particular narrative going*" (54). "In the reflexive project of the self, the narrative of self-identity is inherently fragile. The task of forging a distinct identity . . . is clearly a burden. A self-identity has to be created and more or less continually reordered against the backdrop of shifting experiences of day-to-day life and the fragmenting tendencies of modern institutions" (185).

38. Blind-sightedness is a phenomenon in which individuals with impaired vi-

sual fields (a blind spot) can nevertheless accurately (80 percent success rate) guess the location of an object presented to that blind spot without, however, being conscious of seeing it. Split-brain patients, whose connecting tissue (corpus callosum) between the two hemispheres of the brain has been severed for medical reasons, demonstrate the same capacity in experimental settings: "The right hemisphere is conscious of important distinctions . . . yet if asked about them, consciousness is denied. The right brain makes the correct decisions, but the person cannot consciously explain how that was done." Scientists conclude that our neurological architecture divides the processes that take place in the right hemisphere from the ability of the left hemisphere, the locus of the "interpreter" that "creates our sense of being a whole, rational agent," to consciously and discursively communicate those processes. Restak (1994:129f.) draws the conclusion, widely accepted in cognitive science, that "a distinction must be made here between awareness and consciousness. While consciousness implies awareness, the relationship is not reciprocal. We can respond to something, implying some level of awareness, yet we may remain blithely unconscious of what's happening." See Kihlstrom 1987 on the "cognitive unconscious."

39. Scientific developments, such as we have investigated in this essay, have drastically exasperated this situation. Minsky (1986:306f.), for example, speaks of the untenable predicament brought about by cognitive science: "We each believe that we possess an Ego, Self or Final Center of Control. . . . We're virtually forced to maintain that belief, even though we know it's false."

Gazzaniga (1998:172) colorfully depicts this same predicament:

> "Goddamn it, I am me and I am in control." Whatever it is that brain and mind scientists are finding out, there is no way they can take that feeling away from each and every one of us. Sure, life is a fiction, but it's our fiction and it feels good and we are in charge of it. That is the sentiment we all feel as we listen to tales of the automatic brain. We don't feel like zombies; we feel like in-charge, conscious entities—period. This is the puzzle that brain scientists want to solve . . . the gap between our understanding of the brain and the sensation of our conscious lives.

This is precisely the puzzle that *The Embodied Mind* (Varela, Thompson, and Rosch 1991) addresses utilizing basic Buddhist ideas.

40. By a certain stage of human development, these same dynamics of identity construction at the social and cultural levels became predominant, self-perpetuating evolutionary forces in their own right. Carrithers (1992:49):

> The notion of an evolutionary ratchet is consonant with the idea of co-evolution, which suggests that organisms may produce changes in the environment, changes which redound on themselves, creating a circle of positive feedback. The only peculiarity in human evolution was that human social arrangements

and their unintended consequences became a selective force in themselves.
. . . And with the appearance of these forms there appeared the forms of causation associated with them: not just ecological causation . . . but now distinctly human social, political, and economic causation. These animals were, so
to speak, released into history.

41.   Progress in understanding the complex patterns of interdependence has arguably been hindered by adherence to outmoded, unproductive conceptual
dichotomies. Barkow, Cosmides, Tooby (1992:36f.):

> Thus, the debate on the role of biology in human life has been consistently
> framed as being between optimistic environmentalists who plan for human
> betterment and sorrowful, but realistic nativists who lament the unwelcome in-
> evitability of such things as aggression, or who defend the status quo as in-
> evitable and natural. . . . This morality play . . . has been through innumer-
> able incarnations . . . (rationalism versus empiricism, heredity versus
> environment, instinct versus learning, nature versus nurture, human universals
> versus cultural relativism, human nature versus human culture, innate behav-
> ior versus acquired behavior, Chomsky versus Piaget, biological determinism
> versus social determinism, essentialism versus social construction, modularity
> versus domain-generality, and so on). It is perennial because it is inherent in
> how the issues have been defined in the Standard Social Science Model itself,
> which even governs how the dissidents frame the nature of their dissent.

In many respects, this is largely a matter of searching for conceptual clarity. As Barkow, Cosmides, Tooby (83f.) argue: "Despite the routine use of
such dualistic concepts and terms by large numbers of researchers throughout the social and biological sciences, there is nothing in the real world that
actually corresponds to such concepts as 'genetic determination' or 'environmental determination.' There is nothing in the logic of development to
justify the idea that traits can be divided into genetically versus environmentally controlled sets." Biologist Susan Oyama concurs: "What all this
means is not that genes and environment are necessary for all characteristics, inherited or acquired (the usual enlightened position), but that there is
no intelligible distinction between inherited (biological, genetically based)
and acquired (environmentally mediated) characteristics" (1985:22), cited in
Varela, Thompson, and Rosch 1991:199f.

The models of complex, interdependent causality such as those found in
evolutionary biology, self-organization theory, or Buddhism go far in
avoiding the conundrums created by these unproductive dichotomies by
thinking in terms of patterns of relationship rather than in terms of fixed,
independent entities. These models effectively preclude dichotomization by
encompassing their opposite poles within their basic definitions. See Capra
(1997) for a straightforward introduction to these issues, and Waldron
(2002), "Beyond Nature/Nurture: Buddhism and Biology on Interdependence," for a treatment of them in dialogue with Buddhist perspectives.

42. Behavioral biologists, for example, have long recognized the complexity that these coevolutionary processes require: "The point is . . . that evolutionary processes are inseparable from the behaviour and social organization of animal species. . . . Ethological theory has quite strictly supported the neo-Darwinian view of the interdependence of genetic and behavioral evolution. . . . This is not to argue for 'instinctive determinism,' but to pose a more complex model in which genetic disposition, critical learning, and social environment all interact, even in the simplest and most stereotyped of species" (Nichols 1974:265f.).

Nor is this to suggest that distinct discourses can or should be reduced to a single "master narrative," particularly a biological one. "Darwinian theory," the anthropologist Carrithers (1992:41) argues, "differs from sociological and social anthropological styles of thought: it does not concern humans as persons, humans as realized and accountable agents in a social setting, but only humans as organisms. Evolutionary theory, in other words, does not pretend to explain the full detail of human life in all its dimensions. And because that theory speaks only of humans as organisms, then it can coexist with very different notions of, and practices concerning, human persons constructed in different cultural and social historical circumstances."

43. We must acknowledge yet qualify Geertz's (1979:59) cautionary counsel against uncritically projecting Western notions of "self" onto the world's cultures: "The Western conception of the person as a bounded, unique, more or less integrated motivational and cognitive universe, a dynamic center of awareness, emotion, judgement, and action organized into a distinctive whole and set contrastively both against other such wholes and against a social and natural background is, however incorrigible it may seem to us, a rather peculiar idea within the context of the world's cultures."

First, while the explicit concepts of selfhood, the obvious object of Geertz's remarks, do indeed vary radically from culture to culture, this does not prima facie preclude the possibility of an innate sense of selfhood and psychic organization, such as the evolutionary biologists and cognitive scientists posit and which by definition is a universal, specieswide capacity. Biologists are, after all, discussing humans as organisms, not as socially or culturally defined persons. Some Buddhists, at any rate, make this a similar distinction between a view of self-existence which is innate (and supposedly common to birds and other animals) (*sahajā satkāyadṛṣṭi*) and those views which are conceptual or deliberated (*vikalpita*) and hence unique to the human species (AKBh ad V 19; Shastri 794; Poussin 40. "Kāmadhātau satkāyāntagrāhadṛṣṭī tat samprayuktā ca avidyā avyākṛtāḥ. kiṃ kāraṇam? anādibhir aviruddhatvāt. ahaṃ pretya sukhī bhaviṣyāmi iti dānaṃ dadāti śīlaṃ rakṣati. . . . sahajā satkāyadṛṣṭīr avyākṛtā. yā mṛgapakṣiṇām api vartate. vikalpitā tu akuśala iti pūrvācāryāḥ"). Moreover, one might infer from the history of Buddhism that such "Western" conceptions of self and per-

sonhood are indeed found in other times and places. When Buddhists explicitly argued against a notion of self (*ātman*) that was remarkably similar
to Geertz's "peculiar idea," in classical India for example, they consistently
met with equally explicit, well-argued, and often strident defenses of it.
Moreover, the Buddhist refutation of such a self initially met with puzzled
and often antagonistic responses nearly everywhere Buddhism spread. We
should therefore be as wary of a reconstructed exceptionalism couched in
terms of cultural relativism as we need be of uncritical assumptions of cultural universality.

44. Hans Mol (1976:8f.), a sociologist of religion, elaborates this important
point: "Both in animals and in human beings, security is bound up with order. . . . The need for identity (or for a stable niche in this whole complex
of physiological, psychological, and sociological patterns of interaction) is
very much bound up with continuous regularity. Order means survival;
chaos means extinction. Identity, order, and views of reality are all intertwined. The point is that an interpretation (any interpretation) of reality
is necessary for the wholeness (and wholesomeness) of individual and
society."

45. "There exists convincing fossil evidence that the increased size of our brains
and the development of culture are closely linked. This . . . resulted in very
rapid selection for large brains, and a very finely organized, interdependent
system. Our minds evolved in the context of culture, just as culture has always been produced by the action of our minds" (Barash 1979:221).

46.   There is no such thing as a human independent of culture. . . . As our central
        nervous system—and most particularly its crowning curse and glory, the neo-
        cortex—grew up in great part in interaction with culture, it is incapable of di-
        recting our behavior or organizing our experience without the guidance pro-
        vided by systems of significant symbols. . . . Such symbols are thus not mere
        expressions, instrumentalities, or correlates of our biological, psychological,
        and social existence; they are prerequisites of it. (Geertz 1973:49)

47. This is an important postulate in the sociology of knowledge (Berger and
Luckmann 1966:183): "Man is biologically predestined to construct and to
inhabit a world with others. This world becomes for him the dominant and
definitive reality. Its limits are set by nature, but once constructed, this
world acts back upon nature. In the dialectic between nature and the socially constructed world the human organism itself is transformed. In this
same dialectic man produces reality and thereby produces himself."

48. The psychiatrist Hundert (1989:107) describes these processes: "What is crucial here is the reciprocal nature of the developments of the capacity to have
unitary subjective experiences and the capacity to experience unitary permanent objects. By studying the behaviour of infants, Piaget showed that, in
normal human development, the notion of permanent *objects* and the no-

tion of a separate *self* who is experiencing those objects develop together. From the starting-point of symbiosis, the origins of self and object proceed apace."

49.   This distinction is one of the fundamental themes from traditional religion and philosophy that has been translated into the terminology of theoretical psychology: "Through reflection on the nature of self-consciousness, Kant demonstrated that the notion of 'self' (the 'I') carries with it the notion of 'object' (the external world). Kant said that 'object' necessarily accompanies 'subject' (conceptually). Piaget showed that what is necessarily so, is actually so (epistemologically)! . . . Not only does 'object' always accompany 'subject' (self), but it is our experience with objects that enables us to 'objectify' them!" (Hundert 1989:108–109).

50.   Anthropologist and primatologist Tomasello refers to the multidimensionality of specifically human forms of cognition: "Modern adult cognition of the human kind is the product not only of genetic events taking place over many millions of years in evolutionary time but also of cultural events taking place over many tens of thousands of years in historical time and personal events taking place over many tens of thousands of hours in ontogenetic time" (Tomasello 1999:216).

51.   Berger, Berger, and Kellner (1973:78) make the same point, intertwining themes of selflessness, impermanence, insecurity, and a deluded belief in self-identity: "On the one hand, modern identity is open-ended, transitory, liable to ongoing change. On the other hand, a subjective realm of identity is the individual's main foothold in reality. Something that is constantly changing is supposed to the *ens realissimum*. Consequently it should not be a surprise that modern man is afflicted with a permanent identity crisis, a condition conducive to considerable nervousness."

52.   Although identities imply or require relatively well-defined boundaries, the world is seldom so neatly divided in practice. Identities must be forged, rather, through abstracting shared qualities and categorizing people accordingly. As the anthropologist Mary Douglas (1966:4) argues: "Ideas about separating, purifying, demarcating and punishing transgressions have as their main function to impose system on an inherently untidy experience. It is only by exaggerating the difference between within and without, above and below, male and female, with and against, that a semblance of order is created."

Although we classify, categorize and discriminate for eminently practical reasons, demarcating boundaries between peoples, groups, cultures, etc. by labeling and stereotyping them is almost never neutral. "The maintenance of any strong boundary," Mol (1976:174, 11) observes, "requires emotional attachment to a specific focus of identity," since "it is precisely through emotional fixation that personal and social unity takes place."

53.   Mol (1976:5f.) defines sacralization as "the process by means of which on the

level of symbol-systems certain patterns acquire the . . . taken-for-granted, stable, eternal, quality. . . . Sacralization, then . . . [precludes threats to] the emotional security of personality and the integration of tribe or community. Sacralization protects identity, a system of meaning, a definition of reality, and modifies, obstructs, or (if necessary) legitimates change."

54. "The inherently precarious and transitory constructions of human activity are thus given the semblance of ultimate security and permanence. The institutions are magically lifted above these human, historical contingencies. . . . They transcend the death of individuals and the decay of entire collectivities. . . . In a sense, then, they become immortal. . . . [The modern individual] is whatever society has identified him as by virtue of a cosmic truth, as it were, and his social being becomes rooted in the sacred reality of the universe. Like the institutions, then, roles become endowed with a quality of immortality" (Berger 1967:36f.).

55. Mystification, in the concept of reification, is basic to the sociology of knowledge. Berger and Luckmann (1966:89f.) make a distinction, similar to the Buddhist analysis of self-identity, between two levels of reification, one implicit and unreflective and the other explicit and cultivated. Reification is

> the apprehension of the products of human activity as if they were something else than human products—such as facts of nature, results of cosmic laws, or manifestations of divine will. Reification is possible on both the pretheoretical and theoretical levels of consciousness. It would be an error to limit the concept of reification to the mental construction of intellectuals. Reification exists in the consciousness of the man in the street and, indeed, the latter presence is more practically significant. It would also be a mistake to look at reification as a perversion of an originally nonreified apprehension of the social world, a sort of cognitive fall from grace. On the contrary, the available ethnological and psychological evidence seems to indicate the opposite, namely, that the *original apprehension of the social world is highly reified both phylogenetically and ontogenetically.* (emphasis added)

That is to say, consistent with our main thesis, that we know the world by means of our evolved capacities to reify experience into the categories of language and social and cultural life. These are both fundamental and fundamentally obscuring.

56. As the anthropologist Eric Wolf warns: "By endowing nations, societies or cultures with the qualities of internally homogeneous and externally distinctive and bounded objects, we create a model of the world as a global pool hall in which the entities spin off each other like so many hard and round billiard balls. Thus it becomes easy to sort the world into differently colored balls." Wolf 1982:6, cited in Carrithers 1992:25.

57. "Nationalism depends upon a particular social definition of a situation, that is, upon a collectively agreed-upon entity known as a particular nation. . . . The definition of a particular group of people as constituting a nation

is always an act of social construction of reality. That is, it is always 'artificial'" (Berger, Berger, and Kellner 1973:167). Such reification, Carrithers (1992:19) argues, is "also fully consistent with, indeed necessary to, the notion of cultures or societies as bounded, integral wholes. For once mutability and the vicissitudes of history are allowed, the notion of the integrity and boundedness of cultures and societies begins to waver and melt." See Anderson (1983) for a historical approach to the rise of nations as "Imagined Communities."

58. The reification of processes into entities, a recurrent theme throughout this essay, is a problem for social theory as well, as Norbert Elias (1982 [1939]:228) explains:

> Concepts such as "individual" and "society" do not relate to two objects existing separately but to different yet inseparable aspects of the same human beings. . . . Both have the character of processes. . . . The relation between individual and social structures can only be clarified if both are investigated as changing, evolving entities. . . . The relation between what is referred to conceptually as the "individual" and as "society" will remain incomprehensible so long as these concepts are used as if they represented two separate bodies, and even bodies normally at rest, which only come into contact with one another afterwards as it were.

59. This aspect of nationalism is a modern manifestation of an ancient phenomenon: "The tribe, the race, the nation, and the political state have always been considered sacred by those who shared such collective identities. The nationalism of the nineteenth century gave rise to the sacred adoration of the nation in the twentieth century. Hitler's declaration that the fatherland was sacred left no room for doubt. Mussolini, Stalin and Mao followed suit. Now the nation became the arbiter of morality: anything that furthers the cause of one's country is good; whatever hinders it is evil" (Strivers 1982:26f.).

60. Becker (1975:119): "Each heroic apotheosis is a variation on basic themes. . . . Civilization, the rise of the state, kingship, the universal religions—all are fed by the same psychological dynamic: guilt and the need for redemption. If it is no longer the clan that represents the collective immortality pool, then it is the state, the nation, the revolutionary cell, the corporation, the scientific society, one's own race. Man still gropes for transcendence . . . the individual still gives himself with the same humble trembling as the primitive to his totemic ancestor."

61. Becker (1975:148): "The result is one of the great tragedies of human existence, what we might call the need to 'fetishize evil,' to locate the threat to life in some special places where it can be placated and controlled."

62. Anderson (1983:11) eloquently describes the spiritual vacuum that nationalism came to fill in the early modern period: "In Western Europe the eighteenth century marks not only the dawn of the age of nationalism but the

dusk of religious modes of thought. The century of the Enlightenment, of rationalist secularism, brought with it its own modern darkness. With the ebbing of religious belief, the suffering which belief in part composed did not disappear. Disintegration of paradise: nothing makes fatality more arbitrary. Absurdity of salvation: nothing makes another style of continuity more necessary. What then was required was a secular transformation of fatality into continuity, contingency into meaning." What resulted was the transformation of the modern nation into the sacralized locus of "secular" immortality, a transformation, we might add, that has not gone unchallenged, particularly by fundamentalists around the world. See Bruce Lawrence's incisive treatment of this conflict in *Defenders of God: The Fundamentalist Revolt Against the Modern Age* (1989).

63.   "The hero is, then, the one who accrues power by his acts, and who placates invisible powers by his expiations. He kills those who threaten his group, he incorporates their powers to further protect his group, he sacrifices others to gain immunity for his group. In a word, he becomes a savior through blood" (Becker 1975:150).

64.   Whose desecration or "desacralization," we should remember, is a criminal offense in many countries.

65.   "Since men now hold for dear life onto the self-transcending meanings of the society in which they live, onto the immortality symbols which guarantee them indefinite duration of some kind, a new kind of instability and anxiety are created. And this anxiety is precisely what spills over into the affairs of men. In seeking to avoid evil, man is responsible for bringing more evil into the world than organisms could ever do by exercising their digestive tracts." (Becker 1975:5).

66.   Some qualifications are in order here. First, I have neglected the most obvious and undeniable dimension of this: that ruling elites attain and accumulate personal power and material gain through organized violence. The analysis of these patterns of human interaction and their specific details belong, however, to the political or social sciences; they are not the issues being addressed here. Rather, we are concerned at this point in the essay with understanding the willing participation of the masses of individuals without whom organized violence would be impossible, and for which the "sacralization" of personal, cultural and political identities and the mystifications surrounding the nation seem to be necessary, but certainly not sufficient, conditions. To suggest that the dynamics of these processes are similar in different countries or contexts, however, is by no means to imply either their "moral equivalence" or that they entail equally malicious intent to "deceive the masses." Some things surely are more worth defending than others and some are more compellingly true. They remain, nevertheless, consensual realities defined within our human "life worlds."

67.   This is one reason why Buddhists, for example, do not aim to "destroy" such identities, since, as we have pointed out, these constructs also serve many

useful purposes, both practical and spiritual. Rather, Buddhist traditions emphasize the problems that arise from misconstruing the nature of constructed identities, as if they were unconditioned, permanent, and self-substantially existent. They therefore do not advocate destroying a substantive self, which never existed in the first place, but rather seeing through the illusion that such a constructed self is either substantially real or ultimately dependable and thereafter working with the attachments and desires associated with that sense of self in order to transmute them into more satisfying pathways, i.e., awakened and compassionate activities. A thorough discussion of such transformational practices, important as they are for understanding the relation between Buddhist thought and practice, would take us too far afield from our present focus on the second Noble Truth, the cause of suffering.

68. It is striking to consider the unanimity of opinion concerning the lack of intrinsic or substantial identity—and its concomitant characteristics of being constructed, conflicted, and obscuring—that has been reached in the diverse fields discussed so briefly in these few pages, as the extensive citations in the footnotes aim to demonstrate. Recent progress in many fields has often consisted of deconstructing false or outmoded dichotomies inherited from our earlier "billiard ball" models of life and replacing them with more process-oriented models, such as circular causality and self-organization theory—a change of perspective that relativizes the putative essences, entities, and dichotomies of previous eras into mere heuristic devices that, though sometimes useful in their own way, often serve more as obstacles to deeper understanding, being ultimately misleading at best and deleterious at worst.

References

Abhidharmakośabhāṣya. Ed. S. D. Shastri. 1981. Repr. Varanasi: Bauddha Bharati. Trans. de La Vallée Poussin. 1971. *L'Abhidharmakośa de Vasubandhu*. Bruxelles: Institut Belge des Hautes Etudes Chinoises.

Anderson, B. 1983. *Imagined Communities*. London: Verso.

Barash, D. 1979. *The Whisperings Within: Evolution and the Origin of Human Nature*. New York: Harper and Row.

Barkow, J. 1989. *Darwin, Sex, and Status*. Toronto: University of Toronto Press.

Barkow, J., L. Cosmides, J. Tooby. 1992. *The Adapted Mind: Evolutionary Psychology and the Generation of Culture*. New York: Oxford University.

Barlow, H. 1987. "The Biological Role of Consciousness." In C. Blakemore and S. Greenfield, eds., *Mindwaves*. Oxford: Basil Blackwell.

Bateson, G. 1979. *Mind and Nature: A Necessary Unity*. New York: Bantam.

Becker, E. 1975. *Escape from Evil*. New York: Free Press.

Berger, P. 1967. *The Sacred Canopy: Elements of a Sociological Theory of Religion.* New York: Doubleday.

Berger, P., B. Berger, and H. Kellner. 1973. *The Homeless Mind: Modernization and Consciousness.* New York: Vintage.

Berger, P. and T. Luckmann. 1966. *The Social Construction of Reality: A Treatise in the Sociology of Knowledge.* New York: Anchor.

Bodhi, Bhikkhu, trans. 2000. *The Connected Discourses of the Buddha.* (*Saṃyutta nikāya*). Somerville, Mass.: Wisdom.

Camus, A. 1971. *The Rebel.* London: Penguin.

Capra, F. 1997. *The Web of Life.* New York: Anchor.

Carrithers, M. 1992. *Why Humans Have Cultures.* New York: Oxford University Press.

Collins, S. 1993. "The Discourse on What Is Primary (*Aggañña Sutta*)," *Journal of Indian Philosophy* 21(4): 301–93.

Conze, E. 1973. *Buddhist Thought in India.* Ann Arbor: University of Michigan Press.

Douglas, M. 1966. *Purity and Danger: An Analysis of Concepts of Pollution and Taboo.* London: Routledge and Kegan Paul.

Duncan, H. D. 1962. *Communication and Social Order.* New York: Bedminster.

Elias, N. 1982 [1939]. *The Civilizing Process.* New York: Pantheon.

Gazzaniga, M. 1998. *The Mind's Past.* Berkeley: University of California Press.

Geertz, C. 1973. *The Interpretation of Cultures.* New York: Basic.

——— 1979. *Local Knowledge.* New York: Basic.

Giddens, A. 1991. *Modernity and Self-Identity: Self and Society in the Late Modern Age.* Stanford: Stanford University Press.

Guenther, H. and L. Kawamura. 1975. *Mind in Buddhist Psychology.* Emeryville, Cal.: Dharma.

Hundert, E. 1989. *Philosophy, Psychiatry, and Neuroscience: Three Approaches to the Mind.* Oxford: Clarendon.

Johansson, R. 1979. *The Dynamic Psychology of Early Buddhism.* London: Curzon.

Kegan, R. 1982. *The Evolving Self.* Cambridge: Harvard University Press.

Kihlstrom, J. F. 1987. "The Cognitive Unconscious." *Science* 237:1445–1452.

Lackoff, G. and M. Johnson. 1980. *Metaphors We Live By.* Chicago: University of Chicago Press.

Lawrence, Bruce. 1989. *Defenders of God: The Fundamentalist Revolt Against the Modern Age.* San Francisco: Harper and Row.

Luckmann, T. 1967. *The Invisible Religion: The Problem of Religion in Modern Society.* New York: Macmillan.

Macy, J. 1991. *Mutual Causality in Buddhism and General Systems Theory.* Albany: State University of New York.

Mayr, E. 1988. *Toward a New Philosophy of Biology: Observations of an Evolutionist.* Cambridge: Harvard University Press.

Minsky, M. 1986. *Society of Mind.* New York: Simon and Schuster.

Mitchell, S. 1993. *Hope and Dread in Psychoanalysis.* New York: Basic.

Mol, H. 1976. *Identity and the Sacred: A Sketch for a New Social-Scientific Theory of Religion*. Oxford: Basil Blackwell.

Ñāṇamoli, Bh., trans. 1995. *The Middle Length Discourses of the Buddha: A New Translation of the "Majjhima Nikāya."* Boston: Wisdom.

Nichols, C. 1974. "Darwinism and the Social Sciences." *Philosophy of the Social Sciences* 4:255–257.

Nyanaponika, T. 1986. *The Vision of Dharma*. York Beach, Maine: Weiser.

Nyanatiloka. 1977. *Buddhist Dictionary: Manual of Buddhist Terms and Doctrines*. Colombo: Frewin; rpt. San Francisco: Chinese Materials Center, 1977.

Oyama, S. 1985. *The Ontogeny of Information*. Cambridge: Cambridge University Press.

Ogden, T. 1990. *The Matrix of the Mind*. Northvale, N.J.: Aronson.

Poussin, L. de, trans. 1971. *L'Abhidharmakośa de Vasubandhu*. Bruxelles: Institut Belge des Hautes Etudes Chinoises.

Restak, R. 1994. *The Modular Brain: How New Discoveries in Neuroscience Are Answering Age-old Questions About Memory, Free Will, Consciousness, and Personal Identity*. New York: Touchstone.

Richards, R. 1987. *The Emergence of Evolutionary Theories of Mind and Behavior*. Chicago: University of Chicago Press.

Saul, J. R. 1992. *Voltaire's Bastards: The Dictatorship of Reason in the West*. New York: Vintage.

Shastri, S. D., ed. 1981. *Abhidharmakośabhāṣya*. Varanasi: Bauddha Bharati.

Snellgrove, D. 1987. *Indo-Tibetan Buddhism: Indian Buddhists and Their Tibetan Successors*. Boston: Shambala.

Strivers, R. 1982. *Evil in Modern Myth and Ritual*. Athens: University of Georgia Press.

Tomasello, Michael. 1999. "The Cultural Origins of Human Cognition." Cambridge: Harvard University Press.

Trivers, R. 1976. Forward to R. Dawkins, *The Selfish Gene*. Oxford: Oxford University Press.

Varela, F., E. Thompson, and E. Rosch. 1991. *The Embodied Mind: Cognitive Science and Human Experience*. Cambridge: MIT Press.

von Bertalanffy, Ludwig. 1968. *General System Theory*. New York: Braziller.

Waldron, W. 2000. "Beyond Nature/Nurture: Buddhism and Biology on Interdependence." *Contemporary Buddhism* 1(2): 199–226.

Walshe, M., trans. 1987. *Thus Have I Heard: The Long Discourses of the Buddha*. Boston: Wisdom.

Wolf, E. 1982. *Europe and the People Without History*. London: University of California Press.

Wolf, E. 1991. "Discussion." *Psychoanalytic Dialogues* 1(2): 158–172.

The relation between perception and imagination has been debated by Western philosophers since Greek antiquity, and it has been a central theme of Buddhist contemplative inquiry for the past twenty-five hundred years. Plato emphasized the differences between perception and imagination, whereas Aristotle argued for the continuity between the two. In early Buddhist meditation one finds a strong emphasis on cultivating "bare attention," in which perception is at least relatively purged of conceptual, or imaginative, superimpositions (Nyanaponika Thera 1973). In the Buddhist Madhyamaka view, on the other hand, it is said that all known phenomena arise in one's experienced world (loka) by the power of conceptual designation (Lamrimpa 1999), implying that conceptual superimpositions saturate one's normal, perceptual experience of the world.

In this essay Francisco J. Varela and Natalie Depraz explore this fascinating topic from the perspectives of contemporary neuroscience, the Western tradition of phenomenology, and Tibetan Buddhism. A central strategy of their essay is to use imagination to show the inextricably nondual nature of the material brain basis of mental events and their experiential quality. There is a wide consensus among contemporary neuroscientists that the ability to produce and manipulate imaginary objects stems from the very same neural capacities as those involved in high-level vision and cognition in general. Both require the participation of memory, language, anticipation, and movement, depending on the nature of the cognitive task. This raises the intriguing question: What are the necessary and sufficient causes for the experience of imagination and perception? Are the neural correlates of these mental processes only necessary for the production of such mental events, or are they, together with their interaction with the body and physical environment, causally sufficient? Hasn't the lived experience of imagination a causative power by itself, and what is its specificity? The hypothesis is that perception is a sensorimotor-constrained imagination, and imagination is not an added dimension of perception but belongs to the core of the humanly experienced cognitive life.

According to the Western phenomenological tradition, imagination belongs to the very core of human consciousness, and it is grounded on a prereflexive (or prenoetic, unconscious) level of consciousness from which *it shines forth. On the face of it, this appears very similar to the Buddhist view that all instances of both conceptual and perceptual awareness are grounded in and shine forth from a prenoetic level of consciousness* (bhavaṅga, ālayavijñāna). *That is, they emerge from this immaterial substrate consciousness, while conditioned by the body and its interactions with the physical environment. This essay brings these two views into fascinating juxtaposition.*

Francisco J. Varela and Natalie Depraz

Imagining: Embodiment, Phenomenology, and Transformation

Imagination is one of the quintessential qualities of life and our being. Its central attribute is the manifestation of vivid, lived mental content that does not refer directly to a perceived world but to an absence that it evokes. It is fair to say that imagination is emblematic, in fact, of a *cluster* of human abilities: imagining proper, or mental imagery, remembrance, fantasy, and dreaming. Imagination is an inexhaustible source in all these dimensions, explored and praised by human cultures throughout the world, a witness to its centrality.

Our purpose in this essay is to let imagination be a guiding thread in a journey of exploration of its *inextricably nondual* quality, making it possible to travel from its material-brain basis to its experiential quality without discontinuity. That is, we are not going to propose a "bridge" between a scientific view of imagination and its place in the Buddhist discipline of human transformation. Our purpose is to embrace the entire phenomenon in all its complexity and weave it as a unity with its many dimensions, which need and constrain each other without residue—in the body and brain, in its direct phenomenological examination, and in its pragmatic mobilization for human change. Only such weaving can be called a meeting of Buddhism and neuroscience on a new phenomenological ground.

1. MENTAL IMAGERY: EMBODIMENT

1.1 SEARCHING FOR THE MIND'S EYE

Let us first start from the empirical side, considering what can be said about the explicit embodiment of imagining. This means to start from results in modern cognitive neuroscience, before we expand that to a broader biological framework. Now, as a scientific topic of study, imagination appears very sharply in modern research not as a general topic but in one of its most central aspects: visual mental imagery, the capacity for experiencing, evoking, and examining images in the mind's eye. Mental imagery has a long history that goes back to the Greeks, such is its ease of access and compelling nature. Clearly imagining and perceiving seem to be, at face value, not the same acts. However, most people can make a (more or less sharp) picture of my room, which is not only as vivid as perception but also preserves the spatial properties of the scene they represent, and it is often difficult to separate one from the other. As we shall briefly investigate in this section, with few exceptions, there is a wide consensus in current research that the ability to produce and manipulate imaginary objects can be naturally explained as the endogenous mobilization of the very same neural capacities involved in high-level vision and cognition in general, which requires the participation of memory, language, anticipation, and movement, depending on the nature of the imagery task.

In its time-tested manner, research on the brain basis of imagining thus far has focused particularly on a voluntary mental imagery in carefully controlled laboratory conditions. Only in such conditions can one bring into play the use of modern techniques of global brain study, especially the non-invasive methods of using position emission topography (PET) and functional magnetic resonance (fMRI). Let us turn to a quick tour of some of the most important results and questions (see Kosslyn 1994 and Mellet et al. 1998 for review).

The mind's eye: The debate on primary visual areas   One of the most dramatic questions in this field has been whether the so-called primary visual areas (PVA) are necessarily involved and activated in producing mental imagery as they are during a visual perception. In fact, a major result of modern neuroscience is the discovery that the primary visual areas (area V1 in particular) are topographically organized with respect to the visual field. In other words, its component neurons can be mobilized by showing a stimulus in a small region placed in a precise location of the visual field. Given that the distinctive character of mental images is their topographical verac-ity, a central question is whether the PVA are active to a comparable degree while looking at an image as opposed to imagining it. This has been a thorny debate for science and philosophy alike.

A recent study blocking PVA reversibly by transcranial magnetic stimu-lation (rTMS) has given the best direct response so far (Kosslyn 1999). Blockage of rTMS actually renders subjects unable to visualize striped pat-terns, suggesting very strongly that, at least with regard to tasks of this kind, V1 is indispensable. In this sense perception and imagination can be said to share their spatial characteristics because they also share the same primary basis for the emergence of an image, vividly present to experience. This study is the most recent in a series of other converging evidence. For in-stance Bisiach, Luzzatti, and Perani (1979) found that subjects with hemilat-eral neglect not only ignore objects on one side during perception but also ignore objects on that side during imagery!

But mental imagery research has gone further in trying to study subjects performing a number of different mental tasks. For instance, subjects were presented with a visual map of an island and were compared when imagin-ing this same map. The comparison revealed significant activation under PET in the occipito-temporal regions but not in the PVA. In fact, about half of the mental imagery studies do *not* show an active involvement of PVA (Roland and Gulvas 1994). This might seem paradoxical at first glance. But the answer probably lies in the fact that it is the *type* of mental imagery in-volved that determines an important PVA activation or not. Simply put, vi-sual imagery requires a topographically organized area, whereas spatial mental imagery involving an imagined bodily displacement (such as fol-lowing an island's map with the mind's eye) does not.

But this is still much in exploration, and the precise role of PVA is still open. It is known, for instance, that schizophrenic hallucinations do not in-duce an increase in V1, but that visual association areas were so activated (Silberswieg et al. 1995). Instead of demanding a yes/no answer to this an-cient conundrum, researchers are now focusing on defining more precisely the kind of visual images and acts of imagination that are involved in vari-ous tasks. This again highlights the fact that the neural basis of mental im-agery does not seem to be a network of circuits but rather a *pattern of dy-namic interactions* between multiple candidate subregions subserving various cognitive capacities.

Involvement of multiple areas   In contrast to the heated issue of the primary areas, there is a large consensus that associative areas are constantly present in imagination. Beyond PVA, visual activity is structured, as is well known, in two concurrent streams, the ventral, or occipito-temporal, and the infero-temporal cortex. These are involved in the perception of form and its figurative aspects (such as face recognition) and the intentional content of the percept (the "what?" of seeing; Sergent, Ohta, MacDonald 1992). The occipito-parietal route goes up to the superior parietal region. This subcircuit is quite multifaceted, for it is involved in localization, shift of spatial attention and spatial working memory, and is thus involved in the spatial location (the "where?" of seeing). Several studies have revealed that the dorsal route is active when imagining in the absence of any visual presentation, for example, in subjects who had to mentally navigate a route previously walked. In contrast, the ventral route is easily detected in mental images either visually recalled or named, such as letters and unusual objects. In brief, these results, taken together, clearly underscore the kinship between visual perception and mental images. This was clearly the case with PVA, but the kinship becomes fuzzier for nonprimary regions: a number of brain regions involved in perception are not systematically involved in imagery.

Imagery and language   Interestingly, a mental image is not only a recall of a previous perception but can be generated starting from a *verbal* description.[1] Such language-evoked images are quite comparable to those sensorially induced. In particular, mental scanning and distance comparisons are comparable to those effected on images recalled from previous presentations.

This underscores again the important idea that visual images can be mobilized by extravisual brain circuits. For instance, constructed cube assemblies following auditory instructions activate the dorso-lateral route discussed above. Thus, although visual and verbal activities are quite distinct cognitive entities, there is coherent cross-modal activation that works in imagery just as well as in actual cross-modal perception.

Memory   It stands to reason that episodic memory recall and imagery very closely complement each other. Oftentimes a mental image is generated through a recall, and, conversely, a recall often results in a vivid mental image. This has been observed in various protocols concerning episodic memory. This observation will be especially relevant in light of what will follow later in section 3.

There seems to be a differentiated participation of object and spatial memory and imagery. In fact, it has been shown that at least both types of memories can be differentiated as to the regions they mobilize. The first circuitry is frontalized and seems to be active when the image is dynamic (i.e., spatial transformations of the image). The second is more ventrally located in the middle frontal gyrus and is better related to figurative working memory. This distinction also holds for mental images.

Motor imagery   Finally it should be said here that the mental rehearsal of simple or complex motions is also a human capacity. Although such imagery has some resemblance to visual imagery, a distinction can be made with motor images that allow us to imagine an external or third-person perspective representing actions or an internal or kinesthetic first-person perspective in which we execute the movement ourselves.

Interestingly, in this field the question of whether motor images share the same basis for preparation and execution of actual movements has been hotly debated (Jeannerod 1994) and is still in full development (Berthoz 1998). The functional equivalence between motor imaging and motor preparation has been supported by physiological correlates of motor imagery, which follow closely the activation of areas involved in actual movement or even while seeing someone else executing the same movement: the famous "mirror-neurons" (Decety et al. 1989, 1994). Brain-imaging studies have uncovered a plurality of regions, for instance in finger movement or during saccades (the very rapid eye-movements that accompany normal vision). But, again, not all the regions that are active during overt hand and eye coordination are also involved in mental imagery as well, according to various studies that are sometimes contradictory.

In conclusion, the idea that perception and imagery share common mechanisms has been repeatedly postulated since the time of Aristotle, but the recent evidence just discussed gives a fresh angle on this question. By specifying that this common ground is the cooperative working of a multiplicity of cognitive capacities (including memory, language, and motion), the difference between them is also stressed. Kosslyn (1994:74) lists three such distinctive differences: 1. mental images fade rapidly—in perception the sensory presentation helps to maintain the image, 2. mental images are created from remembrance and association, thus they do not have a veridical relation to their contents, 3. images, unlike perception, are remarkably changeable.

Thus, the same capacities working on an endogenous basis, unconstrained by the sensorimotor embodiment of the organisms, make imagination come to the fore; perception then can be seen as constrained imagination. The far-reaching import of this conclusion should now be examined in more detail.

1.2 IMAGINATION AT THE CORE OF LIFE AND MIND

The organism as an enactive imaginary being   To explore further the consequences of these insights from recent cognitive neuroscience, we wish to pause to place them in the broader studies of the natural history and biology of the brain and mind. In fact, it is still common to regard the cognitive life of an organism as a "representational" coping, where perception is primary and the main source and drive for any valid cognition. A miscognition is thus a misrepresentation, such as mistaking a rope for a snake. However, this view of mind as an accurate or "equate" representation of the world is problematic, and to see why we need to take a broader look at how cognition can be understood.

Varela's overall approach to cognition is based on situated, embodied agents. He has introduced the name *enactive* to designate this approach more precisely. We cannot expand this overall framework extensively (see Varela 1992 [1989]; Varela, Thompson, and Rosch 1991), but its core thesis can be expressed as two complementary aspects:

1. On the one hand, there is the ongoing *coupling* of the cognitive agent, a permanent coping that is fundamentally mediated by *sensorimotor* activities.
2. On the other hand, there are the *autonomous* activities of the agent whose identity is based on emerging, *endogenous* configurations (or self-organizing patterns) of neuronal activity.

Enaction implies that sensorimotor coupling modulates but does not determine an ongoing endogenous activity that it configures into meaningful world items in an unceasing flow. Enaction is naturally framed in the tools derived from dynamic systems, in stark contrast to the cognitivist tradition that finds its natural expression in syntactic information-processing models. The debate pitting embodied-dynamics versus abstract-computational as the basis for cognitive science is very much alive (Port and van Gelder 1997).

From an enactive viewpoint it follows that mental acts are characterized by the concurrent participation of several functionally distinct and topographically distributed regions of the brain and their sensorimotor embodiment. It is the complex task of relating and integrating these different components that is at the root of temporality from the point of view of the neuroscientist. For example, for high-level vision this large-scale integration would draw from all its necessary components, mobilizing not just perceptual abilities but motivation and emotional tonality, attention, memory, and motion. In brain topography this covers a largely distributed set of regions and circuits such as those encountered in the brain-imagining studies related to mental imagery.

A central idea pursued here is that these various components require a *frame or window of simultaneity that corresponds to the duration of the lived present*. This is important for us here, for it places imagination in its factual dimensions: as a transitory nature of an image or a content with a flow of consciousness. In this view the constant stream of sensory activation and motor consequences is incorporated within the framework of an endogenous dynamic, which gives it its depth or incompressibility. This idea is not merely a theoretical abstraction: it is essential for understanding a vast array of evidence and experimental predictions. These endogenously constituted integrative frameworks account for perceived time as discretized and not linear, since the nature of this discreteness is a horizon of integration rather than a string of temporal pulses (Varela, Thompson, and Rosch 1991; Dennett and Kinsbourne 1991; Pöppel and Schill 1995; Varela 1999). Our cursory impression of linearity comes from the fact that in this living present memory will bring a sense of past and continuity.

Within this enactive framework it follows that the self-produced activity from the organism's side is as central to mental/cognitive life as the more traditional idea that the world provides some form of "input." Stated bluntly, the brain mostly relates to its own activity constantly engaged in the organism's maintenance and regulation. This endogenous, self-constituting activity is based on its extensive interconnectivity, but it also occurs because the brain, being part of the organism, never ceases in its self-regulation. This engenders ongoing levels of activity that constantly give rise to dynamic patterns, even in the absence of any world input. And one of the most dramatic manifestations of this fact is the flowery imaginary life that mani-

fests during dreaming (or less naturally when one is put into a state of sensory deprivation). Ordinary perception is, to an essential degree, sensorimotor constrained imagination. Imagination is central to life itself, not a
marginal or epiphenomenal side-effect of perception.

1.3 SELF-EMERGENCE OF IMAGINATION

Large-scale integration and synchrony   What we just said does not make it
clear *how* such large-scale self-organization happens in the brain. Although
cognitive neuroscience knows an enormous amount about the multiplicity
of areas involved and their various specific contributions (cf. Gazzaniga
1999 for review), it knows much less about how these regions can work in
concert together. There are two general principles that we wish to emphasize in this respect that seem to emerge from recent work: *reciprocity* and
synchrony.

Reciprocity refers to the fact that, contrary to the classic idea based on
information processing, cognitive operation can hardly be described as a
linear flow: from raw sensory input, to interim processing, to an output of
action. Both anatomically and physiologically the so-called low level and
high level regions are interconnected in a *reciprocal* fashion. When a visual
image is shown to the eye, it encounters as it enters into the brain in PVA
(i.e., in a bottom-up direction), a highly structured neural context provided
by the multiple regions that connect to PVA (top-down direction). Thus the
sensory flow can modulate but not directly drive the ensuing cognitive
state. Perception is demonstrably constrained and shaped by the concurrent higher cognitive memories, expectations, and preparation for action.
For us, here, this means that what is endogenous (self-activated memories
and predispositions, for example), and hence the manifestation of the
imaginary dimensions, is always a part of perception. Conversely the generation of the imaginary is not a different, or separate, stream but constitutive
of the normal flow of life. It follows that one cannot hope to find a naturalized account of imagination as some sort of cognitive module or brain region. It must necessarily correspond, instead, to a dynamic, emerging global pattern that is able both to integrate the body/brain activity at a large
scale and subside rapidly, for the benefit of the next moment of mental life.

Synchrony refers to the growing evidence that the actual process by
which the reciprocity is carried out is by a back-and-forth fine-tuning of

neural activity throughout the brain (cf. Varela 1995; *Neuron* 1999 for review). It provides the basis for the unified experience during any mental act, instead of being simply a juxtaposition of distinct modules that do not cohere with each other. The basic hypothesis that we follow here is that for every cognitive act there is a singular, specific cell assembly that underlies its emergence and operation. The emergence of a cognitive act, as we have said, requires the coordination of many different regions allowing for different capacities: perception, memory, motivation, and so on. They must be bound together in specific groupings appropriate to the specifics of the current situation the animal is engaged in (and are thus necessarily transient) in order to constitute meaningful contents in meaningful contexts for perception and action.

How are such assemblies transiently self-selected for each specific task? The basic intuition that comes from this problem is that specific cell assemblies emerge through a kind of temporal *resonance* "glue." More specifically, the neural coherency-generating process can be understood as follows: a specific cell assembly (CA) is selected through the fast, transient phase-locking of activated neurons belonging to subthreshold, competing CAs. The key idea here is that ensembles arise because neural activity forms transient aggregates of phase-locked signals coming from multiple regions. Synchrony (via phase locking) must *per force* occur at a rate sufficiently high so that there is enough time for the ensemble to "hold" together within the constraints of transmission times and cognitive frames of a fraction of a second. (For a recent example see Rodriguez et al. 1999.)

Upward causation   Accordingly, when brain-imaging techniques reveal a brain with multiple sites that are lighted during mental imagery tasks, the broader implication of this can now be drawn out. First, to see that imagination is indeed not an added human detail but at the very core of cognitive/mental life altogether. Second, that this imagination works because the autonomous working of the organism operates on the basis of a large-scale integration of multiple concurrent processes. Third, the nature of this nonlinear emergent process (plausibly through nonlinear synchronization) is a dynamic and transient process that occurs in pulses of lived temporality.

Accordingly, mental imagery (like other basic functions of mental life) appears, from the point of view of cognitive neuroscience, as a global dynamic pattern that integrates multiple concurrent activities. This nonlinearity and multiplicity is, we surmise, the very source of the creative and spontaneous nature of imagination. We shall refer to it as the process of *emergent* or *upward causation.* As we shall discuss later on, of equal importance is the converse, or *downward* causation, to which we will return in sections 2 and 3.

2. IMAGINING: THE PHENOMENOLOGICAL EXAMINATION

2.1 IMAGINATION IN THE PHENOMENOLOGICAL TRADITION

At this point in the journey that this essay proposes it is time to pause and return to square one. We have been examining imagination in its brain/bodily basis as a natural phenomenon and finding a number of important observations concerning the commonalties and differences with perception and the embodiment of its emergence. However, the fact remains that imagination is, most strongly and directly, a *lived experience.* People through all times have experienced, used, delighted, and feared what the mind's eye displays, in vivid colors and with the clarity akin of the "real," perceived image. As already said, this concern goes back to the very roots of the Western tradition with Plato and Aristotle, continuing uninterruptedly until the essential contributions of the *phenomenological* approach to mind since Husserl and James, but also with Sartre and Merleau-Ponty.

Phenomenological investigation has brought to the fore some of the basic components of imagination.[2] Two main points that we need to retain here are the following. First, contrary to common sense and the empiricist tradition (namely, Hume), imagination belongs to the very core of human consciousness, in close relation with memory and remembrance, fantasy, dreams, and perception itself. Second, imagination is grounded on a prereflexive (or prenoetic, unconscious) level of consciousness from which it shines forth. Both these points will be important for us in this context, and we need to look at them in some detail.

The intertwining of perception and imagination The ancient quandary that the omnipresence of imagination presented to the new discipline of phenomenology at the end of the nineteenth century is quite direct: is the consciousness of an image that is presented to the eyes comparable, in its essential aspects, to a visualized image or a memory that is recalled in an image? Husserl examined this issue in great detail in various forms. Already in 1904–1905 Husserl came to the conclusion that these represented really two different *kinds* of consciousness. And he drew this conclusion for two interlinking reasons: 1. The categories that are needed to account for the constitution of perception fail when applied to imagination and 2. the discovery that imagination is founded on the temporal character of inner consciousness.

Let us begin with the way we apprehend these acts within our natural attitude (what we can also call common sense) by providing an example coming from a first-person perspective (namely, Depraz's):

> When I am perceiving a pear tree in the garden and its gradually blossoming during early spring, the tree is here in front of me. I can touch it if I stretch out my hand, I can sense its perfume and listen to the noise of the wind in its branches. I am attending to the whole situation in flesh and bone, directly and concretely. If, on the contrary, I close my eyes and try to get a mental image of the tree and its surroundings, I might be able to accurately describe the just-lived scene if I have been quite attentive to its developing. But most probably I will forget some features of the experience and will add some others.

In short, we feel quite spontaneously how different it is to attend to a scene in it immanent immediacy and to recall it by way of mental images, let alone to fancy a similar scene years afterward. Moreover, the difference between imagining a scene based on an initial primary perception and freely fancying a scene that is composed of different features of different fragments of perceptive experiences, but one that has not been lived as such, is crucial.

Varieties of mental images   In his early writings Husserl takes as a motivational lead clue the natural experience we have of such a heterogeneity. He underlines the lived phenomenal contrasts between these different kinds of conscious acts and goes even further: he makes a strong distinction between two main acts of imagination, which he calls on the one side *Bildbewusstsein* (image consciousness) and on the other *Phantasie* (imagination). We will distinguish both in terms of the linguistic distinction between *imaging* (the production of mental images) and *imagining* (the fancying of a radically new world). Thus Husserl offers a more systematized and more gradual differentiation between perception and imagination than that of naive common sense, but he nonetheless follows the trail (if we may say so) of the natural attitude.

Now for the early Husserl (and the early Merleau-Ponty) perception is *the* basic intentional act through which we are able to gain primary access to the world:

> When I am perceiving the pear tree in the garden, I am able to detail its main features. While so observing it, it appears not only as having a meaning but also as having a real and factual existence for me.

Husserl calls perception the primary *"positional* act" (*Setzungsakt*), because it furnishes its intended object with a mode of givenness as being effectively here in front of me.

As far as imaging is concerned, a few paragraphs in *Ideas I* (111–112) present us with a clear account of the distinction between perception and imagination with regard to their reference to factual reality. Whereas perception is a positional act, imagination is defined as an act of consciousness that *neutralizes* every factual existence of the imagined object. Husserl therefore calls imagination a *non*positional act. The imagined pear tree does not have any real existence for me: I am just acting *as if* it had such an existence. In this regard Husserl comes to the same conclusions as recent studies in cognitive neuroscience that took as their starting point the topographical organization of primary visual areas that make it possible to have a (nonpositional) image as an endogenous activation of topographical visual areas that is nevertheless presented topographically but without a compelling facticity. (There is here an interesting convergence between the empirical and the phenomenal analysis that would need to be pursued further).

Imagining and remembering   Now, such a sharp discrepancy between acts of perception and imagination becomes more complex when the act of remembering comes to the fore. Like perception, remembrance is understood by Husserl as being a positional act: the remembered object is endowed with factual reality because, in order to remember it, you *must* have perceived it before, whether just now or a few years ago. Your remembrance will be quite fresh or more diffuse, but it is grounded on a primary perception; unlike perception (and like imagination), remembrance therefore is a founded act. But, unlike imagination, remembrance is a singly founded act (upon perception), whereas the former can be either directly based on perception, based on a remembrance (itself founded on perception), or as a novel emergence, as in the case of free fantasy and daydreaming. In the first two cases we have to do with an act of imaging: we produce a mental image of a perceived object or of the remembered one. In the third case we'll speak of imagining as a relatively unfounded act, because it does not simply follow the trace of a perceived or remembered situation but produces a new synthetic imagining experience based on multifarious perceptual and remembered features.[3]

Next in line in the phenomenological tradition stands the pioneering study of J.-P. Sartre during the war. Sartre strengthens the opposition between these acts, and Husserl's distinctions become sheer dualism. In *L'imaginaire* (1948; originally written in 1936) Sartre criticizes the tendency, common to the natural attitude and to psychologists and philosophers, to confuse the image of an object for the object itself, in the sense of identifying the mental image with an object within consciousness. Now, contrary to such a static apprehension of imagination as a state endowed with internal contents (images), and along with Husserl's analysis of intentionality, Sartre apprehends what he calls the imaging consciousness (*conscience imageante*) as a kind of dynamical and open intentional consciousness, the intended object of which is an image and not a perceived object.

Based on such a justified criticism, Sartre's whole project then lies in strengthening the difference between perception and imagination. Insofar as the confusion between the transcendent object of a perceiving act and the image understood as an immanent thing inside consciousness is his main criticism, he will do his best to avoid any overlapping between perception and imagination. While taking into account Husserl's distinction between position and neutralization, he therefore enlarges the gap between both acts. He even goes further than Husserl does, since he describes the image as enveloping a kind of nothingness (*néant*) and, at the same time, he endows the act of imagining with a radical freedom. Contrasting with perception, which is dependent on the real existence of the object, imagination is totally free. Passivity therefore determines perception, whereas imagination provides consciousness with a complete spontaneous activity.

What can we conclude from Sartre's analysis? We can say that his critical (we could say antinaturalizing) angle is justified: producing mental images does not amount to statically producing immanent things inside consciousness. Imagination is a dynamic intentional act (Husserl) and not an abstract faculty, the objects of which are mental images that we spontaneously describe as "within the mind." Of course, in light of the previous discussion, it is necessary to actualize both Husserl's and Sartre's view of mental images as "inside consciousness," through its naturalization (Petitot et al. 1999), as true but obsolete in its expression. Instead of being "inside," images emerge from a complex underlying network of multiple cognitive dimensions.

Sartre's contention is that perceptive consciousness and imaging consciousness are thoroughly different, so that the gap between them cannot be bridged, for the discrepancy disqualifies perception as passive, in contrast to imagination, which is a totally free and spontaneous consciousness. This gap seems to have been overemphasized in these early phases of phenomenological research up until the 1940s. Even if the Husserl of *Ideas I* makes a distinction between both acts, he acknowledges the dimension of activity that makes of perception a primary act and does not devaluate it as a purely passive experience. In short, Sartre's (nonnaturalized) dualism leaves us still more frustrated with regard to the possibility of understanding what is at stake in the empirical result of an identity (or at least a great commonality) of both neural processes (see also Casey 1976 on this matter).

Nature of the intertwining between perception and imagination   Although Sartre produces an accurate analysis of dreaming and hallucinating consciousness as particular cases of imaging consciousness, such analysis does not lead him to question the strong duality he claims between perception and imagination. Now, we all have had such experiences at least of visual illusions (if not limit experiences of drug-induced hallucinations):

> You are waiting for a friend in a cafe and you are transitorily deluded by the appearance of a person who looks very much like him. You are about to greet him and suddenly discover that he is not the person you are waiting for. (For a more detailed phenomenological account, see Depraz 2001a)

Such limit experiences have been explored as well in great detail in the now classic studies initiated by Perky (1910). This is even more striking in hallucinations people have under varying conditions. In both cases the point is the same: these visual illusions or hallucinations are full perceptions, in flesh and bone, and we experience such delusions as being actually perceived. Still, they lead to imagined objects that are nonexistent, with regard, at least, to the compelling requisite of positionality.

This already allows us to see that one way out of the Sartrean dilemma is to introduce a more detailed examination of imagining, instead of the Sartrean strategy of an a priori rationalistic account. In this sense James's pragmatic contention in his *Principles of Psychology* about visual imagination, mental imagery, and visualizations (vol. 2, chapter 18) opens the possibility for a very close intertwining, and even merging, of perception and imagination. In many accounts he gives of people able to access mental images (very few scientists, according to him!) there seem to be a great continuity between perceiving and imaging. The use of the expressions "visual imagination" and "visualizations" is telling about the potential merging of both acts in James's analysis, even if he refuses to take the imagery too literally. All in all, James's argument is founded on *empirical* psychological accounts, and thus he represents an ideal bridge between the phenomenological approach and modern brain-imagining accounts.

But even the Husserlian advocate of positionality and nonpositionality would be able to reply in two different manners to the contention that, in spite of appearances, imagination and perception do not merge. Such visual paradoxes remain within the realm of perception. However, the appearance of a deluded object is not entirely false merely because it enters in conflict with our habitual perception of objects. As appearance it has its own right to exist, to be real and true. Here Husserl questions the theory of truth underlining classical theories of imagination (Plato, Descartes) as false illusions and claims the truth of images that appear *as* images, endowed with their own intentional mode of givenness that is not the one proper to perception. In that respect the Buddhist Madhyamaka view also concurs on this point.[4]

Thus, the existence of visual illusions requires us to *expand* our concept of perception. In that respect Husserl suggests in many places (*Hua* XVI and *Hua* XXIII) a distinction between *Wahrnehmung* (a narrow positional perception) and *Perzeption* (a perception that includes its own modalizations: doubt, probability, even negation and mental illusions). Such an expanded concept of perception allows us to understand how perception may be permeated by imagination, destroying or at least diminishing the basic difference established in *Ideas I*.

Now, the permeation between the acts of perception and imagination belongs to the late Husserlian investigation (late 1920s, but not published until recently). If one insists upon the enlarged scope of perceptions provided by its inner variants and not on the narrow act of perception reduced to its positionality of the real existence of an object, then perception refers to a far more multiple reality than positionality. It includes our doubts, our confusions, our illusions, and our hallucinations. Perception is not a sheer normative positionality of the object but covers quite different experiences, from very common ones to more liminal ones. In short, perception is a multiform act, not reducible to positionality, which also implies that imagination is in no way reducible to neutrality (cf. mainly Husserl 1939: 20b; see also Depraz 1996a).

It is interesting to notice that among the dimensions that have reemerged into view is the phenomenological (Morley 2000) and psychological (Singer 1964, 1966) study of daydreaming. Daydreaming plays the role of an intermediate condition between dreaming as such and everyday perception, which again indicates the loose boundary between imagination and perception. The unique characteristic of daydreaming is that it manifests as *imagined emotional meaning*. Most of human life in the flow of consciousness is, in fact, such ongoing daydreaming, a point that did not escape Freud's notice (Bernet 1996), and it is also fundamental to the practice of mindfulness and meditative quiescence (*śamatha*) in the Buddhist tradition, as will be discussed. Morley (2000) has recently shown that daydreaming is amenable to a first-person analysis by self-report and interviewing, revealing a complex network of relations between self-world relationships, while contributing a useful example of the application of phenomenological method to the analysis of human consciousness.

Such an extension of the perceptive act on the basis of phenomenological investigation and its consequent mixing with imaging implies that the two acts are not fundamentally different. Now the distinction between positionality and neutralization belongs to the so-called early, or static, phenomenology, which emphasized a stratification of different acts of consciousness. Instead of maintaining a sharp and static opposition between perception and imagination, Husserl's later view offers an account of the dynamic constitution to their relationships. The question then is not, which are the features that distinguish perception from imagination, but rather, how does perception *become* an imaginary act, and, conversely, how does imagination *become* a perceptual one? The emphasis here is on the mutual transformation of one act into the other and vice versa (cf. *Hua* XXIII; see also Depraz 1996b, 1998). In this respect Merleau-Ponty very clearly pointed out the merging between perception and imagination, at best in *The Visible and the Invisible*. All this opens a rich common ground with Buddhism where dialogue will surely prove to be productive.

Through illusions and hallucinations one can analyze how perception can be enlarged to become a kind of imagination, because it goes beyond the common limits of what we usually call perception. Dreams provide us with the reversed process: the dreaming consciousness is an imaginary consciousness that produces images that look very much like perceptions, and are sometimes even more intense. Thus there is a merging of perception and imagination: imagination becomes here a more intense perception. (On this matter, see Depraz 2001b, part 3).

2.2 IMAGINATION AND THE LIVING PRESENT

The intertwining between imagination and perception can now be explored in greater detail, in terms of the dynamic relationship at their *emergence* in any moment of experience. Husserl realized that imagining as the presence of the nonpresent is, in essence, a property of how the living, specious present is constituted. In every moment of now there is surely the just present, which is full of the perceptual content. But one of the subtleties that a careful phenomenology of the present reveals is that together with that perceptual or (as we shall say) *impressional* consciousness of inner time there is also another time consciousness that is proper to imagination, remembrance, and fantasy, which we will refer to as *reproductive* consciousness (in Husserlian terminology this is called *presentificational*[5] consciousness, but the term is awkward for the nonspecialist). In other words, the very core of our temporality is an inseparable mixture of these two modes of apprehension.

The mixture of these two concurrent forms of consciousness means that they are constantly (at every present moment) emerging from a background that is prereflexive or prenoetic, that is, unconscious. From this floating background a constant self-constitution shapes a living present where the impressional and the reproductive coexist. This background's capacity for such recurrent manifestations is reflected in its affective or emotive quality, rather than being a neural or mechanical process. This can be cast also as the performative nature of the memories acquired over a life of habits or intense learning (Squire and Zola-Morgan 1996; Squire and Kandel 1999) and is, as we saw, also intrinsic to the generation of imagery from a neuroscientific point of view.

However, such a dynamic view of the emergence of a lived present should not make us forget the essential ways in which imagination and memory (as reproductions) are also different from perception (as presentations). In memory an object appears in the present but as belonging to the past. It is thus an aspect of inner consciousness that mixes the past and present without collapsing their temporal distance. Thus it is as if consciousness doubles itself, which is why remembrance, or recollection, is very close to reflection altogether. Imagination and visualization are manifestations whose relation as a reproduction of a previous perception is neutralized or suspended, as if presentification never happened. In the same sense that imagination cannot be reduced to perception, perception cannot be derived from pure imagination. But it is fair to say that any perception is *co*determined by the possibility of its imaginary modification.

Thus, while memory and imagination are close cousins, they can be distinguished in inner consciousness. And what is interesting is that both equally express, in an active fashion, the prenoetic background from which they came. In other words, reproductive consciousness is the privileged place for the manifestation of unconscious, sedimented *habitus* and desires. The implications of this observation are very important (see Bernet 1996).

Once again, the conclusions of phenomenological analysis converge with those of cognitive neuroscientific analysis, for both avoid the extreme of ascertaining identity and difference and each discovers in its way their common ground with other mental capacities. This common conclusion is the result of a long history of philosophical analysis from the opposition between Aristotle (who argued for the continuity between perception and imagination) against Plato (who rather emphasized their differences), down to Husserl and Sartre, with the notable exception of Kant. The above discussion points to a converging historical resolution between the work started by Husserl in the 1920s and modern cognitive neuroscience. Casey (1976, especially p. 130) sums up very well the history of this tension.

2.3 RETURN TO NATURALIZATION: DOWNWARD CAUSATION

Global to local   We are now in the position to go back and consider again what we said in section 1.3 on "upward causation." We examined there how the brain/body could arise through large-scale synchronization to a flow of consciousness in a succession of temporal segments, a string of now moments. We discussed how an integrated moment of the present appears as a *transient coherency-generating process* of the organism. But the global nature of this emergence can also be phrased in its reciprocal sense: the large-scale integrative state that underlies a moment of nowness is capable of *accessing any local neural processes.* Stated bluntly, this means that a mental state has agency and causal power over the very substrate that it needs to arise from. In other words, a unitary emergence is, by constitution, a double, or two-way, passage between two levels. This is *key* to the nonreductive type of naturalization we are examining here. This global-to-local action is constitutive because it shows up as order parameters in the dynamics and is mediated by means of the reciprocal and extensive interconnectivities in the brain and the organism itself. No extra ontological ingredients are required for this reciprocal, effective causation (see Varela 2000 for more on this point).

In this sense it is clear that the neural events accompanying any cognitive act are shaped and modified in the context of the rest of the neural events related to, say, limbic and memory activation, bodily posture, and planning. This is what we mean by "neuronal interpretation": the generation of a mental-cognitive state corresponding to the constitution of an assembly, which incorporates or discards into its coherent components other concurrent neural activity generated exogenously or endogenously. In other words, the synchronous glue provides the reference point from which the inevitable multiplicity of concurrent potential assemblies is evaluated, until a single one is transiently stabilized and expressed behaviorally. This is a form of neural hermeneutics, since the neural activity is "seen" or "valuated" from the point of view of the global emergence that is dominant at the time. Dynamically this entire process takes the form of a bifurcation from a noisy background to form a transiently stable, distributed structure bound by synchrony.

The neural events that participate in this process of synthetic interpretation via synchronization are derived indistinctly from sensory coupling and from the intrinsic activity of the nervous system itself. Whatever the mental state thus produced, it will ipso facto have neural consequences at the level of behavior and perception. For instance, if a visual recognition is interpreted in the context of an evasive emotional set and in conjunction with a painful memory association, it can lead to a purposeful plan for avoidance behavior, complete with motor trajectories and attention shifts to certain sensory fields. This illustrates once again the key dimension of the view of mental states we are offering here: there is a level-crossing reciprocity in that a mental state as such (i.e., as a global interdependent pattern) can effectively *act* on neural events (that is, it can have downward causation, as the phrase goes). For this to be more than a simple dualistic rehash, it is essential that the dominant interpretation be itself an emergent neural event, hence the odd-looking part of the theory that requires neural events to be the basis of interpretation of another class of nonsynchronous, less coherent neural events appearing at another level.

Downward causation   By their very nature, mental states make reference both to our own experience (and thus require a phenomenological account) and to our biological makeup (and thus require a fully scientific account). Now we are in a position to ask the central question that animates our inquiry here: How are these two accounts related to each other? What is the specific nature of their circulation?

Cognition is not only enactively embodied but is *enactively emergent*, in that technical sense that we just tried to sketch. Some people might call that by various names: self-organization, complexity, or nonlinear dynamics. The core principle is the same: the passage from the local to the global. It is a codetermination of neural elements and a global cognitive subject. The global cognitive subject belongs to that emergent level, and it has that mode of existence.

Now this principle of emergence is normally interpreted with a rather reductionist twist underlining only its upward causation (section 2.2). What we mean is that many will accept that the self is an emergent property arising from a neural/bodily base. However, as we have been arguing, the *reverse* statement is typically missed. If the neural components and circuits act as local agents that can emergently give rise to a self, then it follows that this global level, the self, has direct *efficacious actions* over the local components. It is a two-way street: the local components give rise to this emergent mind, but, vice versa, the emergent mind constrains and affects directly these local components.

To avoid thinking this is merely descriptive, let me provide an example. We have been working with epileptic patients who have electrodes implanted in their brains for future surgery. Thus, we have access to very detailed electrical signals of the brain of a waking human. This makes it possible to also analyze the moments that precede the crisis and, in fact, to predict its occurrence some minutes before it takes place (Martinerie et al. 1998). This is of course a good example of local properties (the local currents) leading to a global state (the crisis) in a lawful manner. But we were also able to find evidence for the converse: if a patient engages in purposeful, cognitive activity (such as recognizing a visual form), we could see changes in the detailed attributes of the local epileptic dynamics. This means the global state has downward effects over local electrical activity in a very precise fashion (Le van Quyen et al. 1997).

In brief, cognition is enactively emergent and is the codetermination of local elements and the global, emergent cognitive subject. Mind is pervaded by imagination and is not just about representing an "external world." The mind is about constantly generating a coherent reality that constitutes a world through the dynamics of local-global transitions. Perception is as imaginary as imagination is perception based, by now a familiar theme that we recover from its dynamic grounds as nonlinear causality.

3. THE TIBETAN TRADITION OF MENTAL IMAGING AS TRAINING AND THE PHENOMENOLOGICAL IMAGINATIVE SELF-TRANSPOSAL

We are finally ready to address our last point where the contributions of the Buddhist tradition are highlighted. The intertwining of the neurobiological accounts of the living present and imagination and the phenomenological discoveries set the background from the Western tradition for the rich terrain of imagining. A missing element in both science and phenomenology, however, is a thorough exploration of the *pragmatic* consequences of such observations, that is, on how such human capacities are also a means for human change and *transformation*.

It can be said that this dimension of imagining does not really need the Buddhist tradition, since there is significant literature on imagery and learning, for example, in sports training. Since memory is integral to imagery from neuroscience's point of view, it follows that learning can be brought to bear on stabilizing imagined contents, and thus to produce a desired learning by repetition and coaching. It is also well known that many structures (including the limbic system/mesial temporal lobe) participate in so-called procedural memory. These are memories manifested in performance and not conscious recall, as is the case of declarative memory (Squire and Zola-Morgan 1996; Squire and Kandel 1999). Procedural memory is at the center of acquired habits and sedimented ways of life, particularly in the emotional domain. This context makes it even more plausible to follow the Buddhist tradition in its discoveries for learning through imagery that do not attempt a specific intentional result but rather a shift in human traits in the entire range of social and individual life. It is precisely this hands-on, broad-based approach where one can learn from the Buddhist tradition because it cannot be conceived apart from an effort directed toward human transformation to unfold its full potential. For centuries we know they have excelled in these pragmatic efforts and sustained a treasure chest of methods and experience pertaining to human change. This is the topic of this last section.

The role of mental imagining in the practices of the Buddhist tradition, and especially in the Tibetan tradition of mental development upon which we focus here, is all-pervasive. In fact, the entire tradition can hardly be understood at all unless one carefully analyzes the multiple effects and sources of imagination as symbolization and of visualization as active imagining. Recalling the discussion above, this tradition has thus exploited in great detail the downward changes made possibly by the very constitution of the living body as a unity of global and local influences, as something that is both conscious and organic. Although at first the elaborate visualizations and techniques might appear as an idiosyncratic or folkloric content, in view of what the scientific and phenomenological analysis reveals this is a very superficial understanding. Before we return to this point let us briefly provide a sketch of practices and methods where explicit imagery figures explicitly.

A variety of visualization methods   The varieties of visualization can be basically described in the traditional three-fold distinction between Theravāda, Mahāyāna, and Vajrayāna approaches of Tibetan Buddhism. This roughly corresponds to the basic foundation practices dealing with

1. cultivating the basic skill of mindfulness and nondistraction as an antidote to the core ignorance of human life (Theravāda),
2. the extension of a renewed awareness extended to the concerns for others in our intersubjective constitution (Mahāyāna),
3a. the so-called preliminary practices (*ngöndro*), intended to reconfigure one's psychophysical constitution, preparing or purifying one's psychophysical ground for Vajrayāna, and
3b. the "advanced" methods of the Vajrayāna tradition dealing with a radical transformation of one's psychophysical reality.

Here we will concentrate on the Mahāyāna practice of exchanging self-for-others, but for the sake of context I will briefly touch on other imagery practices in the Hīnayāna and Vajrayāna schools.

Śamatha sitting meditation   The very basis of the training of mind is, first and foremost, grounded on cultivating the stability of attention. The exercise of *śamatha* (pronounced "sha-ma-ta"), figuratively rendered as mindfulness practice (Skt. *śamatha*; Tib. *zhi gnas*; Eng. "quiescence"),[6] is based on an examination of the nature of our mind and the origin of habitual patterns by paying meticulous attention to every moment of appearance. In other words, using the activity of mind to go beyond mind, looking at the givenness of experience with a fresh, inquiring glance.

A *śamatha* meditation session is a highly structured event. We focus here on what can be called the daily routine of cultivation for *śamatha*. This is done regularly, during more or less prolonged sessions and over a long period of time (at least a few years). The practice is carried out according to an explicit method or technique, which has variants over different schools in the Buddhist tradition, but for our purposes here I will deal as well as I can with the common core they share. More precisely, I follow here the kind of training I have received from various Kagyü-Nyingma schools. For a very detailed description in a classical setting see Tashi Namgyal (1984), and for a succinct modern presentation Trungpa (1995). Although unique in its own way, this background shares much common ground with most other Buddhist traditions.

The practice is, first of all, based on an attitude of *nondoing*, embodied in a dignified sitting (on the ground or a chair). The posture is centered on the straightness of the spine, the relaxed alignments of neck and arms, and the hands resting on the knees or over one another. The eyes are open or half-open, and the breathing is done through both nostrils and mouth. Once settled into the basic posture, one follows the injunction to "merely" follow what is going on, without engaging in it. Since breathing is ongoing, breath typically becomes a guideline, as an attentional track (in other variants, a mental image is used as support for attention). Although this does not mean that all other sensations, thoughts, and emotions stop, they are considered as if from a distance, from a position of an abstract observer, as clouds on the background of the foreground of the breathing followed into the lungs and out the nostrils.

The cultivation of this mindful presence is done with or without an explicit support of visualization but always with an attentive following of the breath. Some schools use active visualization for *śamatha*. Tsongkhapa, the founder of the Gelug school, wrote extensively on this method in *The Great Treatise on the Stages of the Path to Enlightenment*.[7] As he explains, the object to be used as support for visualization can be classified as 1. a mental image (most typically, a Buddha image), the attention being focused without further analysis, or 2. a mental image on the basis of which one cultivates insight beyond stable attention, and thus it is accompanied with conceptual analysis (Wallace 1998:chapter 2). Tsongkhapa goes on to discuss the way different mental images should be used depending on the individuals' abilities and obstacles.

This is the concise manifestation of the *capacity* being cultivated: mindfulness to what is happening in the present and the breath as the point of focus. As all kinds of experiences appear within this attentive space, we explicitly redirect our attention "inwardly," *without* engaging in the examination of their contents, their arising, emerging in full form, and then their subsiding into the background again.

As distracting thoughts, emotions or bodily feelings arise, against the background of sustained attention to breathing, we can become aware of how much we waver from this focusing center. We realize that we are *not* simply following our breath but have gone elsewhere in our experience, wandering along in a chain of thoughts, fantasies, and daydreaming. As soon as we note the sudden jolt of realizing we had not been following the instruction, we simply let go of the distraction and come back to the breathing, our engaged object of attention. This calls for the necessary faculty of introspective monitoring used to see if the mind has fallen into such distraction. The Indian Buddhist philosopher Asaṅga asserts: "What is mindfulness? The non-forgetfulness of the mind with respect to a familiar object, having the function of non-distraction" (1971:6).

This practice of *śamatha* entails an intelligent, active examination and monitoring of the awareness of the breathing, not only as object of attention but as an assessment of our mental state, whether it is actually engaged or not and in what quality. It is this reciprocal engagement of mindfulness and introspection that provides the efficacy of the learning. In fact, the quality of engagement changes constantly through a practice session. It can typically become very excited (full of ideation) or very lax and drowsy. There is rich and abundant literature concerning the skillful methods, obstacles, and antidotes for this training in order that the practitioner eventually finds a balance between excitation and laxity, to a relaxed equipoise (Trungpa 1980; Wallace 1998). Through this kind of sustained training, guided by the experience of others over time, along with the accompanying methods that provide the learning path, one achieves a degree of stability in *śamatha.*

The Vajrayāna tradition   The depth and richness of the ways in which mental imagery is put into action reach an extraordinary degree of refinement in Vajrayāna, the tantric tradition of Tibetan Buddhism. This long-standing tradition contains an accumulation of carefully selected visualizations that are claimed to touch on the most resistant core of people's obstacles to realization. In other words, Vajrayāna exploits to its full extent the dynamic self-organization of the mind and imagination that was laid out in the beginning of this essay to encompass the local changes not only in specific brain functions but also to the full extent of the phenomenologically integrated brain/lived body. The fruition is an opening to a direct experience of the open nature of being alive.

Tantric visualization embodies the marriage of a vivid yet nonexistent presence. In spite of its imagined quality, such visualization is said to be closer to our basic nature than so-called real perception. Thus, Vajrayāna deals with a symbolic reconstruction of one's self. It follows that the mental images chosen are not arbitrarily selected; the tradition has concentrated on their detailed efficacy for inducing a transformation in the individual. This know-how goes back to the tradition of Indian *mahāsiddhas*, and its detailed sources are hard to establish with historical accuracy. However, as pragmatic tools, they are available to examination by all those who are willing to engage with the discipline. Each visualization, directed to a specific mode of transformation, corresponds to a particular imagined gestalt, typically with a central figure, or *yidam*. The visualization takes place in the setting of an entire set of procedures, or *sādhana*. At the core of each *sādhana* there is an initial phase of establishing the visualization, or "development stage" (*utpattikrama*), always followed by a "dissolution stage" (*sampannakrama*) into the open background. The development stage is understood as an instrumental approach, in that everything is included in the *sādhana's* practice: attitudes, gesture enunciation, and actions. One becomes a totality that is embodied in the *yidam's* character (Kontrul 1999).

The reader familiar with the Vajrayāna tradition will understand that we are referring to an enormous domain only pointed at by the preceding sketch for the purposes of this presentation. Given its depth and diversity, I will not even attempt to address it further here; a detailed presentation of the relation between imagination and the Vajrayāna tradition is a major undertaking for the future. However, the overall background developed here for a renewed understanding of imagination and its role in human transformations can serve as a first step to more ambitious studies. Here, I will stay closer to a more basic practice of visualization: the tradition of mind training and the meditative technique called *tonglen*.

Mind training and tonglen   The visualization training I will examine here in more detail is traditionally referred to as mind training (*lojong*). The origins of this teaching date back to the coming to Tibet of a remarkable Indian teacher, Atīśa (982–1054) during the eleventh century, the period of Buddhist renaissance in Tibet. He was known at the time as Khedrup Nyiden, one who has accomplished both scholarship and realization by practice. He evidently was also a brilliant teacher who left behind a very active lineage, preserved by oral transmission. Atīśa's core text was made widely available in a compilation by his follower Geshe Chekawa Yeshe Dorje (1101–1175) as a brief root text entitled *The Seven Points of Mind Training*, a pithy summary for a path to develop the aspiration of awakened heart, or *bodhicitta*, by *cultivating* one's compassion and sensitivity to others with the releasing and letting go of self-centeredness as an automatic habit. The teaching took the form of "grandfatherly" advice: aphorisms to be applied at every moment of life and specific techniques for more formal periods for the cultivation of *bodhicitta* (Trungpa 1993; Wallace 2001). The teaching spread to many other schools in Tibet and became known as the Kadampa tradition. It is actively taught today, in particular in the Gelug order and the Kagyü order. This latter tradition has been inspired largely by the remarkable commentary written in the nineteenth century by Jamgon Kontrul the Great, know as *The Great Path of Awakening (Changchup Shunglam)* (Kontrul 1987). The lineage of transmission of these Kadampa teachings was received in the eleventh century by Gampopa, the founder of the Dakpo Kagyü lineage, and transmitted through a succession of practitioners of that order, to the contemporary teacher Chögyam Trungpa. It was from him that Varela received the oral instructions for this practice that animate what is discussed here (Trungpa 1980; Depraz received such instructions from Varela himself).

Tonglen as practice: "Imagine all the people"   Atīśa's root texts begins with very evocative lines:

All phenomena should be regarded as dreams.
Contemplate the nature of unborn insight.
Self-liberate the antidote.
Rest in the nature of basic cognition (*ālaya*)
In postmeditation one should consider all phenomena as illusions.

Giving and taking should be practiced alternately. That alternation should be put on the medium of the breath.

The last lines are a condensed reference to an explicit practice called *tonglen* (Tib. *gtong len*). *Tong* means sending out, letting go, and *len* means taking in. So sending and taking is the basis of this *bodhicitta* field training. Like *śamatha* and *yidam* visualization, *tonglen* is an actual practice one should conduct regularly and intensify its meaning by periods of intensive retreat. It is only after such familiarization over the years that its fruits can be recognized. A cursory exploration or a weekend trial would not do, as is true in any training in sports, music, and so on.

The practice should be done in formal sessions, lasting about thirty minutes. Here is a procedural description, in three steps.

THIS DESCRIPTION constitutes one formulation of what has been transmitted through the Kagyü lineage and should not be taken dogmatically. Such descriptions are inseparable from detailed oral instruction, and each person should pursue the practice according to one's own individuality. It is highly recommended not to engage in such practices unless the context for its refinement and progress is available.

Mobilizing imagination in tonglen   This traditional and celebrated practice has been cultivated, as we said, for centuries by a multitude of practitioners. These accumulated experiences provide telling evidence that such practices (done repeatedly) do lead to a progressive softening or weakening of the automatic position of the "me-first" characteristic of our cognitive ego, or self. The habit of self-interest is gradually replaced by an automatic inversion of one's position so the welfare of others spontaneously takes precedence. Needless to say, in the practice itself the visualizations have a quality of being discursive and fictitious, which they clearly are. But the key point is to regard the imagined situation as if it were real and effective; the exercise then seems to bring about an actual transformation in one's consti-

TONGLEN

STEP 1: STARTING GROUND
The ground for this practice is an attitude of letting go and a light touch to one's experience, whatever it may be, as a reminder of the emptiness of phenomena as the ground.

STEP 2
Two-stage visualization:

STEP 2A
Begin the practice proper by closing the eyes and, in a sort of free association, just allow any painful or emotionally charged recollections to come to mind. One can to some extent trigger, or evoke, a situation that is pressing, or focalize on some specific contents such as someone's illness and suffering or a recent personal painful event. This pain need not necessarily be physical, bodily pain. It could be moral or psychic pain such as depression, neurotic blockages, or external obstacles. The content is visualized in whatever form this comes and then stabilized and sharpened into an image. It is essential this image be very singular and precise; a "general sense" will not do. Typically this visualization is accompanied by an enhanced emotional tone that might vary in each case.

Once the situation is visualized, begin the process of *tonglen* itself, by breathing in the pain, darkness, sorrow, and heaviness of the chosen scene and breathing out from one's core openness, warmth, and release back into the person or situation. In other words "exchange" means to replace oneself in the position of the person who is suffering to provide space and relief to the other. Practice this exchange on the medium of the breath for some time so that the specific situation evoked is felt to the core.

STEP 2B
As the visualization in Step 2a become more or less established, proceed to extend the same exchange and felt presence beyond the singular situation to a larger field touching many other people (or sentient beings) who are in a similar predicament. Make this extension literally by visualizing the multitude of such beings, known or unknown, so that they populate the space before your mind's eye in front of you. Continue to provide release and comfort coming from your open core and absorb into that core the quality of pain and suffering before you.

When this extension has become too abstract and diffuse, interrupt the practice, make a fresh start, and cycle back to step 2a, perhaps with a different event or situation.

STEP 3: CONCLUSION
When it is time to finish the session of *tonglen*, one dissolves the visualization into its ground and rests one's mind in free-flowing mindfulness again.

tution (physical and psychological) into further openness. Stages 2a-b play explicitly on the interdependency between memory and imagination; the distinction between the two is not obliterated but kept in active contrast. Given the findings both in cognitive neuroscience and phenomenology summarized above, the effectiveness and skillfulness of *tonglen* become much more intelligible. The road of such Bodhisattva mind training is surely long, but what's important is that it can be taken at all, and that this can be mediated by explicit practices. Thus, *tonglen* thoroughly exemplifies the skillfulness of imagining, an emotional training and moral transformation based on know-how rather than on abstract moral injunctions.

It is essential to remark that *tonglen* is eminently a practice based on the existing intersubjective nature of one's experience. The exchange is possible only because humans are already immersed in a network of empathic relations. One's cognitive identity is inseparable from this foundation, as modern research is making more and more clear (cf. Thompson 1999), and phenomenology has explored this also in great detail (cf. Depraz 1995). Thus, we are dealing neither with a private self-absorption nor with visualization akin to elaborate *yidam* symbols, as in Vajrayāna. *Tonglen* seems to exploit explicitly the fact that each person's individual life is like a hologram of human social life, with its bonds and interpersonal circulation. Through this training, which initially goes against the river of our phylogenetic heritage of self-preservation, the opposite of a "private" thing, the true nature of experience comes to the fore. In this respect the *tonglen* practice meets very closely the Husserlian imaginative self-transposal at work in empathy (*Hua I* 1950a and *Hua* XV, no. 18) that Spiegelberg developed still more concretely in his *Doing Phenomenology* (see Depraz, Varela, and Vermersch 2001; Depraz and Varela 2001).

4. BREAKING NEW GROUND

Drawing consequences   It is now time to draw some conclusions from the admittedly complex road we have followed. Perhaps the most important conclusion we want to emphasize here is that as one brings together the empirical and the experiential as corresponding mutual constraints, old dualisms disappear. The dualism of mind and matter as forever apart merges into a new conceptual space where we see that, if one gives the local to global *and* the global to local their proper role, mind and experience reveal without any mysterious residue an effective or efficacious potential. Our minds are enmeshed in multilevel causalities in the material basis of our bodies, just as much as this organic basis is the substrate from which our mind can be said to emerge. A purely one-sided emergence view that deprives experience of its active dimensions is bound to negate its understanding as a mere epiphenomenon.

Imagination is a privileged, detailed example for such a new framework. It provides us with a unique case study where we can put into place (in an unfinished form, to be sure) all the ingredients present in this important entire phenomenon rather than a lopsided view. Let us summarize this itinerary, which is entirely cyclical:

- Brain-imaging studies of mental imagery reveal the following:
- Imagination is at the crossroads of many other mental capacities: language, memory, motor actions.
- The physiological basis of imagining can be traced to a network of many distributed circuits and sites, typically all those that are active during high-level vision in active life. The specific networks that are active are highly dependent on the imagining task being examined and the individual's style.
- The neurodynamical study of such large-scale phenomena reveals that the large-scale integration of this multiplicity of brain/body sites appears as a dynamic signature, or fleeting emergence, via synchronization for the duration of moment of experience (upward causality).
- The globally emergent configuration of the organism, however, can be reflected down as a local constraint on detailed physiological and even genetic processes (downward causality).

The phenomenology of imagination reveals that:

- Imagination is clearly different from perception, but they appear to be closely related.
- Imagination is part of a family of mental events that include memory, fantasy, daydreaming, and dreaming.
- Perception and imagination work as complementary, or codefined, modes of consciousness in any moment of the present lived moment. They modify and condition each other.
- The transition points between the purely empirical and the experiential, however, are crucial and must be analyzed at their appropriate levels of dynamic patterns that provide a passage between the experiential and the

observations from a third-person point of view. In other words, the gap between neurons and experience remains forever.

- Imagination arises out a background of prereflexive, or unconscious, sedimented habits, what paves the way for the intersubjective imaginative self-transposal.

The Buddhist tradition implicitly shares most of these conclusions, but it takes them into the realm of pragmatic implications for human transformation.

- *Tonglen* constitutes a precise case of a skill to achieve change in one's spontaneous reactions and attitudes toward the other and the world.
- The transformation is induced by carefully selected visualizations that are designed to induce changes in one's associations and emotional responses.
- This transformation can be seen as based on the pervasive interlinking between what appears in our experience in visualization (global) and the basis of the appearing (our body/brain), which is revealed by studies on brain imaging.
- The efficacy of such downward causation is evident in a long history of practitioners and their transformation (and also echoed in a number of recent studies on the brain's plasticity, as, for example, in sports training and child emotional development).

This cycle, we repeat, reveals the *entire* range and coherence of the phenomenon of imagining rather than merely one or another of its dimensions. The three dimensions of the empirical, the experiential description, and transformation practices form a coherent whole, not contradictory views. They illuminate, rather than exclude, one another. Here we have attempted to trace the phenomenon of imagining from the side of the material, never abandoning its material support, and yet explore it in such a way that it leads to the global level that manifests as first-person experience. On the contrary, one could start from the full-blown efficacy of Buddhist visualization practice and, by tracing the phenomenon without ever leaving its specificity, open up, as it were, into its most detailed empirical level. This is what can appropriately be called a *neurophenomenological* analysis, as Varela has described elsewhere (Varela 1999).

Thus imagination is a perfect example of what we wish to call (with Bruno Latour) a *mixed* object, like an alloy where the notion of "ridges" becomes irrelevant. There is only *one* phenomenon, and one can traverse it from one to another of its qualities, from experiential or organic without rest or jump. There is no gap to bridge, only traces to follow, as we have done in this essay. In other words, once the constitution of the natural object is adequately understood in the phenomenological realm, pure experiences can also be considered as belonging to a psychological consciousness and hence belong to an organism. In this precise sense data rooted in *lived firsthand experiences are intrinsically open to a nonreductive naturalization.* This is a central thesis that animates the neurophenomenological research project, which is possible only if the central issues of embodiment are of central concern for cognitive science (such as the enactive approach), for phenomenology, and with regard to first-person methods dealing with human transformation, where Buddhism excels. In fact, it is in the lived body, broadly conceived, that one finds the "the close relationship" between experience and its grounding, both as lived body (*Leib*) and biological body (*Körper*; see Depraz 1997). It is in this realm of events that we are given access to both the constitutive natural elements familiar to cognitive science as well as the required phenomenological data.

The notion of reciprocal constraints between the brain and experience can now be more precisely presented by exploring the nature of mixed objects as such. This means that in the study of mind, any phenomenon is understood from the beginning as a mixed object, as if the real is also in delicate balance between two avenues of discourse. On the one hand, we have the avenue that seeks to naturalize phenomena (i.e., imagination) and that leads directly to the account we can glean from science. On the other hand we have the avenue that seems to make experiential, or phenomenalize the empirical (i.e., the emergent patterns), by discovering in them one's entire experience (including our social history and language), which is always already present. This balancing act of traversing the route of naturalizing and the route of experientially phenomenologizing is both possible and productive. It requires the hard work of exploring with precision and discipline the potentials in specific domains.

Notes

1. For a Buddhist account of the distinction between experience-evoked images and language-evoked images see Gen Lamrimpa 1999:32–38. This passage also explains the distinction between the basis of conceptual designation and the designated object.—Ed.

2. *Imagination* takes its roots from Latin *imaginari*, to copy. A mimetic quality

is thus intrinsic to words that derive from this term (image, imagining). It is unfortunate that the series of words stemming from the Greek *phantasia* are not available in English, except in a narrow sense (fancy, fantasy), but present in German as *Phantasie* or in French as *fanstasme*. This last group is particulaly interesting for us since it is close to the root *phainos*, which gives rise to *phenomena*.

3. Such a basic distinction dates back to Kant's third *Critique* and is also mentioned by James 1890:44: "The imagination is called "reproductive" when the copies are literal; "productive" when elements from different originals are recombined so as to make new wholes"; and p. 45: "When the mental pictures are of data freely combined, and reproducing no past combination exactly, we have acts of imagination properly so called." As for Husserl, he details such a distinction between *Bildbewusstsein* and *Phantasie* in many texts from *Hua* XXIII. See, for example, nos. 19 and 20.

4. See the discussion in Wallace 2000 that is implicitly written from the Madhyamaka perspective.

5. This term has been introduced to translate the German *Vergegenwärtigung,* a key term for the analysis of present time consciousness (cf. *Hua* X). The connotation is to focus on the qualitative jump between the immediate present (*Gegenwärtigung*), presentations, and the past brought to bear in the maintenance of a mental content, whence the reproductive nature of this temporal mode.

6. This expressive rendering was first introduced by C. Trungpa (cf. *Meditation in Action* [Boulder: Shambhala, 1972]) and is widely used today. It is not, however, a strict translation that gives terms such as *quiescence, peace, attention,* and *recollection.* The issue is not purely technical, as every inflection chosen is also a choice on the style of practice. For a recent presentation of *śamatha,* see Wallace 1998.

7. This text has been translated into English and is being published in three volumes by Snow Lion Publications. The reader can also consult his more concise, but equally authoritative, presentation of this topic in Tsongkhapa, *The Medium Treatise on the Stage of the Path to Enlightenment*, translated and copiously annotated in Wallace 1998.

References

Asaṅga. 1971. *"Abhidharmasamuccaya," Le Compendium de la Super-doctrine d'Asanga.* Trans. W. Rahula. Paris: Publications de l'Ecole Francaise d'Extrême-Orient.

Bernet, R. 1996. "L'analyse husserlienne de l'imagination." *Alter: Revue de phénoménologie* no. 5, pp. 43–69.

Berthoz, A. 1998. *Le sens du mouvement.* Paris: Jacob.

Bisiac, E., C. Luzzatti, and D. Perani. 1979. "Unilateral Neglect, Representational Scheme, and Consciousness." *Brain* 102:609–618.

Casey, E. 1976. *Imagining: A Phenomenological Study.* Indiana: Indiana Univ.Press.

Decety, J., M. Jeannerod, M. Germain, and J. Pastene. 1989. "The Timing of Mentally Represented Actions." *Behavioral Brain Research* 34:35–42.

Decety, J. et al. 1994. "Mapping Motor Representation with PET." *Nature* 371:600–602.

Dennett, D. and M. Kinsbourne. 1991. "Time and the Observer: The Where and When of Time in the Brain." *Behavioral Brain Sciences* 15:183–247.

Depraz, N. 1995. *Husserl: Transcendance et Incarnation. Le statut de l'intersubjectivité comme altérité à soi chez Husserl.* Paris: Vrin.

———— 1996a. "Puissance individuante de l'imagination et métamorphose du logique dans *Expérience et jugement.*" *Phenomenologische Forschungen*, Sonderdruck, pp. 163–180.

———— 1996b. "Comment l'imagination 'réduit'-elle l'espace?" *Alter: Revue de phénoménologie* no. 4, pp. 179–211.

———— 1997. "La traduction de *Leib*, une crux phaenomenologica." *Etudes phénoménologiques.* Louvain.

———— 1998. "Imagination and Passivity: Husserl and Kant, a Cross-Relationship." In N. Depraz and D. Zahavi, eds., *Alterity and Facticity: New Perspectives on Husserl.* Phaenomenologica 148. Dordrecht: Kluwer.

———— 2001a. "A la source de l'hallucination: L'expérience du conflit." In G. Charbonneau, ed., *Phénoménologie des hallucinations.* Paris: L'art du comprendre.

———— 2001b. *Lucidité du corps. De l'empirisme transcendantal en phénoménologie.* Dordrecht: Kluwer.

Depraz N. and F. J. Varela. 2003. "Empathy and Compassion: Confronting Experiential Praxis and Buddhist Teachings." In D. Carr and Chang-Fai eds., *Space, Time, Culture.* Dordrecht: Kluwer.

Depraz N., F. J. Varela, and P. Vermersch. 2001. *On Becoming Aware: An Experiential Pragmatics.* Amsterdam: Benjamins.

Gazzaniga, M. 1999. *Cognitive Neuroscience.* Cambridge: MIT Press.

Husserl, E. 1939. *Erfahrung und Urteil.* Prague: Govert and Claasen.

———— 1950a. *Cartesianische Meditationen. Hua* I. The Hague: Nijhoff.

———— 1950b. *Ideen zur einer reinen Phänomenologie* I. *Hua* II. The Hague: Nijhoff.

———— 1966. *Phänomenologie des inneren Zeitbewusstseins. Hua* X. The Hague: Nijhoff.

———— 1973. *Zur Intersubjektivität (1929–1935). Hua* XV. The Hague: Nijhoff.

———— 1976. *Ding und Raum. Hua* XVI. The Hague: Nijhoff.

———— 1980. *Phantasie, Bildbewusstein, Erinnerungen. Hua* XXIII. Dordrecht: Kluwer.

James, W. 1890. *Principles of Psychology.* London: Dover.

Jeannerod, M. 1994. "The Representing Brain: Neural Correlates of Motor Intention and Imagery." *Behavioral Brain Science* 17:187–245.

Kontrul, J. 1987. *The Great Path of Awakening*. Boston: Shambhala.
———— 1999. *Creation and Completion*. Ithaca: Snow Lion.
Kosslyn, S. 1994. *Image and Brain: The Resolution of the Imagery Debate*. Cambridge: MIT Press.
Kosslyn, S. et al. 1999. "The Role of Area 17 in Visual Imagery: Converging Evidence from PET and RTMS." *Science* 284:167–170.
Lamrimpa, G. 1999. *Realizing Emptiness: Madhyamaka Insight Meditation*. Trans. B. Alan Wallace. Ithaca: Snow Lion.
Le Van Quyen, M., J. Martinerie, C. Adam, H. Schuster, and F. Varela. 1997. "Unstable Periodic Orbits in Human Epileptic Activity." *Physica* E56:3401–3411.
Martinerie, J. C. Adam, M. Le van Quyen, M. Baulac, B. Renault, and F. J. Varela. 1998. "Epileptic Crisis Can Be Anticipated by Non-linear Analysis." *Nature Medicine* 4:1173–1176.
Mellet, E., L. Petit, B. Mazoyer, M. Denis, and N. Tzorio. 1988. "Reopening the Mental Imagery Debate: Lessons from Functional Anatomy." *Neuroimage* 8:129–139.
Merleau-Ponty, M. 1966. *The Visible and the Invisible*. Paris: Gallimard.
Morley, J. M. 2000. "The Private Theater: A Phenomenological Investigation of Day Dreaming." *Journal of Phenomenological Psychology*.
Namgyal, T. 1984. *Mahāmudrā: The Quintessence of Mind and Meditation*. Boston: Shambhala.
Nyanaponika, T. 1973. *The Heart of Buddhist Meditation*. New York: Weiser.
Petitot J., F. J. Varela, B. Pachoud, and J.-M. Roy, eds. 1999. *Naturalizing Phenomenology: Contemporary Issues in Phenomenology and Cognitive Science*. Stanford: Stanford University Press.
Pöppel, E. and K. Schill. 1995. "Time Perception: Problems of Representation and Processing." In M. A. Arbib, ed., *Handbook of Brain Theory and Neural Networks*, pp. 987–990. Cambridge: MIT Press.
Port, E. and T. van Gelder. 1997. *Mind in Motion*. Cambridge: MIT Press.
Rodriguez, E. N. George, J. P. Lachaux, J. Martinerie, B. Renault, and F. Varela. 1999. "Perception's Shadow: Long-Distance Synchronization in the Human Brain." *Nature* 397:340–343.
Roland, P. E. and B. Gulvas. 1994. "Visual Imagery and Visual Representation." *Trends in Neuroscience* 17:281–286.
Saraiva J. 1970. *L'imagination chez Sartre*. The Hague: Nijhoff.
Sartre, J.-P. 1948 [1936]. *L'imaginaire*. Paris: Gallimard.
Sergent, J., S. Ohta, B. MacDonald. 1992. "Functional Neuranatomy of Face and Object Processing." *Brain* 115:15–36.
Silberswieg, D. A. et al. 1995. "A Functional Neuroanatomy of Hallucinations in Schizophrenia." *Nature* 378:176–179.
Singer, J. L. 1964. "Daydreaming and the Stream of Thought." *American Scientist* 4:417–425.
———— 1966. *Daydreaming: An Introduction to the Experimental Study of Inner Experience*. New York: Random House.

Spiegelberg H. 1972. *Doing Phenomenology*. The Hague: Nijhoff.

Squire, L. R. and S. Zola-Morgan. 1996. "The Medial Temporal Lobe Memory System." *Science* 253:1380–1386.

Squire, L. R. and E. R. Kandel. 1999. *Memory: From Molecules to Memory*. New York: Freeman.

Thompson, E. 1999. "Human Consciousness: From Intersubjectivity to Interbeing." A report to the Fetzer Institute.

Trungpa, C. 1980. *Hinayana-Mahayana Seminary 1979*. Boulder: Vajradhathu.

———— 1993. *Training the Mind and Cultivating Loving-Kindness*. Boston: Shambhala.

———— 1995. *The Path Is the Goal: A Handbook of Meditation*. Boston: Shambhala.

Varela, F. 1992 [1989]. *Invitation aux Sciences Cognitive*. Paris: Seuil.

———— 1995. "Resonant Cell Assemblies: A New Approach to Cognitive Functioning and Neuronal Synchrony." *Biological Research* 28:81–95. [*Neuron* 1999 for review.]

———— 1999. "The Specious Present: The Neurophenomenology of Time Consciousness." In J. Petitot, F. J. Varela, B. Pachoud, and J. M. Roy, eds., *Naturalizing Phenomenology*, pp. 266–314. Stanford: Stanford University Press.

———— 2000. "Upwards and Downwards Causation in the Brain: Case Studies on the Emergence and Efficacy of Consciousness." In K. Yasue and M. Jibu, eds., *Towards a Science of Consciousness*. Amsterdam: Benjamin.

Varela, F., E. Thompson, and E. Rosch. 1991. *The Embodied Mind: Cognitive Science and Human Experience*. Cambridge: MIT Press.

Wallace, B. A. 1998. *The Bridge of Quiescence: Experiencing Tibetan Buddhist Meditation*. Chicago: Open Court.

———— 2000. *The Taboo of Subjectivity: Toward a New Science of Consciousness*. New York: Oxford University Press.

———— 2001. *Buddhism with an Attitude: The Tibetan Seven-Point Mind-Training*. Ithaca: Snow Lion.

In this comparative study of the traditional Tibetan Buddhist discipline of dream yoga and the modern psychological discipline of lucid dreaming, Stephen LaBerge juxtaposes two profoundly different worldviews. Dream yoga is usually practiced within the context of Madhyamaka philosophy and Buddhist Vajrayāna, whereas the modern cognitive sciences are usually practiced within the context of scientific materialism, which is rooted in Cartesian dualisms. This worldview promotes an absolute dichotomy between mind and matter and between subjective and objective phenomena. But the lack of parity in development of the physical sciences and the cognitive sciences in the West has been largely responsible for the current, widespread marginalization of subjective experience in the natural world and reduction of all mental events to states of matter.

In fairness, the same cannot be said of LaBerge's psychophysiological approach, which considers subjective experience to be the primary source of information about consciousness and uses physiological data as convergent evidence validating the testimony of introspection. The basic assumption of psychophysiology, that mind interacts with physiology, undermines the absolute distinction of subjective and objective phenomena. As a lucid dreamer, LaBerge knows that the "realness" in experience is not in the "external world" but in the mind. Nor does he assume that if it is true some mental states are states of matter (i.e., brain states) the same must be true of all.

A fundamental difference between dream yoga and modern techniques of lucid dreaming is that the former encourages the trainee to view all waking events as a dream, whereas the latter encourages "state checks" during the daytime to see whether or not one is dreaming. Thus, Tibetan Buddhism challenges the status of waking reality, suggesting that it is not ontologically more real than the dream state. Although the lucid dreaming approach views dreaming and waking experience both as mental constructions, it assumes the essential difference between dreaming and waking perception is whether sensory input from a presumed part of reality is absent or present.

When a Tibetan dream yogin claims to have had an out-of-body experience while dreaming, witnessing events in our communal "real" world, is he experiencing a subjective illusion or is he actually moving about within the objective world? While this question makes perfect sense within the framework of Cartesian dualism, it is considered fundamentally flawed from the perspective of the Madhyamaka. For in this context, the so-called objective world, of which our experience in the normal waking state is allegedly "direct and true," is but an intersubjective illusion. A dream yogin, nevertheless, may be deluded. He may think he is having an authentic out-of-body experience, witnessing the intersubjective world of other people in their waking state. But in fact he may be experiencing a singularly subjective state, in which there is no subjective participation other than his own. Thus claims of out-of-body experiences can be studied scientifically not with reference to a purely objective world but rather with reference to the intersubjective world of scientists and the general public.

The issue of subjective versus objective realities is raised again in this essay with respect to cakras and "psychic energy channels." Are these physical or mental, objective realities or subjective realities? From a materialist perspective the answer to these Cartesian questions seems quite obvious: they are in all likelihood illusory, nonphysical, subjective constructs. But from a Buddhist perspective, once again, the flawed nature of the questions obscures the actual status of these hypothetical phenomena. While it is quite true that Buddhism has no theory of the brain and presents no third-person, objective account of the nervous system, its elaborate theories concerning the cakras may be viewed as first-person, theory-laden, phenomenological accounts of elements of the nervous system as viewed with refined states of mental perception. From this perspective these cakras do not fit into the reified categories of mind and matter or subjective and objective. Nevertheless, the nature and functions of the cakras may be tested pragmatically, as the yogin applies the relevant Buddhist theories in his contemplative practice. In this way it may turn out that the third-person, scientific account of the brain and nervous system is complementary to the first-person, contemplative account of the cakras and the psychic energies that course through them. This field of dream yoga and lucid dreaming is an especially fertile one for interdisciplinary, cross-cultural research.

Here we are, all of us: in a dream-caravan.
A caravan, but a dream—a dream, but a caravan.
And we know which are the dreams.
Therein lies the hope.
                                            —Bahaudin

May I awaken within this dream,
And grasp the fact that I am dreaming
So that all dreamlike beings may likewise awaken
From the nightmare of illusory suffering and confusion.
                                            —Surya Das

Stephen LaBerge

Lucid Dreaming and the Yoga of the Dream State: A Psychophysiological Perspective

Until very recently Western science regarded "lucid dreaming"—dreaming while knowing that one is dreaming—as no more than a curiosity: at best a metaphorical unicorn, rare to the point of being mythical, at worst, an oxymoron (i.e., "How can one be *conscious* while *asleep*?"). Indeed, before eye-movement signaling (LaBerge et al. 1981; see LaBerge 1990 for more references) provided objective proof of its existence, few sleep and dream researchers were willing to credit subjective reports of lucid dreaming. Probably the main reason was a widespread theoretical assumption that being asleep meant being unconscious; thus, claiming to be conscious of anything at all during sleep, including the fact that one is dreaming, seemed a contradiction in terms (LaBerge 1985, 1990). At least one philosopher went so far as to contend that to say one is dreaming is "unintelligible; nonsense— . . . an inherently absurd form of words" (Malcolm 1959:50).

In contrast, for more than a thousand years Tibetan Buddhists have believed that it is possible to maintain the functional equivalence of full waking consciousness during sleep. This belief is not based on anything as tenuous as theoretical grounds but upon firsthand experience with a sophisticated set of lucid dreaming techniques collectively known as the *Doctrine of Dreams* or *dream yoga* (Evans-Wentz 1958; Norbu 1992). I will present in what follows a commentary on the currently available literature on dream yoga. Thus the scope of this article is narrower than a review of lucid dreaming (for which see LaBerge 1985, 1990; LaBerge and Rheingold 1990).

On one hand, the twenty-five years or so of recent Western experimental and experiential research has yielded much more extensive knowledge about the lucid dreaming state, including its psychophysiological characteristics (LaBerge 1998), methods of induction (LaBerge and Rheingold 1990), and theoretical basis (LaBerge 1998). On the other hand, the more than twelve centuries of history accumulated in the Tibetan dream yoga tradition of intensive experiential observation has correspondingly greater depth of meaning. As a result, it will be seen in this commentary that while Western knowledge of the topic may surpass the Eastern teachings in a number of technical details the most important aspect of dream yoga, namely, its transcendental purpose—"enlightenment"—is difficult to even frame in the terms of Western science.

Given that *Buddha* means "the awakened one," it is not surprising that lucid dreaming arises very naturally in Buddhism, providing both a metaphor for enlightenment and a method for attaining the goal of awakening:

> The whole purpose of the Doctrine of Dreams is to stimulate the *yogin* to arise from the Sleep of Delusion, from the Nightmare of Existence, to break the shackles in which *maya* thus has held him prisoner throughout the eons, and so attain spiritual peace and joy of Freedom, even as did the Fully Awakened One, Gautama the Buddha. (Evans-Wentz 1958:167)

The Tibetan Buddhist point of view reverses the order of valuation of the waking and dreaming states: while most Westerners consider the waking state the only reality and dreams to be unreal and unimportant, in the East the dream state is considered to have greater potential for understanding and spiritual progress than the so-called waking state, and both states are considered to be equally real or unreal. From this point of view a relativistic metaphysics seems compelling:

> If the feeling of reality in dreams is perfectly credible and if this feeling disappears on waking, then there is no reason for not imagining life as a dream from which one could also wake up. Already, one can dream that one is dreaming and awake from one's dream within the dream: thus a metaphysics

of degrees of the real becomes necessary, an approach which it seems it will be hard to avoid in the future. (De Becker 1965:402–403)

The original sources of the methods and doctrines on dream yoga are lost in antiquity; according to some scholars, the teachings can be traced back twenty-five hundred years to the Buddha himself (Mullin 1997). It is like-wise claimed that a shamanistic pre-Buddhist lucid dream tradition existed in Tibet (Norbu 1992; Tarab Tulku 1991). Be that as it may, the oldest certain source of detailed teachings on Tibetan dream yoga is the work known as the *Six Doctrines* or *Yogas of Naropa* (Mullin 1997). The six yogas include 1. the doctrine of psychic heat, 2. the doctrine of the illusory body, 3. the doc-trine of the dream state, 4. the doctrine of the clear light, 5. the doctrine of the after-death state, and 6. the doctrine of consciousness transference (Evans-Wentz 1958).

Naropa's teacher, Tilopa (988–1069), is regarded as the compiler of the six yogas (Mullin 1997). Most of Tilopa's teachings were orally transmitted; in the only extant text of certain authenticity Tilopa specifically attributes the teachings on dream yoga to Lawapa of Oddiyana:

Know dreams as dreams, and constantly
Meditate on their profound significance.
Visualize the seed syllables of the five natures
With the drop, the *nada* and so forth.
One perceives buddhas and Buddha fields.
The time of sleep is the time for the method
That brings realization of great bliss.
This is the instruction of Lawapa. (Tilopa 1997:28)

Lawapa is said to have been born a prince in the eighth century in Oddiyana but renounced his kingdom to travel the path of spirituality as a wandering monk (Mullin 1997). According to tradition, he acquired his name, which means "the blanket master," by sleeping wrapped in his blanket in front of the local king's palace for twelve years without rising. It is said that all who touched his sleeping form were miraculously cured of any of their illnesses. In any case, we know that Lawapa passed on his teachings to Jalandhara, who was the guru of Krishnacharya, a teacher of Naropa.

The earliest written formulations of the dream yoga are extremely con-cise and even aphoristic; mostly they seem to be reminders of the much more extensive teachings transmitted orally from guru to disciple. The account from Tilopa (1997) just quoted is typical. However, over the many centuries more and more details of the dream yoga appear in print. The main source is "The Doctrine of the Dream-State," a part of the volume edited by Evans-Wentz (1958) deriving from a compilation by Padma Karpo in the seventeenth century.

"The Doctrine of the Dream-State" consists of four parts or stages (Evans-Wentz 1958:215):

1. "Comprehending the nature of the dream-state" (i.e., that it is a dream, and thus, a construction of the mind).
2. "Transmuting the dream-content" (practicing the transformation of dream content until one experientially understands that all the contents of dreaming consciousness can be changed by will and that dreams are essentially unstable.)
3. "Realizing the dream-state, or dream-content to be *Maya*" (i.e., that everything that appears in the dream is a mental construction.)
4. "Meditating on the thatness of the dream-state" (Realizing that the sensory experiences of waking consciousness are just as illusory as dreams and that, in a sense, "it's all a dream," resulting in union with the Clear Light.)

Normally the dream yoga is preceded by a number of preparatory exercises. These include the basic exercises practiced by those undergoing Buddhist training as well as a specific preparation designed to "purify the mind": essentially a series of meditations designed to set one's intention to a pure aspiration for enlightenment (Gyatrul 1993). The essential frame of mind with which to approach dream yoga is described as follows:

> Then when you go to bed in the evening, cultivate the spirit of enlightenment, thinking, "For the sake of all sentient beings throughout space, I shall practice the illusion-like *samādhi*, and I shall achieve perfect Buddhahood. For that purpose, I shall train in dreaming." (Padmasambhava 1998:151).

The inner heat yoga is regarded as a prerequisite by some (e.g., Gyatso 1997; Tsongkhapa 1997) but not all (e.g., Gyatrul 1993; Norbu 1992) commentators. This is an important issue that needs to be addressed by anyone seeking to adapt the dream yoga teachings to a Western context: what are the recommended and minimal necessary prerequisites for these practices? It is notable that the Dalai Lama recently stated that dream yoga can be practiced without a great deal of preparation. Dream yoga could be practiced by non-Buddhists as well as Buddhists. If a Buddhist practices dream yoga, he or she brings a special motivation and purpose to it. In the Buddhist context the practice is aimed at the realization of emptiness. But the same practice could be done by non-Buddhists. (Varela 1997:45).

COMPREHENDING THE NATURE OF THE DREAM STATE

The doctrine of the dream state describes three types of practice aimed at "comprehending the dream-state" (i.e., producing lucid dreams): the power of resolution, the power of breath, and the power of visualization (Evans-Wentz 1958).

THE POWER OF RESOLUTION

The power of resolution refers to resolving to maintain unbroken continuity of consciousness throughout both the waking state and the dream state:

> Under all conditions during the day hold to the concept that all things are of the substance of dreams and that thou must realize their true nature. Then, at night, when about to sleep, pray to the *guru* that thou mayest be enabled to comprehend the dream-state; and firmly resolve that thou wilt comprehend it. By meditating thus, one is certain to comprehend it. (Evans-Wentz 1958:216)

The daytime training, "sustaining mindfulness without distraction during the daytime experience" is the opposite of the Western lucid dreaming induction technique of reality testing in which one repeatedly questions whether or not one is dreaming, with the aim of creating a habit that will recur during the dream state (LaBerge and Rheingold 1990). Here is Gyatrul Rinpoche's description of the Tibetan technique:

> During the daytime, one must sustain mindfulness without distraction. This mindfulness is to constantly remind oneself that all daytime appearances are nothing other than a dream. Throughout the different experiences during the daytime reality, you just keep on mindfully sustaining the awareness, "This is a dream, this is a dream, I'm asleep and I'm dreaming," and this will create a habit. (1993:104)

An experiment comparing the effectiveness of these two opposed yet related techniques should certainly be done: is it better to focus on the differences between waking and dreaming, as in the Western approach, or on the similarities, as in the Tibetan approach?

THE POWER OF BREATH

In the second practice, the power of breath, the procedure is as follows:

> Sleep on the right side, as a lion doth. With the thumb and ring-finger of the right hand press the pulsation of the throat-arteries; stop the nostrils with the fingers [of the left hand]; and let the saliva collect in the throat. (Evans-Wentz 1958:216)

Falling asleep while engaged in this rather elaborate practice is likely to present a challenge for most people; accordingly, Norbu (1992) recommends that it be used only by those who fall asleep easily. However, a major part of the desired effect is to maintain a certain degree of consciousness as one falls asleep, making it necessarily more difficult (but not impossible) to fall asleep, so in practice the requisite balance may be delicate.

The Evans-Wentz account of the procedure departs from the typical practice: one is usually advised to close only the right nostril (Norbu 1992) instead of both, as described above. Most commentators agree that sleeping on the right side is optimal for lucid dreaming, although several claim this is only true for men (e.g., Norbu 1992; Surya Das 2000; Wangyal 1998). For example, Norbu asserts that "a woman should lie on her left side and try to block her left nostril. . . . The reason that the positions are reversed for men and women has to do with the solar and lunar channels" (1992:52–53). Wangyal (1998:44) provides more explicit detail, explaining that according to the Tibetan tradition there are three main channels through which "pranic energy" flows: the "blue central channel . . . the channel of non-duality" in which moves "the energy of primordial awareness" (Tib. *rigpa*), the "white channel (the right in men and the left in women) is the channel through which energies of negative emotions move," while the "red channel (the left in men and the right in women) is the conduit for positive or wisdom energies." "Therefore," Wangyal writes,

> in dream practice, men sleep on their right side and women on their left in order to put pressure on the white channel and thus close it slightly while open-

ing the red wisdom channel. This contributes to better experiences of dream, involving a more positive emotional experience and greater clarity. (44)

Considering the reasoning given, one suspects that the technique has not actually been empirically tested in women, although it would not be the first time that men and women have been shown to have typically opposite organization of neuropsychological systems.

This practice might seem somewhat bizarre to Western scientists, and the explanations even more so. However, it may be that most or all components of the procedure have specific effects on the nervous system. For example, pressing the throat arteries stimulates the baroceptors, lowering the heart rate and perhaps facilitating rapid onset of REM sleep (Puizillout and Foutz 1976) and hence "Wake-Initiated Lucid Dreams" (WILDs; LaBerge 1980). Mouth breathing may also favor WILD initiation by requiring a higher level of the central nervous system control and perhaps a greater degree of consciousness than nose breathing, given the fact that newborn human infants cannot breath through the mouth. Sleeping on the right side is likely to stimulate a neural reflex evoked by pressure between the fifth and sixth intercostal spaces that causes the dilation of the contralateral (left) blood vessels and possibly also changes in cerebral laterality (Werntz et al. 1983). If this is the mechanism by which sleeping on the right side facilitates lucid dreaming, one would expect that lucid dreams will be correlated with the relative dilation of the left nostril.

Indeed, in a pilot study of nasal dilation laterality, sleeping posture, and lucid dreaming, with seven men and eight women, all right handed, LaBerge and Levitan (1991) found that, for women, lucid dreams were three times as likely to be accompanied by left nasal dilation than right nasal dilation. For both men and women, lucid dreams were three times more frequent when sleeping on the right side than on the left. These results confirm the basic Tibetan claim that sleeping on the right side is more favorable for lucid dreaming than sleeping on the left side, for men and women alike. Although this study found several gender differences (for example, women reported relatively clearer thinking in dreams with the left nostril open, while men reported relatively greater emotional intensity with the right nostril open), it did not confirm the claim that women should sleep on the left side. We are currently engaged in a replication study with a larger number of right- and left-handed men and women.

I believe the relationship between the results of empirical study and Tibetan method and theory illustrate a pattern that will be found in other studies: namely, that the methods are more likely to be confirmed than the theoretical explanations. The reason for that seems clear. The traditional methods were probably derived mainly from actual experiences—in some cases, by serendipitous observations—of what procedures produced the desired effect. In other words, the methods were empirically based. Moreover, ineffective methods are relatively easily recognized: if a sufficient number of practitioners follow a procedure without the expected results, the suspicion must arise that there is something wrong with the technique. In contrast, the validity of the theoretical explanation of how the method is thought to work is not likely to come into question as the result of observations of the effectiveness of the method, especially not if the technique is effective. Tests of the theory itself take place in a context outside the accepted paradigm (Kuhn 1970).

The effectiveness of a psychophysiological technique can be tested by careful observers of the contents of consciousness without the need of any technology other than a well-trained mind and disciplined body. In contrast, testing the validity of an explanation of that technique may require the extremely sophisticated technology needed for the visualization and measurement of neural activity. As a result, the Tibetan Buddhist tradition almost completely lacks any notion of the role the brain plays in the activities of local mind, including perception, emotion, cognition, and dreaming (Houshmand, Livingston, and Wallace 1999). In consequence, Tibetan theorizing about, for example, the types of dreams and how they arise (e.g., Norbu 1992; Wangyal 1998) sometimes exhibit a distinctly prescientific flavor and naiveté in striking contrast to the usual high degree of sophistication of Buddhist thinking. The absence of a role for the brain or nervous system casts doubt on the status of the entire system of *cakras* and "psychic energy channels." Is the *cakra* system in the body or in the brain? Claims that the channels are nonphysical (e.g., according to Wangyal 1998, "The channels that carry this very subtle energy cannot be located in the physical dimension but we can become aware of them"; 44) are contradicted by claims (e.g., Wangyal's paragraph quoted above) that particular physical postures compress them. A possible alternative explanation is that they represent structures in the central and peripheral nervous system, which may explain why "we can become aware of them."

THE POWER OF VISUALIZATION

The third practice, the power of visualization, consists of the following:

Thinking that thou art the deity Vajra-Yogini [the feminine aspect of primordial wisdom], visualize in the throat psychic-center the syllable AH, red of
color and vividly radiant, as being the real embodiment of Divine Speech.

By mentally concentrating upon the radiance of the AH, and recognizing
every phenomenal thing to be in essence like forms reflected in a mirror,
which, though apparent, have no real existence of themselves, one compre-
hendeth the dream.

. . . At nightfall, [strive to] comprehend the nature of the dream-state by
means of the visualization just described. At dawn, practice "pot-shaped"
breathing seven times. Resolve [or try] eleven times to comprehend the nature of the dream-state. Then concentrate the mind upon a dot, like unto a
bony substance, white of color, situated between the eyebrows. (Evans-Wentz
1958:217–218)

The text recommends that the dot should be visualized as red if one is too
sleepy or as blue if one is too vigilant. If these means are insufficient to induce lucidity, one is advised to increase one's morning practice to twenty-
one pot-shaped breathings and twenty-one resolutions to comprehend the
nature of the dream state. "Then, by concentrating the mind on a black dot,
the size of an ordinary pill, as being situated at the base of the generative or-
gan, one will be enabled to comprehend the nature of the dream-state."
(Evans-Wentz 1958:218)

There are many variations on what is visualized and where it is visualized. Another text recommends,

Lie down to sleep with the resolve to apprehend your dreams. Let your behavior be unhurried and calm. As for the mind, in your heart imagine a white,
stainless AH sending forth varicolored rays of light which melt samsara and
nirvana into light and dissolve them into the AH. Fall asleep with the sense of
a clear vision, like the moon rising in a stainless sky. (Gyatrul 1993:105)

Another source recommends

Visualize at your throat a four-petaled lotus with Oṃ in its center, Aḥ in
front, Nu on the right, Ta in back, and Ra on the left. . . . First direct your

interest to the Oṃ in the center, then when you become sleepily dazed, focus your awareness on the Aḥ in front. As you are falling asleep, attend to the Nu on the right. When you are more soundly asleep, focus on the Ta in back. When you have fallen fast asleep, focus on the Ra on the left. . . . While sleeping, focus your interest on Oṃ, and with the anticipation of dreaming, without being interrupted by other thoughts, apprehend the dream-state with your sleeping awareness. . . . If the seed syllables are unclear and you still do not apprehend the dream-state in that way, focus your attention clearly and vividly on a *bindu* of light at your throat; and with the anticipation of dreaming, fall asleep and thereby apprehend the dream-state. (Padmasambhava 1998:153)

Head, heart, throat, or lower? Four-petaled lotus with one to six letters, a flame, a deity, or point of light (white, red, blue, or rainbow colored)? Clearly there are more variations here than in the *Kāmasūtra*. One wonders whether the particular imagery assigned to specific areas of the body has any objective reality. The same issue arises when one compares the Hindu and Buddhist descriptions of the *cakra* system. The two systems generally agree on placing psychic energy centers or *cakras* in the region of the head, throat, heart, navel, and genitals, but they disagree on almost all the details, including the number of *cakras* and the colors, syllables, number of petals, elements, and functions associated with each (Mann and Short 1990).

OVERCOMING OBSTACLES TO THE PRACTICE

The Evans-Wentz text next presents guidance in preventing "the spreading-out of dream content" (i.e., loss of the dream-state or awareness due to premature awakening, poor recall, etc.). The clearest instructions are in Padmasambhava (1998).

DISPERSAL THROUGH WAKING

"Dispersal through waking" refers to waking up immediately after becoming lucid, a typical problem for beginning lucid dreamers (LaBerge 1985).

To dispel that, maintain your attention at the level of the heart and below, and focus your mind on a black *bindu*, the size of a pea, called the "syllable of darkness," on the soles of both feet. That will dispel it. (Padmasambhava 1998:157)

The basis of this technique is apparently derived from the meditative practice of raising the gaze when too sleepy and lowering the gaze when too vigilant (Lamrimpa 1995:83). Other sources recommend eating nutritious food and performing bodily work or exercise until fatigued, resulting in deeper sleep (Evans-Wentz 1958). As for the visualization described above, it should be experimentally compared to the Western techniques for preventing premature awakening from lucid dreams already proven effective (i.e., "dream spinning"; LaBerge 1980, 1985, 1993).

LaBerge (1980) serendipitously discovered that spinning one's dream body when the dream begins to fade causes a return of stability to the dream state. Presumably, this sensory engagement with the dream discourages the brain from changing state from dreaming to waking. Research (LaBerge 1993) has proven the effectiveness of spinning: the odds in favor of continuing the lucid dream were about 22 to 1 after spinning, 13 to 1 after hand rubbing (another technique designed to prevent awakening), and 1 to 2 after "going with the flow" (a control task). That makes the relative odds favoring spinning over going with the flow 48 to 1, and for rubbing over going with the flow 27 to 1.

DISPERSAL THROUGH FORGETFULNESS

"Dispersal through forgetfulness" refers to initially becoming lucid but then forgetting that one is dreaming and letting the dream go on as usual.

> To dispel that, train in the illusory body during the day, and accustom yourself to envisioning the dream-state. Then as you are about to go to sleep, do so with the yearning, "May I know the dream-state as the dream-state, and not become confused." Also cultivate mindfulness, thinking, "Also, when I am apprehending the dream-state, may I not become confused." That will dispel it. (Padmasambhava 1998:157)

The tendency to lose lucidity by forgetfulness or false awakenings is common with beginning lucid dreamers. In contrast to the practice above, LaBerge (1985) recommends reminding oneself *in the dream* that one is dreaming: repeating phrases like "This is a dream" or "I'm dreaming" until one has had sufficient experience so that lucidity is no longer likely to be lost (LaBerge and DeGracia 2000).

DISPERSAL THROUGH CONFUSION

"Dispersal through confusion" means never becoming lucid at all because of a confused state of mind and resultant diffused awareness. In this case,

> during the daytime powerfully envision dreaming, and strongly emphasize the illusory body. Apply yourself to purifying obscurations, practicing fulfillment and confession, and performing the *gaṇacakra* offering. By forcefully practicing *prāṇāyāma* with the vital energies, and continuing in all this, the problem will be dispelled. (Padmasambhava 1998:158)

DISPERSAL THROUGH INSOMNIA

"Dispersal through insomnia" means failing to become lucid as a result of failing to sleep (an obvious prerequisite!).

> If sleep is dispersed due to powerful anticipation, and you become diffused as your consciousness simply does not go to sleep, counteract this by imagining a black *bindu* in the center of your heart. Bring forth the anticipation not forcefully and just for an instant, and by releasing your awareness, without meditating on sleep, you will fall asleep and apprehend the dream-state. (Padmasambhava 1998:158)

TRANSMUTING THE DREAM CONTENT

After developing the ability to induce and remember frequent and stable lucid dreams, the next stage is to practice "transmuting the dream-content" in a wide variety of ways:

> Practice moving gross and subtle appearances of sentient beings and the environment back and forth; increase one to many; gradually reduce many to one; transform pillars and pots and so on into living beings, both human and animal; within the environment and its inhabitants change living beings into pillars, pots and so forth just as you please; transform the peaceful into the wrathful and the wrathful into the peaceful and so on. Increase and transform

in various ways whatever you like in whatever way you like. At night, recognize the dream as a dream, and with your previous imagination and objects, increase things as much as you wish, and change them in any way you like. (Gyatrul 1993:82–83)

After gaining proficiency in control of dream figures and objects, the dream yogi practices "visionary travel," visiting any dream scene desired, typically a Buddha field:

When about to sleep, visualize a red dot as being within the throat psychic-center, and firmly believe that thereby thou shalt see whichever of the Realms thou desirest to see, with all of its characteristics, most vividly. (Evans-Wentz 1958:220)

It is sometimes claimed that under certain circumstances one actually does visit the place dreamed about. This is in spite of the general principle that everything experienced in the dream is illusory.

By engaging in the technique of illusory dream deeds one projects oneself in the dream state to a buddhafield and meditates there. But is this a reliable experience? Can one really project oneself to a buddhafield? [Perhaps.] . . . For example, Lama Wonton Kyergangwa projected himself into the presence of Guru Padmasambhava and received direct teachings on the Hayagriva Tantra. Nonetheless such instances are rare and their validity is difficult to ascertain.

With dream experiences that arise as a result of the technique being applied through energy control, it seems that sometimes they are . . . valid and sometimes not. When the method is conscious resolution, they usually are not valid. (Gyatso 1997:65)

H. H. the Dalai Lama himself appears to believe in the reality of actual out-of-body travel:

But it's also said that there is such a thing as a "special dream state." In that state, the "special dream body" is created from the mind and from vital energy (known in Sanskrit as *prana*) within the body. This special dream body is able to disassociate from the gross physical body and travel elsewhere. (Varela 1997:38–39)

Here as elsewhere, it is not always clear which body is being referred to. How, by direct experience, does one distinguish between the body one experiences, i.e., the phenomenal body or "body image" from that theoretical entity, "the gross physical body"? Does "travel elsewhere" really mean travel in physical space or travel in mental space? How does Gyatso distinguish valid and invalid projections, as mentioned above? These kinds of questions are ordinarily answered by the aid of the convergent corroboration of independent instruments or observers. How can introspection alone ever distinguish external reality from internal, phenomenal reality when the latter is all that by definition can be observed?

In any case, with my colleagues (Levitan et al. 1999) I argue that the concept of "out-of-body experience" is best understood, without the unproven and unnecessary hypothesis of a dream body disassociated from the physical body, as a purely mental experience:

> The worlds we create in dreams and OBEs are as real as *this* one, and, further, they are unfettered by the constraints of the physical universe. In dreams, we have the potential to explore the true powers of the mind without the limitations imposed in the "real world" by the need to survive in a hostile environment. How much more exhilarating it must be to be "out-of-body" in a world where the only limit is the imagination, than to be loose in the physical world in a powerless body of ether! Freed of the constraints imposed by the physical, expanded by the knowledge that we can transcend all previously known limitations, who knows what we could be, or become? (p. 194)

REALIZING THE DREAM STATE TO BE ILLUSION

The next level of practice in dream control is to become fearless in the dream, realizing that everything that happens in the dream is illusory:

> Whenever anything of a threatening or traumatic nature occurs in a dream, such as drowning in water or being burned by fire, recognize the dream as a dream and ask yourself, "How can dream water or dream fire possibly harm me?" Make yourself jump or fall into the water or fire in the dream. Examine the water, stones or fire, and remind yourself of how even though that phenomenon appears to the mind, it does not exist in the nature of its appearance. Similarly, all dream phenomena appear to the mind but are empty of an inherently existent self-nature. Meditate on all dream objects in this way. (Tsongkhapa 1997:127)

Very similar practices are described in the Western literature on lucid dreaming (LaBerge 1985; Tholey 1988) with special emphasis on trusting one's lucidity by being willing to put the dream body "at risk," since if one is indeed dreaming what harm can come? The suggestion in LaBerge (1985) that if there is nothing threatening in the dream, as one finds it, one can (and should) go looking for trouble is also found in the Tibetan tradition:

> Apprehend the dream-state and go to the bank of a great river. Consider, "Since I am a mental-body of a dream, there is nothing for the river to carry away." By jumping into the river, you will be carried away by a current of bliss and emptiness. (Padmasambhava 1998:156)

After becoming "thoroughly proficient" in the art of transforming dream content, the yogi turns his attention to his own dream body: this he now sees as just as illusory as any other element of his lucid dream. He now visualizes his own dream body in the form of a deity, and likewise all other bodies in the dream are seen in their divine form (Evans-Wentz 1958:221).

> While apprehending the dream-state, consider, "Since this is now a dream-body, it can be transformed in any way." Whatever arises in the dream, be it demonic apparitions, monkeys, people, dogs, and so on, meditatively transform them into your chosen deity. Practice multiplying them by emanation and changing them into anything you like. (Padmasambhava 1998:155)

This transmutation of base metal to dream gold prepares the ground for the final stage of dream yoga: transcending form altogether in mystical union with the formless.

MEDITATING ON THE THATNESS OF THE DREAM-STATE

The fourth and final stage of dream yoga training is meditating upon the "thatness" (voidness, primordial awareness, transcendent source of being) of the dream state. The text tells us that by means of this meditation "the dream propensities, whence arise whatever is seen in dreams as appearances of deities, are purified" (Evans-Wentz 1958:222). The yogi is instructed to concentrate in the lucid dream state, focusing on the previously visualized divine forms, and to keep his mind free of thoughts. In the undisturbed quiet of this mental state, the divine forms are said to be "attuned to the non-thought condition of mind; and thereby dawneth the Clear Light, of which the essence is of the voidness."

The last stages of the process of realization are summarized by Evans-Wentz in a footnote as follows:

[Thus, one realizes that the appearance of form] . . . is entirely subject to one's will when the mental powers have been efficiently developed by . . . [the dream yogic practices]. In other words, the *yogin* learns by actual experience, resulting from psychic experimentation, that the character of any dream can be changed or transformed by willing that it shall be. A step further and he learns that form, in the dream-state, and all the multitudinous content of dreams, are merely playthings of mind, and, therefore, as unstable as mirage. A further step leads him to the knowledge that the essential nature of form and of all things perceived by the senses in the waking state are equally as unreal as their reflexes in the dream-state, both states alike being *sangsaric*. The final step leads to the Great Realization, that nothing within the *Sangsara* is or can be other than unreal like dreams. The Universal Creation . . . and every phenomenal thing therein . . . are but the content of the Supreme Dream. With the dawning of this Divine Wisdom, the microcosmic aspect of the Macrocosm becomes fully awakened; the dew-drop slips back into the Shining Sea, in *Nirvanic* Blissfulness and At-one-ment, possessed of All Possessions, Knower of the All-Knowledge, Creator of All Creations—the One Mind, Reality Itself. (Evans-Wentz 1958:221–222)

The Eastern "Great Realization" that all phenomena are dreamlike has a clear parallel in recent Western psychological and neuroscientific conceptualizing of consciousness as a world model. According to this conception, the intrinsic nature of consciousness does not vary from one state of consciousness to another. Whether awake or asleep, our consciousness functions as a model of the world constructed by the brain from the best available sources of information. During waking conditions this model is derived primarily from sensory input, which provides the most current information about present circumstances, and secondarily from contextual and motivational information. While we sleep very little sensory input is available, so the world model we experience is constructed from what remains, contextual information from our lives, that is, expectations derived from past experience and motivations (e.g., wishes, as Freud observed, but also fears). As a result, the content of our dreams is largely determined by what we fear, hope for, and expect (LaBerge 1985, 1994, 1998; LaBerge and Rheingold 1990).

From this perspective, dreaming can be viewed as the special case of perception without the constraints of external sensory input. Conversely, perception can be viewed as the special case of dreaming constrained by sensory input (Llinas and Pare 1991). According to this model, dreaming should be phenomenologically more similar to than different from waking perception, as the results of a series of studies directly comparing the contents of consciousness in dreaming and waking (Kahan and LaBerge 1994, 1996; Kahan et al. 1997; LaBerge, Kahan, and Levitan 1995).

APPLICATIONS OF LUCID DREAMING FROM THE PERSPECTIVE OF DREAM YOGA

It should be clear by now that the principal application of dream yoga is nothing more nor less than enlightenment. This ought to be reason enough to pursue lucid dreaming. Still, from the perspective of dream yoga there are additional benefits of lucid dreaming described in the Tibetan tradition.

Tarthang Tulku explains one of the beneficial effects of lucid dreaming on personal outlook as follows:

> Experiences we gain from practices we do during our dream time can then be brought into our daytime experience. For example, we can learn to change the frightening images we see in our dreams into peaceful forms. Using the same process, we can transmute the negative emotions we feel during the daytime into increased awareness. Thus we can use our dream experiences to develop a more flexible life. (Tarthang 1978:77)

"With continuing practice," Tarthang Tulku explains,

> we see less and less difference between the waking and the dream state. Our experiences in waking life become more vivid and varied, the result of a lighter and more refined awareness. . . . This kind of awareness, based on dream practice, can help create an inner balance. Awareness nourishes the mind in a way that nurtures the whole living organism. Awareness illuminates previously unseen facets of the mind, and lights the way for us to explore ever-new dimensions of reality. (90)

As we have seen above, the practice of dream control techniques is considered by many authorities in the Tibetan tradition to lead to the capacity to dream anything imaginable. "Advanced yogis are able to do just about anything in their dreams. They can become dragons or mythical birds, become larger or smaller or disappear, go back into childhood and relive experiences, or even fly through space" (Tulku 1978:76). Here is mentioned in passing a fascinating possibility that raises several questions: to what extent is it possible to relive childhood experiences? How does this method compare with hypnotic age regression?

Western studies of lucid dream control have so far established rather more modest claims. For example, a study (Levitan and LaBerge 1993) investigated lucid dreamers' abilities to carry out a variety of simple dream control tasks ranging from looking at one's hand, front and back, turning on and off a light, checking one's reflection in a mirror, entering the mirror ("passing through the looking glass" as in *Alice in Wonderland*). All these tasks could be performed by some of the lucid dreamers some of the time. Perhaps the extensive visualization practice undergone by traditional practitioners of dream yoga leads to much higher ability to manipulate the dream imagery.

The wish-fulfillment possibilities of this degree of dream control may seem compelling, but it may be that the more valuable applications of lucid dreaming are elsewhere: "Dreams are a reservoir of knowledge and experience, yet they are often overlooked as a vehicle for exploring reality" (Tulku 1978:74). The lucid dream represents "a vehicle for exploring reality," an opportunity to experiment with and realize the subjective nature of the dream state and, by extension, *waking* experience as well.

One of the most interesting aspects of reality that can be explored is the phenomenology of embodiment and various forms of subjective relationships between "self" and "other." As discussed by Wallace (2001), when we dream nonlucidly our relationship with dream figures is intersubjective, i.e., as if they exist separately from the self. But in lucid dreams one can enter into an intrasubjective relationship to other dream figures by understanding that self and other are two as-if constructions within a single mind. Tholey (1988, 1991) has described a number of fascinating accounts in which lucid dreamers enter the bodies of other dream characters via a mobile "ego-core." The ego core is defined by Tholey as that location in the phenomenal field from which one seems to experience the world. It is usually localized within the phenomenal body, typically at the experienced origin of the visual field, the "cyclopean eye." It is possible to liberate this ego core from its usual location within the phenomenal body by a variety of means (e.g., destroying the dream body by fire, splitting, etc.) and then to enter into the body of another dream character, apparently "taking control" of that host body.

One might expect that entering deeply into the being of another in such a manner ought to have considerable potential for enhancing empathy, and

Tholey cites at least one case in which a lucid dream of this sort had exactly the expected effect. A teenaged woman was in love with a young man who was both friendly and pleasant, yet reserved with her. Before going to sleep one night she spent some time wondering why he was so reserved towards her. That night she dreamt that she was talking with the young man and suddenly realized she was dreaming. She asked herself again why he didn't return her feelings and wanted to get an answer in the dream. At that moment she became aware of her spirit detaching itself from her body and floating across the room and entering his body. It felt like she had taken over all his bodily functions without him being aware of it. At first it felt very strange and awkward, like driving a new car. But soon she got used to being in his body and saw with his eyes, felt with his hands, talked with his voice, and so on. From his perspective she saw herself standing in front of him and, what is more, how he perceived her, the effect she had on him, and the feelings he had for her. She saw that he was conflicted because, although he was very fond of her and had noticed her feelings toward him, he did not want to become romantically involved. She knew exactly what he was thinking and why he had been so reserved with her. She realized he would never return her feelings and then awoke. With her feelings sorted out by the experience, she was satisfied with being the young man's friend and felt relief because the tension that had previously existed between them vanished completely following the dream (Tholey 1988).

Realizing that our experience of reality is subjective, rather than direct and true, may have practical implications. According to Tarthang Tulku, when we think of all of our experiences as being subjective, and therefore like a dream, "the concepts and self-identities which have boxed us in begin to fall away. As our self-identity becomes less rigid, our problems become lighter. At the same time, a much deeper level of awareness develops" (1978:78). As a result, "even the hardest things become enjoyable and easy. When you realize that everything is like a dream, you attain pure awareness. And the way to attain this awareness is to realize that all experience is like a dream" (86).

Another aspect of reality that the lucid dream may allow us to explore is those unexplained phenomena currently referred to as paranormal. Dreams seem the ideal sphere in which to test how intentionality alters reality. Also, if verified, the claims described above of veridical projections to locations distant in space and time would have profound implications for our understanding of the world.

Finally, according to Tibetan lore, practice of the dream yoga provides essential preparation for the dreamlike after-death state, allowing the yogi to become illuminated at the point of death or to choose a favorable rebirth.

> It is said that by training in this transitional process of dreaming, as the transitional process of reality-itself and of cyclic existence are like the dream-state, those transitional process will be apprehended. Moreover, it is said that if the dream-state is apprehended seven times, the transitional process will be recognized. (Padmasambhava 1998:160)

The transition from waking to sleeping is regarded in the Tibetan tradition as very closely analogous to the experience of death, and the dream-state is considered the closest parallel to the after-death *bardo* state (Norbu 1992). Thus, practice in apprehending the dream-state can make recognition of the *bardo* state possible, something that may one day be of the greatest possible importance to all of us. At the least, it might make possible an experimental phenomenology of the ego-death process.

DISCUSSION: PROSPECTS FOR FUTURE WORK

Some scholars seem pessimistic about the possibility that Westerners can understand anything at all about the practices of dream yoga. For example, Wendy Doniger writes that "LaBerge attempts to apply the Oriental dream-control techniques of yogis and shamans to Western goals of improving one's life, or even, indeed, one's lifestyle. But this cannot be done" (1996:172). Why not? the reader may well ask. Unfortunately Doniger doesn't tell us. Instead, she switches to arguing that medieval yogis would not be interested in pursuing Western goals. This seems a non sequitur, given that no one is suggesting that they should do so! What is at issue is whether we can in the twenty-first century learn from practitioners of, say, the tenth century. I believe that the answer is yes, because even though we may widely differ in cultural heritage, we are all human, with similar enough brains, and possess a Rosetta stone—*experience*. A similar issue is discussed in LaBerge and DeGracia (2000) in which out-of-body experiences and lucid dreams are considered to share a common essential criterion, namely, a "reference to state," while differing in semantic interpretation.

Be that as it may, to take up again Doniger's point about goals, one typically has many goals in life, including getting enough food and air, raising a family, contributing to society, *and* obtaining whatever degree of enlightenment one can. Only in a limited view of spirituality are the goals of everyday life contradictory to the goal of spiritual development.

Serinity Young (1999) also seems to misunderstand this important point. In an appendix to her otherwise excellent review of the Buddhist tradition of dreams, she attempts to show (in my view, unconvincingly) that dream yoga and lucid dreaming are almost "totally different": "The context, content, method, and aim of these two practices [dream yoga and lucid dreaming] remain totally different, however. And they certainly have very different histories" (167). Young then goes on to present a limited and rather fanciful history of lucid dreaming in the West, starting with Kilton Stewart and ending with Patricia Garfield, apparently using Domhoff's *Mystique of Dreams* (1985) as her nearly exclusive source. The subtext? Western lucid dreaming is the fruit of the claims of a liar (Stewart, according to Young) being accepted by the credulous and self-deceived Human Potential Movement.

> *Context.* Dream Yoga: spiritual advancement takes place in a culturally and religiously supportive environment with at least a thousand-year history of such practice. Lucid dreaming: practitioners work in isolation or in recently formed dream groups, sometimes under the supervision of a trained psychologist, often not. (Young 1999:169)

The claim about lucid dreaming practitioners, if true, sounds like the sort of knowledge that would come from an anthropological study. If such a study exists, Young does not give the reference, and the claim appears to be wholly unsupported. The statement about the context of dream yoga makes one wonder how the context of Tibet one thousand years ago could be identical to the context of Tibetan Buddhists in the West today. Still, it is likely to be true that classical dream yoga is a less variable context than the diverse collection of modern Western practices falling under the rubric of *lucid dream work.*

> *Content.* Dream Yoga: practitioners share Buddhist imagery with very specific meaning. Lucid dreaming: dream content and meaning are often highly individualized. (Young 1999:169)

Certainly, the world of a Buddhist monk of the tenth or twenty-first century is likely to differ in many ways from the world of the archetypical Western lucid dreamer. But so, too, are the worlds of the scientist exploring lucid dreaming experientially and experimentally and the scholar studying ancient manuscripts (in translation or otherwise) *about* lucid dreaming or dream yoga.

> *Method*: Dream Yoga: practitioners work with a guru and radically alter their lifestyle by taking religious vows, forming an intention to achieve the religious goal of enlightenment, and, often, living apart from others for years. Lucid dreaming: practitioners work with an experienced lucid dreamer or simply read a book on the subject. (Young 1999:169–170)

> Two final methodological differences are that lucid dreaming may begin when awake, and in Buddhism the time of nights [*sic*] one dreams is important. (Young 1999:170)

These alleged "final differences" are in fact similarities. Some "Western" lucid dreaming techniques (see LaBerge and Rheingold 1990:chapter 3) for producing wake-initiated lucid dreams (WILDs) in fact derive from the dream yoga tradition (e.g. Tarthang 1978). Moreover, time of night is a widely established determinant of lucid dreaming frequency (LaBerge 1985; LaBerge et al. 1986). Again one gets the impression that for some reason Young wants to believe, and is grasping at straws in attempting to prove that lucid dreaming and dream yoga are "totally different" (Young 1999:167).

> *Aim*. Dream Yoga: the goal is spiritual advancement, reduction of attachment to earthly pleasures, and, ultimately, dissolution of the notion of an enduring self or world. Lucid dreaming: the goal is realizing earthly pleasures and maintaining attachment to them—for instance, achieving sexual orgasm—or attaining other psychological or practical benefits that enhance the sense of self. Although lucid dreaming can be used for spiritual practice, its broad usage and its existential base set it apart from Dream Yoga. (Young 1999:170)

Young cites no research on the usage and "existential base" of lucid dreams in the West, so it is difficult to take her claim seriously. Moreover, she blithely ignores the fact that in some Western writings (e.g., Bogzaran 1990; Kelzer 1987; LaBerge 1985; LaBerge and Rheingold 1990; Sparrow 1976) the spiritual dimensions of lucid dreaming are treated as the culmination of lucid dreaming practice.

While it should be clear from the preceding that the findings of the Eastern tradition are of great interest and value to Western researchers studying lucid dreaming, it may be less certain that Buddhists have a corresponding interest in and valuation of the results of Western science. That at least one very influential representative of Tibetan Buddhism, H. H. the Dalai Lama values science is made unmistakably explicit in the following quotation:

> For example, scientific investigation of the existence of a particular subject may reveal a multitude of logical fallacies. If we then persisted in accepting its existence, it would contradict reason. If it can be clearly proved that something that should be findable if it exists cannot be found under investigation, then from a Buddhist point of view we accept that it does not exist. If this somehow contradicts some aspect of Buddhist doctrine as contained in the scriptures, we have no other choice but to accept that that teaching is in need of interpretation. Thus, we cannot accept it literally simply because it has been taught by the Buddha; we have to examine whether it is contradicted by reason or not. If it does not stand up to reason, we cannot accept it literally. We have to analyze such teachings to discover the intention and purpose behind them and regard them as subject to interpretation. Therefore, in Buddhism great emphasis is laid on the importance of investigation. (H. H. the Dalai Lama 1997:169)

And so, too, is it in science. The fact that both traditions highly value careful observation and lucid discrimination (cf. "and we know which are the dreams") provides the middle ground prerequisite to fruitful cooperation. "Therein lies the hope."

Note

I am grateful to Kenny Felder, the Fetzer Institute, and the Institute of Noetic Sciences for financial support, to Alan Wallace for many illuminating discussions and great patience as an editor, and to Mushkil Gusha for the usual reasons.

References

Bogzaran, F. 1990. "Experiencing the Divine in the Lucid Dream State." *Lucidity Letter* 9(1): 169–176.

De Becker, R. 1965. *The Understanding of Dreams.* New York: Bell.

Domhoff, W. 1985. *The Mystique of Dreams: A Search for Utopia Through Senoi Dream Theory.* Berkeley: University of California Press.

Doniger, W. 1996. "Western Dreams About Eastern Dreams." In K. Bulkeley, ed., *Among All These Dreamers: Essays on Dreaming and Modern Society,* pp. 169–176. Albany: SUNY Press.

Evans-Wentz, W. Y. 1958. *Tibetan Yoga and Secret Doctrines.* London: Oxford University Press.

Gillespie, G. 1988. "Lucid Dreams in Tibetan Buddhism." In J. I. Gackenbach and S. LaBerge, eds., *Conscious Mind, Sleeping Brain*, pp. 27–35. Ithaca: Plenum.

Gyatrul Rinpoche. 1993. *Ancient Wisdom: Nyingma Teachings on Dream Yoga, Meditation, and Transformation.* Trans. B. Alan Wallace and Sangye Khandro. Ithaca: Snow Lion.

Gyatso, Jey Sherab. 1997. "Notes on a Book of Three Inspirations." In G. H. Mullin, ed., *Readings on the Six Yogas of Naropa*, pp. 43–70. Ithaca: Snow Lion.

H. H. the Dalai Lama. 1997. *The Joy of Living and Dying in Peace.* San Francisco: HarperCollins.

Houshmand, Z., R. Livingston, and B. A. Wallace, eds. 1999. *Consciousness at the Crossroads: Conversations with the Dalai Lama on Brain Sciences and Buddhism.* Ithaca: Snow Lion.

Kahan, T. L. and S. LaBerge. 1994. "Lucid Dreaming As Metacognition: Implications for Cognitive Science." *Consciousness and Cognition* 3:246–264.

——— 1996. "Cognition and Metacognition in Dreaming and Waking: Comparisons of First- and Third-Person Ratings." *Dreaming* 6:235–249.

Kahan, T. L., S. LaBerge, L. Levitan, and P. Zimbardo. 1997. Similarities and Differences Between Dreaming and Waking: An Exploratory Study. *Consciousness and Cognition* 6:132–147.

Kelzer, K. 1987. *The Sun and the Shadow: My Experiment with Lucid Dreaming.* Virginia Beach: ARE.

Kuhn, T. S. 1970. *The Structure of Scientific Revolutions.* 2d ed. Chicago: University of Chicago Press.

LaBerge, S. 1980. *Lucid Dreaming: A Study of Consciousness During Sleep.* Ph.D. diss., Stanford University, 1980. University Microfilms no. 80–24,691.

——— 1985. *Lucid Dreaming.* Los Angeles: Tarcher.

——— 1990. "Lucid Dreaming: Psychophysiological Studies of Consciousness During REM Sleep." In R. R. Bootsen, J. F. Kihlstrom, and D. L. Schacter, eds., *Sleep and Cognition,* pp. 109–126. Washington, D.C.: APA.

——— 1993. "Prolonging Lucid Dreams." *Night Light* 7(3–4). http://www.lucidity.com/NL734SpinFlowRub.html

——— 1994. "The Stuff of Dreams." *Anthropology of Consciousness*, 5:28–30.

——— 1998. "Dreaming and Consciousness." In S. R. Hameroff, A. W. Kaszniak, and A. C. Scott, eds., *Toward a Science of Consciousness* 2:495–504. Cambridge: MIT Press.

LaBerge, S. and D. J. DeGracia. 2000. "Varieties of Lucid Dreaming Experience. In R. G. Kunzendorf and B. Wallace, eds., *Individual Differences in Conscious Experience*, pp. 269–307. Amsterdam: John Benjamins.

LaBerge, S., T. Kahan, and L. Levitan. 1995. "Cognition in Dreaming and Waking." *Sleep Research* 24A:239.

LaBerge, S. and L. Levitan. 1991. "Sleep on the Right Side, As a Lion Doth . . ." "Tibetan Dream Lore Still True After Ten Centuries." *Night Light* 3(3): 4–11.

LaBerge, S., L. Levitan, and W. Dement. 1986. "Lucid Dreaming: Physiological Correlates of Consciousness During REM Sleep." *Journal of Mind and Behavior* 7:251–258.

LaBerge, S., L. Nagel, W. Dement, and V. Zarcone. 1981. "Lucid Dreaming Verified by Volitional Communication During REM Sleep." *Perception and Motor Skills* 52:727–732.

LaBerge, S. and H. Rheingold. 1990. *Exploring the World of Lucid Dreaming.* New York: Ballantine.

Lamrimpa, Gen. 1995. *Calming the Mind: Tibetan Buddhist Teachings on the Cultivation of Meditative Quiescence,* p. 83. Trans. B. Alan Wallace. Ithaca: Snow Lion.

Levitan, L. and S. LaBerge. 1993. "Testing the Limits of Dream Control: The Light and Mirror Experiment." *Night Light* 5(2). http://www.lucidity.com/NL52LightandMirror.html

Levitan, L., S. LaBerge, D. J. DeGracia, and P. G. Zimbardo. 1999. "Out-of-Body Experiences, Dreams, and REM Sleep." *Sleep and Hypnosis* 1(3): 186–196.

Llinas, R. and D. Pare. 1991. "Of Dreaming and Wakefulness." *Neuroscience* 44:521–535.

Malcolm, N. 1959. *Dreaming.* London: Routledge.

Mann, J. and L. Short. 1990. *The Body of Light.* Boston: Tuttle.

Mullin, G. H., ed. 1997. *Readings on the Six Yogas of Naropa.* Ithaca: Snow Lion.

Norbu, N. 1992. *Dream Yoga and the Practice of the Natural Light.* Ithaca: Snow Lion.

Padmasambhava. 1998. *Natural Liberation: Padmasambhava's Teachings on the Six Bardos.* Trans. B. Alan Wallace. Boston: Wisdom.

Puizillout, J. J. and A. S. Foutz. 1976. "Vago-aortic Nerve Stimulation and REM Sleep: Evidence for an REM-triggering and a REM-maintenance Factor. *Brain Research* 111:181–184.

Shah, I. 1968. *Caravan of Dreams.* London: Octagon.

Sparrow, G. S. 1976. *Lucid Dreaming: Dawning of the Clear Light.* Virginia Beach: ARE.

Surya Das, Lama. 2000. *Tibetan Dream Yoga.* Audio cassette. Boulder: Sounds-True.

Tarab Tulku. 1991. "A Buddhist Perspective on Lucid Dreaming." *Lucidity* 10(1/2): 143–152.

Tarthang Tulku. 1978. *Openness Mind.* Berkeley: Dharma.

Tholey, P. 1988. "A Model for Lucidity Training As a Means of Self-Healing and Psychological Growth. In J. I. Gackenbach and S. LaBerge, eds., *Conscious Mind, Sleeping Brain,* pp. 263–287. Ithaca: Plenum.

——— 1991. "Overview of the Development of Lucid Dream Research in Germany. *Lucidity Letter* 10(1, 2): 340–360.

Tilopa. 1997. "The Oral Instruction of the Six Yogas." In G. H. Mullin, ed., *Readings on the Six Yogas of Naropa,* pp. 23–29. Ithaca: Snow Lion.

Tsongkhapa, Lama Jey. 1997. "A Practice Manual on the Six Yogas of Naropa: Tak-

258   *Lucid Dreaming and the Yoga of Dream State*

ing the Practice in Hand." In G. H. Mullin, ed., *Readings on the Six Yogas of Naropa*, pp. 93–135. Ithaca: Snow Lion.

Varela, F., ed. 1997. *Sleeping, Dreaming, and Dying: An Exploration of Consciousness with the Dalai Lama.* Boston: Wisdom.

Young, S. 1999. *Dreaming in the Lotus.* Boston: Wisdom.

Wallace, B. A. 1989. *Choosing Reality: A Buddhist View of Physics and the Mind.* Ithaca: Snow Lion.

——— 2001. Intersubjectivity in Indo-Tibetan Buddhism. *Journal of Consciousness Studies* 8:209–230.

Wangyal, T. 1998. *The Tibetan Yogas of Dream and Sleep.* Ithaca: Snow Lion.

Werntz, D., R. G. Bickford, F. E. Bloom, and D. S. Shannahoff-Khalsa. 1983. "Alternating Cerebral Hemisphere Activity and the Lateralization of Autonomic Nervous Function." *Human Neurobiology* 2:39–43.

Much has been made in recent years about the extraordinary progress in technology that has enabled brain scientists to study the living brain as never before. This rapidly increasing sophistication in the instruments to study very specific brain processes is indeed remarkable and enables scientists to gain fresh insights into the neural correlates of a wide range of mental processes. But, to a crucial extent, such discoveries of mind/brain correlations depend not only on third-person observations of the brain but also upon first-person observation of mental processes. It is in this juxtaposition of third-person and first-person modalities that one finds a glaring lack of parity: while the third-person modes of observation and experimentation are highly professional, rigorous, and increasingly sophisticated, first-person modes of observation and experimentation have been left to amateurs—untrained "subjects" who are expected to have no more expertise than a laboratory animal.

In this essay Matthieu Ricard presents the outlines for a contemplative science that would rectify this problem by introducing sophisticated means of exploring and transforming the mind firsthand through sustained contemplative training. Taking the Buddhist contemplative tradition as a model for such a science, he points out that the fundamental aim of contemplative science is to understand the mind through direct experience. And the function of such knowledge is to purify the mind of its "afflictions," such as craving, hatred, and deluded self-centeredness, and thereby discover a state of genuine well-being. Such happiness is not a stimulus-driven pleasure, not even an intellectual or aesthetic joy, but a way of flourishing that stems from our deepest nature as human beings. That nature, according to Buddhism, is "pure awareness," the experience of which enables one to transcend self-centeredness and open to a deep sense of altruism.

Ricard addresses a wide range of issues that may be explored by way of a contemplative science, including the origins of consciousness, free will, the relation between mental afflictions and genuine happiness, and the false dichotomy of reified concepts of mind and matter. A central theme of his essay is that a contemplative science can be—and for centuries in Tibet actually has been—every bit as rigorous as the physical and life sciences. The Buddhist tradition in particular presents precise modes of observation and a wide array of experiments that have been conducted for *generations and have yielded repeatable results. Such contemplative science is presented not as a substitute for the modern cognitive sciences but as a crucial counterpart to them, fulfilling aims that the methods of the physical sciences were never designed to achieve.*

Matthieu Ricard

On the Relevance of a Contemplative Science

What is it that might best fulfill human needs? Science? Spirituality? Money? Power? Fame? Pleasure? No one can answer such questions without asking themselves what mankind aspires to most deeply, and what the very purpose of life might be. Buddhism's answer to that question is to point out that, finally, what we all seek in life is happiness. But it is important not to misunderstand the apparent simplicity of that observation. Happiness, here, is not just some agreeable sensation but the fulfillment of living in a way that wholly matches the deepest nature of our being. Happiness is knowing we have been able to spend our lives actualizing the potential that we all have within us and have understood the true and ultimate nature of the mind. For someone who knows how to give meaning to life, every instant is like an arrow flying toward its target. Not to know how to give meaning to life leads to discouragement and a sense of futility that may even lead to despair and the ultimate failure—suicide.

Happiness necessarily implies wisdom. Without wisdom it would be impossible to put right the principal cause of what we perceive as unhappiness—that is, persistent dissatisfaction dominating the mind. Such dissatisfaction comes from an inability to overcome the mental poisons of hatred, jealousy, attachment, greed, and pride, which arise from an ego-centered vision of the world and from the attachment to the idea of a self that is so powerful within us.

The other essential component of happiness is summed up in three words: altruism, love, and compassion. Draw a line, on one side of which you put yourself and on the other all living beings—the asymmetry is obvious. Just so, whatever happiness and suffering I experience is insignificant compared to the happiness and suffering of others. And anyway, whatever happens, my own happiness is intimately linked to that of others.

What is the need for a contemplative science? Is it not enough to try to relieve all our problems materially? The conditions the external world affords may well be vital to us in terms of our well-being, our comfort, our health, our longevity—our very lives, even. The techniques and remedies that work through material, external circumstances are important in bringing us certain kinds of happiness. But none of them can bring us true, inner well-being. Here it is the mind that counts, for the mind plays the essential role in satisfaction and dissatisfaction, happiness and suffering, fulfillment and failure. The mind is behind every experience in life. It is also what determines the way we see the world. The mind is the window through which we look out at "our" world. It takes only the slightest change in our minds, in how we perceive people and things, for that world to turn completely upside down.

Someone in the throes of an overpowering feeling of hatred sees the whole world as hostile. But his hatred only has to fade away for his view of the world to change completely. "Where could I ever find enough leather to cover the whole earth? But my leather sandal is enough to cover it wherever I go," says the famous Buddhist text *The Way of the Bodhisattva*.[1] If we sow the thorns of our own hostility everywhere, we will never see the last of our enemies. But, if we know how to protect the mind from hatred, all perception of those enemies will vanish of its own accord. It is at the inner causes of poverty and conflict that contemplative science takes its aim.

It is certainly legitimate to look into the nature of reality, the universe, the origins of life, how the body and the brain work—but it is impossible to access a universe independent of our perception, our consciousness, or our concepts. In fact, according to Buddhism, there is no such "parallel" world that would be totally independent of our consciousness. To posit the existence of such a world is a gratuitous, unverifiable assumption. Interdependence means that everything is interconnected in a global, mutual causality that naturally includes the animate and the inanimate.

As Heisenberg said, "What we observe is not nature in itself, but nature exposed to our methods of questioning."[2] We can only perceive the world through our senses and our thoughts. It seems only natural, therefore, to look into what the nature of mind might be. So it is curious that the cognitive sciences, as they have developed over the last century or two in the West, have found it very hard to accept that in the final analysis the mind is the only thing that could know itself. That distrust of the mind has led them to conclude that no information apprehended and evaluated by the mind alone could ever form the basis of any science.[3] In 1913 the American behaviorist J. B. Watson declared: "The time has come for psychology to get rid of all reference to consciousness. . . . Its sole task is to predict and control behavior; introspection can take no part whatsoever in its methods."[4]

The problem arises from the methods used. The first requirement for the practice of a contemplative science is to have a suitable tool at hand with which to work. An unstable mind, perpetually in motion, or a mind weighed down by torpor, is of little use in such an undertaking. It is essential to acquire mental stability and clarity, for without those qualities the mind is quite inadequate as an instrument with which to investigate its own nature. That is the obstacle confounding those few psychologists who, in the late nineteenth century, attempted to study the mind introspectively. They lacked the indispensable prerequisite of having mastered the mind, an achievement brought about only through long, sustained effort. Too little time spent on personal experience was what led one of the founders of modern psychology, William James, to declare that it was impossible to still the flow of discursive thoughts.

When we observe our own minds, we can see that from morning to night our thoughts never stop following one another in an uninterrupted flow. It takes time to calm that chaos down. No contemplative would expect to be able to master his own mind in the space of a few days, nor to experience its nature just through discursive thinking.

This indicates the need for training, the long and patient work that a contemplative must undertake over periods counted in months and years. This is not a question of blocking all the thoughts that might arise, or of anesthetizing the mind, but of gradually freeing it from the perpetual chain reaction of discursive thoughts. Inner calm is not a goal in itself but simply makes it possible to examine in depth the nature of the mind, which is the only way we can ever liberate ourselves from the yoke of the ego and ignorance.

How do we proceed? If we want to experience the ultimate nature of mind directly, we have to be able to "watch" the mind. But when we try to watch it those continually arising thoughts create interference, in the same way that stirring up a pond with a stick spoils the clarity of the water because of the fine particles suspended in it, making it very difficult to see the bottom. So the first stage is to begin to calm down the flow of our thoughts.

When we first try to gain some mastery over what happens in the mind, our initial impression is that it gets more agitated, rather than calming down. Our thoughts look like the water of a waterfall and seem to be more numerous than usual. In fact, there are no more of them than before; we have simply become aware of how many there are. Usually, whenever a thought arises, another one follows on, and then a third, and so on, building up a whole chain of thoughts. For example, if we say to ourselves, "That person was wrong to do that to me," this thought will multiply itself many times over, invading the mind and eventually leading us to say something hurtful or carry out some act of violence. Like clouds building up rapidly to form a huge, dark, threatening mass, those small thoughts gather more and more strength and end up making us think that the world around us—sometimes hostile, sometimes pleasant, sometimes neutral—is something very solid.

If, instead of fanning those thoughts, we simply let them cross our mind, the waterfall becomes a torrent, sometimes flowing over rapids, sometimes through more tranquil stretches. The torrent then becomes a peacefully flowing river, interrupted here and there by falls or eddies and swirls in the current. This corresponds to a state in which the mind remains calm, unless it is stimulated by the perception of external events. Finally, the mind becomes like the sea in calm weather. Ripples of discursive thoughts occasionally run over its surface, but in the depths it is never disturbed. In this way we can reach a state of consciousness called "clear consciousness" in which the mind is perfectly lucid, without being the plaything of any thoughts.

That state of consciousness, reached only after months or years of perseverance, can subsequently be entered as often as we wish and becomes habitual. Consciousness devoid of any mental activity can then be directly observed. This is not a new state of consciousness that we have created. We have simply recognized the possibility of directly contemplating "pure awareness," the mind's most fundamental component. It is something that usually escapes us, as it is almost always blotted out by mental images arising from imagination, memory, or perceptions of the external world. But a seasoned contemplative is capable of observing consciousness dissociated from any object and remaining in a state of inner simplicity. This is naturally accompanied by a lasting serenity and finds its expression in a great openness toward others.

Such an exercise makes it possible to understand that consciousness is not a separate entity, endowed with its own existence, that might correspond to an "ego," or a "soul," and thus to a "person." That discovery has profound repercussions on our perception of the world. Consciousness is seen to be a perpetually changing flow whose ultimate nature, enlightenment, transcends any conceptual activity.

Clearly, everything within us changes. We move on from youth to old age, and our minds change and develop along with the experiences we undergo. We nevertheless have the impression that there is something constant in us that characterizes us as persons and lasts until we die. That impression gives rise to the deep-rooted feeling of an independent ego, creating a gulf between ourselves and others. We have a natural tendency to want to satisfy and protect this "me," while similar attitudes in others are perceived as threatening—or at best as tools, insofar as we can use them to satisfy our own desires. Perceiving this solid "me" engenders a multitude of mental events that wreak havoc on our inner peace. We get wrapped up in the game of attraction and repulsion, from which arise desire, anger, pride, jealousy, and lack of discernment. These mental poisons destroy our serenity and prevent us from opening ourselves to others, enclosing us in a prison of self-centeredness.

The contemplative must therefore analyze this idea of a self. It quickly becomes apparent that the self cannot be identified with the body, being present in neither the brain, nor the heart, nor the other organs of the body, nor even in the atoms of which the body is composed. Could the idea of "me" arise from the perception of the body and mind as a combined whole? If so, it would be a mere label applied to an ever changing flow. Given that a self with its own true existence cannot be identified in either mind or body, nor in both together, nor as something distinct from both of them, it would be logical to conclude that it does not exist. As Alan Wallace puts it in *Choosing Reality*,

> Such empirical insights can have an extremely profound effect upon one's intuitive sense of personal identity. This in turn makes a major impact on one's

emotions, one's way of regarding other people, and the manner in which one responds to them. . . . There remains a difference between self and other, but this distinction is of a conventional, not an absolute, nature. . . . The sense of absolute isolation from the rest of the world is banished. The arising of joy and sorrow may also be seen as dependently arising events, and this insight leads to the cessation of selfish craving and aggression. In the absence of these mental distortions one can explore ways of life that are in accord with reality rather than at odds with it. If grasping onto intrinsic existence is the fundamental mental affliction, realization of emptiness is the fundamental cure.[5]

However, that mistaken identification with a nonexistent self is the result of a long-standing habituation and could never be swept away for good in a few minutes of meditation. We have to persevere with our analysis until this new understanding becomes an integral part of us.

Let us go back to our examination of the nature of consciousness. According to Buddhism, consciousness is simply a function, a "mere appearance" devoid of intrinsic reality, not an entity. However, on the relative level, there is a difference between a conscious unreality (mind considered as a stream of conscious instants) and an unconscious unreality (the inanimate world it apprehends).

Buddhism also distinguishes several levels of consciousness: gross, subtle, and extremely subtle. The first one corresponds to the working of the brain and the second one to what we intuitively call consciousness from the point of view of our own experience: the faculty the mind possesses to investigate its own nature and exert free will. This aspect also includes the expression of tendencies accumulated in the past. The third and most essential level is called "fundamental luminosity of the mind." It is a pure knowledge that does not operate on the dual mode of subject and object and does not involve discursive thoughts.

Consciousness is not simply a question of matter becoming more and more sophisticated. Something physical by nature cannot turn into a cognitive phenomenon. Between those two things there is a difference in kind, and, in order to function, cause and effect must have a common nature. If something could be born from another thing of a fundamentally different nature, then everything could be born from anything.

Science explains clearly how, in terms of matter, increasing complexity appeared in the living world. The simplest organisms developed the ability to move, which allowed them to approach a source of nourishment and to retreat from danger. These simple tropisms gradually evolved into more and more complex capacities for interaction with the external environment, leading to the development of a nervous system and culminating in the evolution of intelligence. But this does not solve the problem of consciousness itself, of which intelligence is only one aspect.

Neurobiologists believe that the meaning we make out of our experience of the world emerges from the continued activity of the body in a given environment, since the brain exists in a body that interacts with the world around it. From their point of view, consciousness arises from this constant interaction with the outer world, and it could thus emerge from inanimate matter without requiring any additional ingredient.

If one considers emergence as involving both an upward and a downward causation, as in Francisco Varela's concept of *"embodied cognition,"*[6] one can interpret emergence within the Buddhist perspective of an interaction between the world, the body, and the various levels of consciousness. The upward causation does not suffice for the body to create consciousness, but it allows it to influence mental events. The downward causation allows consciousness to influence the body.[7] Downward causation also allows consciousness to mold what constitutes "our" world (the world we perceive as a particular type of living being) according to the tendencies accumulated by our consciousness during countless lifetimes.

The world is not, as idealists claim, a nonexistent projection of an existing mind. Yet, "our world" has been slowly shaped by the mind as a vase is formed by a potter. This shaping corresponds to what Buddhism calls "collective karma," which makes human beings, for instance, perceive the world more or less in a similar way, while our varied experiences within this world correspond to our individual karma.

But what could distinguish consciousness from a mechanism that could fully imitate the behavior of a conscious being? According to Buddhism, only the subtlest aspect of consciousness possesses the inherent faculty of nondual self-awareness. One calls this faculty the luminous aspect of mind, because it can "illuminate," or know, both itself and outer phenomena.

The difference between a computer and a purely biological consciousness does not seem to be fundamental. In the case of a computer, whose sole ability is to calculate, no one would consider attributing a consciousness to it. It is obvious that it has no more consciousness than a heap of scrap iron or an abacus of the kind still used in China to make calculations. In the case of neurobiological man, what reductionist biologists call consciousness is nothing more than the mechanical functioning the brain.

The computer Deep Blue's victory over Kasparov at chess made it obvious that such a combat between man and machine had little real significance, for Kasparov's defeat was not at all the defeat of human intelligence; it was simply evidence of the limits in his calculating capacity. And it was clearer than ever that this is one of the least important aspects of human consciousness. As one American commentator said at the close of the tournament, "Let's hope Deep Blue would have been capable of celebrating his victory over a bottle of champagne with Deep Pink!"

If consciousness were no more than the outcome of a physico-chemical process, could it ever question its own nature? Specialists in cognitive science say that the brain works in a much more flexible and complex way than a computer functioning on a binary system. They offer a more dynamic vision, according to which the interaction between the various biological components of the brain causes the emergence of global states that can be identified with consciousness.

Systems of artificial intelligence are able to learn and display self-organizing faculties, but would they worry about their future or rejoice at functioning well in the present? To this question Francisco Varela answers: "For the same reasons that life does."[8] Where does the emergence of "meaning" come from? According to Varela, once the organization of neurons in the brain has reached a certain level of complexity, the meaning of the world can emerge, the definition of an emerging principle being that it can manifest qualities different from and superior to the sum of its components. The emergence of meaning is possible only when the organism is inserted into a particular environment. This meaning could therefore emerge from robotic systems if they could learn by interacting with the environment similar to the way that meaning has appeared during the evolutionary process of living beings. This is the hypothesis of Rodney Brooks and others.[9]

It seems, however, that artificial intelligence can play but know nothing of the spirit of playfulness. It can calculate the future but can never worry about it; it can record the past but can never feel joy or sadness about it. It does not know how to laugh or cry, be sensitive to beauty or ugliness, or feel friendship or compassion. Above all, a mechanism without any consciousness, whether a computer or a network of neurons, has no reason to question its own nature. Why should a system of artificial or neuronal intelligence wonder what it itself might be, or what will happen to it after death or when its batteries run out? Why should an "intelligent" robot spend hours or days and months simply looking introspectively at consciousness to find out what is its ultimate nature? A mechanical system, whether made of metal or of flesh and blood, has no reason to begin asking such questions, let alone find answers to them. Is the very fact that consciousness is capable of wondering about its own existence not some indication that it cannot be exclusively a mechanism, however sophisticated?

The question of free will is also one of the trickiest points of the reductionist approach. According to the neurobiological model, consciousness can no longer have any power of decision. Anything that might look like a decision is determined by a complex set of interactions between neurons; free will is just an illusion. What we take for a conscious decision is the result of an evaluation made by the nervous system of the optimal response to an external situation in terms of maximum contribution to the survival of the individual or the species. It could be an "egocentric" response, when it favors the survival of the individual, or an "altruistic" response, when it favors the survival of the species, sometimes to the individual's detriment (and let us not forget that all this is supposed to happen without any consciousness!).

According to this theory, we have the illusion of acting, but what we usually call consciousness is really only a spectator, a passive witness that takes no active part in the workings of the brain and cannot influence the final result of any calculations made by the neurons. In short, consciousness has no power to make decisions.

Some people explain that we believe ourselves to have free will because that is an illusion that has given our species an evolutionary advantage. Such arguments leave unanswered the problem of who or what it might be that experiences this illusory feeling of freedom. But as Christian de Duve, winner of a Nobel Prize for medicine, says in *Vital Dust*, "If free will does not exist, there can be no responsibility, and the structure of human societies must be revised. Very few among even the most uncompromising materialists are willing to drive this argument to its logical conclusion." And he sums up with the qualified opinion, "We still know too little about the human mind to affirm categorically that it is a mere emanation of neuronal activity lacking the power to affect this activity."[10]

Finally, there is the conclusion of my former boss, the Nobel Price winner François Jacob, in his inaugural lecture at the Collège de France:

Can knowledge of the relevant structures and the intelligence of the mechanisms involved ever be enough to describe processes as complex as those of the mind? Is there any chance that one day we will be able to specify in terms of physics and chemistry the sum total of all the interactions giving rise to a thought, a feeling, or a decision? Some doubt is surely permissible.

If consciousness took no part in decisions, what on earth would it be for? Why does it observe us, and who does the observing? A bunch of robots would manage just as well, and the existence of consciousness would just be a completely free luxury, a bonus granted to man so that he can enjoy the show.

If man is no more than his neurons, it is hard to understand how sudden events of deep reflection and the discovery of inner truths could lead us to completely change the way we see the world or influence the direction our lives take. There are criminals whose lives are dominated by hatred, and who even kill each other in prison, but who suddenly realize one day how monstrous their way of life has become. They say that the aberrant state they have been in until that moment was something akin to madness. Any such major upheaval would have to be accompanied by an equally deep, sudden, major restructuring of the complex circuits of neurons that determine our habits and behavior. If, on the other hand, the continuum of consciousness has a nonmaterial component, there is no reason why it should not be able to undergo major changes quite easily—and much more flexibly than a network of physical connections formed during a slow and complex process.

The next question is, "Can the mind know itself?" This is a problem that has preoccupied everyone who has tackled introspection, notably within Buddhism, where it has given rise to frequent argument and debate. Common sense tells us that the mind can know itself. We are quite clearly capable of observing our own mental activity, of knowing whether our minds are calm or agitated, and of identifying the thoughts crossing them. But logic tells us that the mind cannot know itself—the mind, seen as an autonomous entity, can no more observe itself than a sword can cut its own blade or an eye see itself. So we would need a second consciousness to observe our thoughts, a third to observe the second, and so on ad infinitum. This is the problem of the observer, which has led neurobiologists to the conclusion that consciousness is nothing other than the very workings of the brain and nervous system. That is a valid argument, if the mind is considered to be an independent entity endowed with true existence. On the other hand, if thought is seen as a function, as a continuing, interdependent process, its faculty of knowing itself would be inherent in its very nature. It is self-aware in a nondual way that does not involve a subject-object relationship, like a lamp flame that illuminates itself without needing any external light source. If that is the case, self-awareness is only one instant of consciousness out of many others, because in one individual there could not be several coexisting streams of consciousness.

To come back to the purpose of contemplation, spending days and years watching one's mind and trying to determine its nature might appear to be a pointless exercise. It might also seem pretentious to call such an approach scientific. But it all depends on the goal one has in mind. If our aim is to dissolve the negative emotions that constantly destroy our inner peace, if our working hypothesis is that awareness of the ultimate nature of the mind, and how thoughts arise, will allow us to dissolve those emotions, and if experience shows that this is indeed what happens, it is certainly worth the effort. If contemplative science can cure us of hatred and attachment as effectively as aspirin can cure us of a headache, that would surely be a more than worthwhile result.

Let me repeat, this is a long-term project, like acquiring any knowledge or experience. The mind has gotten used to working in ways that maintain it in confusion from which it is difficult to disengage. When a sheet of paper has been left rolled up for a long time, it is not enough to flatten it out for a second or two. As soon as you let go of it, it rolls itself up again. Perseverance is required. The effort is neither disordered nor arbitrary; it has its own hypotheses, its own laws, and its own results. And for someone who takes the trouble to put it into practice, certainty dawns little by little and becomes no less compelling than the certainty derived from the discovery of the physical laws governing external phenomena. The rigor of this inner approach certainly justifies the term *contemplative science.* In the West, although it has always been present in the Christian contemplative tradition, it is not seen as one of life's priorities. This is not the case in Tibet, where it was the inspiration of an entire society for more than a thousand years. Until the Chinese communist invasion in 1949, a fifth of the Tibetan population lived in monasteries, retreat centers, and philosophical colleges and made the spiritual life their principal activity. This is a fact probably without parallel in human history. So it is not surprising that out of this large number of people there emerged a few remarkable beings who progressed far along this path and attained unusual degrees of perfection. These sages reached the certainty that it is possible to have direct experience of the state of pure awareness and its continuity and that death is not the end of this stream of consciousness but a transition from one state of existence to another. Their certainty did not arise from a stubborn belief in any dogma, nor from some unwarranted hypothesis, but from their own inner experience. It is a conclusion difficult to accept from a commonsense point of view, as our materialist tendencies make us want to see things with our own eyes. We find it easy to accept that a trained athlete can pole vault up to six meters, an achievement totally beyond our ordinary capacities. But it is much harder to accept that training the mind can also be taken very far, and produce astonishing results, because it all happens in a domain invisible to us.

Buddhism suggests three criteria for determining whether a statement can be accepted or not. The first is direct proof supplied by the immediate observation of a phenomenon. The second is indirect, inferential proof of the sort that makes it possible to assert, for example, that if there is smoke there must be a fire. The third is valid testimony, like that which allows a court of law to deliver a judgment, even in the absence of physical proof.

In the case of contemplative science, direct proof is afforded by the transformation that occurs in a person who undertakes spiritual training with sufficient perseverance. Inferential proof, for a Buddhist, is the way external perfection reflects internal perfection. Valid testimony is supplied by spiritual teachers, those unusually accomplished contemplatives who in all aspects of their character and behavior are manifestly worthy of confidence, of whose words everything that can be verified through our own experience is found to be true. With a little reflection it can be seen that this is not a leap of faith. We often call on deduction of this kind. For example, when an astronomer tells us what he has discovered, or a physicist explains the nature of elementary particles, we feel confidence in their assertions and consider their testimony valid, since a large number of scientists whose honesty is not in doubt have reached identical conclusions. We nevertheless have no way of verifying what they say, whether by their reasoning, which may well be beyond us, or by their calculations, which are inaccessible to us. We would have to undertake a long and arduous program of study to verify their claims through our own experience; yet it seems much more likely that they are telling the truth than otherwise.

Contemplatives follow a similar process. Their methodology is rigorous, and their findings corroborate those of others and stand up with just as much strength as any mathematical reasoning. The texts of Buddhist contemplative science are precise, clear, and coherent. Finally, cognitive phenomena are just as real as physical phenomena. The contemplative life is not a matter of familiarizing oneself with complex equations but of progressing with perseverance and coherence through the different stages of inner experience, which make it possible to clarify knowledge of the nature of mind. Anyone who makes the right kind of effort with the right attitude will reach identical conclusions. They also show, outwardly, irreproachable human qualities such as uprightness, lucidity, altruism, serenity, and inner joy. In short, they are beings like the Buddha, any chink in whose armor is impossible to find. It therefore seems natural to accept their testimony as valid. When they say that the stream of consciousness can act on the brain and nervous system without being reduced to mere neuronal activity, they should not be taken lightly.

According to Buddhism, the conflict between materialist and idealist points of view, between mind and matter, is a false problem. Most philosophers and scientists, in fact, have seen things in terms of a Cartesian dualism of "solid" matter and "nonmaterial" mind in opposition to each other. But the dominant idea today among scientists is that such dualism infringes the laws of conservation of energy by supposing that a nonmaterial object can influence a material system. Such a view of things does indeed raise insoluble problems.

There are several ways of resolving that paradox. First of all, it could be thought that consciousness is just a particular property of matter. It could also be thought—and this is the approach that Buddhism adopts—that no such duality exists, for neither consciousness nor the world of material phenomena have any intrinsic reality.

It might therefore be useful to investigate the solid characteristics that we invest matter with at first sight, for it is actually in reifying matter that materialism comes up against its own failure to understand the nature of mind. The history of ideas tells us that the first atomic theory was posited by Democritus, who regarded his "hooked atoms" as the building blocks of the universe. In fact, however, it was much earlier, in the sixth century B.C.E., that Buddhism had already set out an analysis of the notion of atoms that was much more elegant. It was an analysis whose purpose was not to establish a physical science but to investigate the tangible nature of reality. Buddhist logicians asked themselves if there could be indivisible particles ("atoms" in the etymological sense of the word) that are permanent, autonomous, and endowed with true existence and might, in some way, be the substance of the real world. They reasoned that an indivisible particle would have to be the equivalent of a mathematical point, without any dimension. If they had any dimension, and hence sides or directions, they would no longer be indivisible. Now, such particles could never be in contact with other particles except by merging with them, because if they were to retain their definition of being indivisible one particle would have to be in simultaneous contact with the entirety of another, and that would mean merging completely with it. A whole mountain of particles would therefore be able to dissolve into a single one. Reasoning of this kind is essentially aimed at breaking down our stubborn belief in the substantiality of things. So, twenty-five hundred years ago, Buddhism set out the hypothesis that elementary particles are neither solid nor endowed with independent existence but exist only in dependence on one another. Without making too much of the parallels with modern physics, it is difficult not to be reminded of Heisenberg, who wrote, "Atoms and the elementary particles themselves are not real; they form a world of potentialities or possibilities rather than one of things or facts."[11] Modern physics has yielded data that also lead us to doubt their solid reality. For example, we now accept that the phenomenon we call an electron can appear both as a particle and as a wave. While it is undeniable that the so-called electron has properties of a particulate nature, as well as others more characteristic of waves, it would nevertheless be hard to conceive of two more antinomic possibilities. Particles and waves are phenomena of totally different kinds. A particle is a localized object with a certain mass, while a wave is just the opposite.

Certain interference phenomena can only be explained by supposing that the same particle passes through two separate holes at the very same instant. This is an observation that defies our common sense, which takes particles as having a localized material existence. The imprecision that surrounds the constituents of the phenomenal world led Heisenberg to say that "atoms are not things."[12] If one can thus question the solidity of the physical world at the microscopic level, this should lead one to question as well its reality at the macroscopic level.

Such an understanding of the nature of phenomena bridges the apparently irreducible gulf between a so-called immaterial consciousness and a so-called material physical world. Consciousness and phenomena are constantly changing streams, interdependent and devoid of true existence, neither one being any more or less real than the other. It is from this angle that the Cartesian duality of physical and mental phenomena vanishes. This does not mean, as we pointed out earlier, that, from a conventional and relative point of view, there isn't a clear qualitative difference between the animate and inanimate and consciousness cannot be reduced to matter. In a dream, for instance, although nothing is real, a difference remains between the animate and the inanimate. Stones do not think, even in our dreams.

An irreducible Cartesian dualism would not allow the existence of an interface between consciousness and the world, for those two would have nothing in common. Consciousness would not be able to apprehend the world, since in order to act on it and be influenced by it, there has to be a dynamic relationship between the two. In brief, the duality of mind and matter is not irreducible because both are devoid of independent reality and partake of the interdependent globality of phenomena.

It is a principle of modern physics that mass-energy never completely disappears in the material world—"nothing is lost, nothing is created." What exists cannot entirely cease to exist; nothingness cannot modified, however many billions of causes might be brought to bear on it. For this last reason, Buddhism considers that the world can only be beginningless and that an apparent beginning, such as the Big Bang of our present universe, can only be an episode in a vaster process. Buddhism applies a similar reasoning to consciousness. According to Buddhist contemplatives, the flow of consciousness of any sentient being, human or otherwise, cannot simply appear out of nothing or disappear into nothing. Just as transformations of matter always take place through causes and conditions coming from matter's constituents and the forces that govern it, transformations of the stream of consciousness take place through causes that share its nature, too—that is to say, conscious ones. Consequently, the stream of consciousness, being a function that perpetuates itself, must also be considered to have neither beginning nor end.

This might seem to contradict the gradual evolution of intelligence in the history of the cosmos. But that depends on what is meant by consciousness, or by animate as the opposite of inanimate. When consciousness has a physical basis (Buddhism also envisages states of consciousness that are not associated with any form) and that basis is endowed with a complex structure such as that of the human nervous system, consciousness can be expressed in the form that we call intelligence. When the basis is more rudimentary, consciousness is expressed in much less sophisticated ways.

There is therefore no reason why the gross aspect of consciousness should not be correlated with the brain's chemical reactions, and give rise to physical processes that act on the body, nor that such processes should not in return influence consciousness. Such interactions last as long as consciousness is associated with a body. Neurologists who deny that this could be so are not speaking in terms of scientific investigation; they are expressing a metaphysical opinion. As Alan Wallace says, for the "hard" sciences,

> all of reality essentially boils down to matter and energy subject to the mindless, immutable laws of nature. Life is reduced to an ephiphenomenal byproduct of complex configurations of chemicals; and mind is a coemergent property of the organization of the neural system. Such physicalist reductionism is not simply a conclusion based upon scientific research. Rather, it provides the metaphysical context in which such research and theorizing are pursued.[13]

For Galileo, all that did not concern the study of measurable and quantifiable properties of material bodies (forms, numbers, and movements) was not science. By its very nature the subtlest aspect of consciousness thus escapes direct detection by the methods of investigation defined and used by the physical sciences. Their mechanical, quantitative techniques were never designed to be applied to unsubstantial phenomena. But not to be able to find something using those parameters of physical measurement is no proof of its nonexistence. We should not forget what Einstein said: "On principle, it is quite wrong to try founding a theory on observable magnitudes alone. In reality the very opposite happens. It is the theory which decides what we can observe."[14] That said, although the subtle or nonmaterial consciousness is without substance, it is not nonexistent, being capable of fulfilling a function. Consciousness carries within it the capacity to interact with the body. There is therefore no reason why one should not be able to study the interactions of subtle consciousness with its grosser aspects and with the body.

Buddhist scriptures tell the story of two blind men who wanted to have it explained to them what colors were. One of them was told that white was the color of snow. He concluded that white was cold. The other blind man was told that white was the color of swans. He concluded that white was the swishing of wings. The story is a good illustration of how we usually perceive the outside world. Light can be perceived as heat or as a source of colors; it can be measured as a wave as well as being detected as a particle. But what is there to tell us that one of those characteristics represents the true nature of light? To reflect in this way calls into question the possibility of any independent reality or, in other words, any phenomenon independent of other phenomena and the consciousness observing it. According to Buddhism, the key to understanding the world is the idea of the interdependence of all phenomena, whether inanimate or animate. There is no reality independent of all perception, no reality that defines itself by itself, as those who believe in realism would hold. Nor are phenomena a projection of the mind, as idealists think. Buddhism does not take an anthropocentric position, according to which the universe was formed *in order that* life could develop.[15] It observes that there is interdependence between our world and the consciousness perceiving it. This is what is called the "middle way," which avoids falling into either materialism or nihilism.

The fact that all phenomena are related to one another, whether in the fields of consciousness or matter, naturally implies a certain harmony between them. It is therefore possible to understand that the universe's physical constants (Planck's constant, gravitational force, etc.) are not incompatible with the conditions that make it possible for life and consciousness to exist. From this point of view, it would not be said that consciousness *had* to appear to observe the universe and its beauty, nor that the universe is as it is because it *had* to culminate in life and in humankind (a very anthropocentric idea). Nor does it mean that life automatically *had* to arise within the physical conditions defined by those constants, but rather that the universe and consciousness are two "coemergent" continua. Both continua are necessary, since although they are interdependent and are part of one globality, matter cannot produce the fundamental aspect of consciousness.

Can science fulfill human needs? Natural sciences do not pretend, cannot and will not give meaning to human existence. There is nothing wrong with this, since such meaning is purely a matter of inner discovery. In fact, science became immensely successful after it stopped trying to be an all-encompassing field of knowledge that would explain everything about everything. It has focused instead on the study of natural phenomena and became extremely efficient in finding, measuring, describing, and acting upon such phenomena. The amount of knowledge that came out of this effort is so immense that, in many people's minds, it has eclipsed the idea that science is not meant to answer a number of fundamental questions about the meaning of our existence. This is not a failure of modern science, for it has never been its purpose to help one find happiness and inner peace.

A doctor's prescription is not enough to cure a sick person, and, in just the same way, a purely theoretical approach will not make us into better human beings. Any spiritual move we make should have two purposes: to perfect ourselves and to contribute something to others. The modalities of spiritual life vary greatly from one tradition to another, but both of these criteria need to be fulfilled. Returning to the definition of happiness as the purpose of life, I would say that it results from the feeling of having fully actualized the potential for perfection that we all have within us, the potential for wisdom, personal fulfillment, and altruism. That is how we can bring about the union of wisdom and compassion, which is the very essence of Buddhism.

Scientific and contemplative knowledge are not antagonistic to one another. However, there is a hierarchy not to be overlooked. Science gives priority to understanding outer phenomena and acting on the world, while contemplative traditions emphasize inner peace, the elimination of mental suffering, and making the mind lucid, serene, and altruistic. One experiences with things, the other with consciousness.

Once we are committed to a spiritual path, it is essential to check that over the months and years we are actually freeing ourselves from hatred, grasping, pride, jealousy, and, above all, ignorance. That is the result that counts. The discipline that brings it about deserves to be called a science, in the sense that it is a form of knowledge and—far from being useless information—constitutes true wisdom.

Notes

1. Shantideva, *The Way of the Bodhisattva* (Boston: Shambhala, 1996), chapter 4, verse 13.
2. Werner Heisenberg, *Physics and Philosophy: The Revolution of Modern Science* (New York: Harper and Row, 1962), p. 58.
3. B. Alan Wallace, *The Taboo of Subjectivity: Toward a New Science of Consciousness* (New York: Oxford University Press, 2000).
4. Cited in Arthur Koestler, *The Ghost in the Machine* (New York: MacMillan, 1967), p. 5.
5. B. Alan Wallace, *Choosing Reality: A Buddhist View of Physics and the Mind* (Ithaca: Snow Lion, 1996), pp. 191, 160.
6. Francisco Varela, Evan Thompson, and Eleanor Rosch, *The Embodied Mind: Cognitive Science and Human Experience* (Cambridge: MIT Press, 1991).
7. This has been shown, for instance, in the case of the expression of certain

genes that become inactivated in young primates when they suffer from a lack of motherly care. See M. J. Meany et al., "Early Environmental Regulation of Forebrain Glucocorticoid Receptor Gene Expression: Implications for Adrenocortical Responses to Stress," *Developmental Neuroscience* 18(1996): 49–72.

8. Franscico Varela, personal communication.

9. R. A. Brooks, "Intelligence Without Reason," in *Proceedings of the 1991 International Joint Conference on Artificial Intelligence* (San Mateo, Cal.: Morgan Kaufmann, 1991), pp. 569–595; R. A. Brooks, "Intelligence Without Representation." *Artificial Intelligence Journal* 47:139–160; R. A. Brooks, C. Breazeal, R. Irie, C. C. Kemp, M. Majanovic, B. Scassellati, and M. M. Williamson, "Alternative Essences of Intelligence," *Proceedings of American Association of Artificial Intelligence* (1998).

10. Christian de Duve, *Vital Dust: Life As a Cosmic Imperative* (New York: Basic, 1995).

11. Heisenberg, *Physics and Philosophy*, p. 181.

12. Werner Heisenberg, "On the History of the Physical Interpretation of Nature," *Philosophical Problems of Quantum Physics* (Woodbridge, Conn.: Ox Bow, 1979 [1932]).

13. Wallace, *Choosing Reality*, p. 12.

14. Cited in Werner Heisenberg, *Physics and Beyond: Encounters and Conversations* (New York: Harper and Row, 1971), p. 63.

15. This is known as the "strong" anthropic principle.

PART 3

Buddhism and the Physical Sciences

In this wonderfully succinct essay William Ames presents a comparative analysis of Buddhism and physics, from the philosophical realism of the Abhidharma to the ontological relativity of the Madhyamaka, and from the classical physics of Newton and Maxwell to the breakthroughs of relativity theory and quantum mechanics. While the philosophical premises of early Buddhist thought and classical physics are remarkably similar in some respects, they are crucially different in others. Perhaps most important, Buddhism is concerned with qualitatively *understanding the world of experience (loka), including both mental and physical phenomena, as a means to eradicating the sources of suffering and achieving spiritual awakening. Physics, on the other hand,* quantitatively *probes the nature of the nature of physical phenomena as they ostensibly exist in the purely objective world, independent of subjective experience, and this knowledge is put to the use of technology and the mastery of the external world. Moreover, while Buddhism is primarily concerned with the investigation of mental phenomena, physics is solely concerned with physical phenomena. Insofar as the whole of modern science is based on the paradigm of physics, in the scientific view of the universe subjective mental phenomena are marginalized, relegated to the status of emergent phenomena arising out of objective configurations of matter.*

Underlying both early Buddhism and classical physics is the assumption of the radical duality between subject and object, between the system of measurement and the measured phenomena. Both assume that the physical world consists of elementary building blocks that exist by their own inherent nature, independent of observation or measurement. But in the Madhyamaka school of philosophy, regarded by many as the pinnacle of Buddhist philosophical thought, the independent existence of both subjective and objective phenomena is challenged. All phenomena are found to exist as dependently related events, "empty" of any intrinsic identity of their own. In this view both subjective and objective phenomena have only a conventional existence, relative to the mind that perceives or conceives of them. The only invariant truth among all cognitive frames of reference is that all phenomena are empty of an inherent nature, and even emptiness itself is empty of an inherent nature. For this reason it can be said that the essential nature (Tib. ngo bo*) of all phenomena is their very lack of an essential nature.*

In a manner that invites comparison with the Madhyamaka view, experimental evidence in quantum mechanics indicates that the fundamental building blocks of the physical universe—quanta of matter and energy—do not intrinsically bear either the qualities of particles or of waves. Rather, the characteristics they exhibit depend on the system of measurement by which they are detected. Moreover, their very existence prior to measurement is highly abstract, presented as a probability function. Only when a measurement is made does a quantum phenomenon come into "real" existence, a discovery that profoundly challenges the assumption that the objective world of physics is simply waiting out there to be discovered. Rather than viewing the elementary particles of the physical universe as discrete, independent, local entities, they may better be characterized as nonlocal, dependently related events, arising in relation to the system of measurement by which they are detected.

William L. Ames

Emptiness and Quantum Theory

Emptiness (*śūnyatā*) is a key concept in Buddhism, especially in Ma-hāyāna Buddhism; quantum theory is the heart of contemporary physics. Emptiness does not mean "nonexistence" but rather that all entities, in-cluding ourselves, lack the independent identity we tend to assume that they possess. Quantum theory has replaced the mechanistic worldview of nineteenth-century physics with a view that offers far less support to naive realism. To see what these ideas have in common, we begin by considering Buddhism.

SOME BASIC BUDDHISM

The Buddha's lifetime (c. 560–480 B.C.E.) was a period of great social and political change in northern India. The use of iron led to an increase in pop-ulation and wealth as people used iron axes to clear the forest and iron plows to put more land under cultivation. Cities grew, and large kingdoms re-placed earlier small kingdoms and aristocratic republics. Urbanization and social change led to alienation and discontent and a questioning of older re-ligious formulations. In this atmosphere new religious and philosophical teachings proliferated. The doctrines of the *Upaniṣads* profoundly transformed what we now call Hinduism; and from among the various new schools that arose at that time, Buddhism was to become a major world religion.

The major elements of the Buddha's teaching were an analysis of the pervasive suffering in life and a proposal for a path leading to the end of suffering. The Buddhist First Noble Truth of suffering has often been misinterpreted as meaning that all of life is overtly painful, an assertion that contradicts common experience. Instead, the Buddha claimed that all conditioned phenomena are impermanent, and thus even our pleasure and happiness are unsatisfactory in that they are subject to change and loss. Even neutral experiences, which provoke neither pain nor pleasure, are unsatisfactory because of their conditionedness, which involves our supposed selves in a complex web of causes and conditions whenever we try to act.

Another famous Buddhist teaching is that all phenomena are "not a self" or "without a self" (*anātman*). In the case of a human being, this means that neither one's body nor one's mind is a self or under the control of a self. The notion of a self is in fact rather slippery. Although we sometimes identify with our bodies or our thoughts or feelings or other mental phenomena, at other times we speak of "my" body, thoughts, feelings, etc., implying that the self is something separate from these, something that possesses them. We can imagine a self without any one of these kinds of phenomena, though not without all of them.

In Buddhism body and mind are seen as a stream of impermanent physical and mental events. The body constantly changes and ages, while the mind, with its perceptions, thoughts, feelings, etc., changes at each moment. Body and mind are like a river. The flow goes on, but there is no enduring entity that we can point to. The flow itself is not a thing, and it is constantly changing.

Body and mind are not under the control of a self because they are impermanent and because their changes depend on various causes and conditions, many of which are independent of what the self may want. Moreover, most Buddhists hold that there is no self apart from the body or mind that could control them. When we observe ourselves, we find only the physical and mental phenomena we call body and mind. Thus there is no evidence for a self different from mind and body, and a self with neither a body nor a mind makes no sense.

DEPENDENT ORIGINATION

According to Buddhism, our minds and bodies and the external world are all made up of impermanent, changing phenomena. This idea is closely related to the Buddhist emphasis on causality. On the one hand, the fact that things are dependent on causes and conditions drives the process of change, because the causal factors are themselves constantly changing. On the other hand, causality provides a source of order that prevents impermanence from becoming arbitrary and chaotic. For example, given the necessary conditions, such as soil and water and warmth, a rice seed will give rise to a rice sprout, but never to a barley sprout.

Causality in Buddhism is expressed in the principle of dependent origination (*pratityasamutpāda*). This principle is often formulated as a succession of twelve factors beginning with ignorance and leading, via craving, to suffering such as old age and death. In its most general form, though, the idea of dependent origination is stated in the following way: "When this exists, that comes to be; because of the origination of this, that originates. When this does not exist, that does not come about; because of the cessation of this, that ceases."

In other words, if A is one of B's necessary conditions, then the occurrence of B is dependent on or conditioned by the occurrence of A. When A does not exist, B cannot exist. (We should note that in Buddhism multiple causal factors are always necessary to produce any given result.) To go back to the traditional example of a rice sprout, the rice seed is considered the cause (*hetu*) of the sprout, while the other necessary factors, such as soil, moisture, warmth, etc., are causal conditions (*pratyaya*). A rice sprout originates only when all the causal factors necessary to its existence occur. When the conditions necessary to maintain it are not all present, it ceases to exist.

Dependent origination is related to the idea of no self. The world is composed of impermanent phenomena, but past and present phenomena are related by the principle of dependent origination. The causal relation between past and present provides for a degree of continuity and the possibility of relative permanence. That is, because of causal regularity, things may endure for some time with only minor changes, changes that we may be able to ignore when carrying out the activities of our daily lives. Thus, thanks to dependent origination, strict impermanence can be reconciled with the sort of regularity and persistence of things that we observe in our ordinary experience. Buddhists generally do not think that we need to invoke selves or essences to account for the continuity in our experience. In fact, they emphasize that our tendency to try to hold on to things as if they had a permanent essence and to regard ourselves as real, enduring entities is at the root of our suffering.

Dependent origination is also an impersonal process. Phenomena condition other phenomena without there being an agent of actions or an experiencer of results, though there may be mental and physical phenomena that we are accustomed to think of as being someone's action or someone's experience. Thus Buddhists have argued that our actions (karma means "action") can have results, including rebirth through successive lifetimes, without there being any permanent self. What we call a self is a stream of impermanent phenomena, causally interrelated by the fact that each mental or physical event originates in dependence on other events.

ABHIDHARMA

The preceding survey of basic Buddhist teachings is very incomplete. For example, I have said nothing about liberation or the path leading to it. Nevertheless, the ideas mentioned so far will provide enough of a foundation for the purpose of comparing emptiness and quantum theory.

Within a few centuries after the death of the historical Buddha, various early Buddhist schools had developed elaborate descriptions of the mind and the world, called Abhidharma. Abhidharma takes an analytical approach to the world. Living beings and physical objects are not considered to be unitary entities. Instead, they are made up of more fundamental units, called dharmas. The Sanskrit word *dharma* has many meanings, including "law" and "teaching of the Buddha." Here dharma is used in the sense that I have been translating as "mental or physical phenomenon." Dharmas are impersonal phenomena, which may be either mental or physical.[1]

This use of the term *dharma* preceded the Abhidharma systems. The *sūtra*s, or recorded discourses of the Buddha, as they have come down to us, speak of dharmas, perhaps because the doctrine of no self made it necessary to account for human experience in a way that did not appeal to the notion of a self. As time went on, there would have been a natural tendency to want to elaborate the teachings contained in the Buddha's discourses in order to extend the description and classification of things in terms of dharmas. More ambitiously, there would have been a desire to explain everything in terms of dharmas and the relations among them. This desire to make the Buddhist picture of the world complete and systematic probably accounts for the rise of Abhidharma.

Thus Abhidharma can be thought of as an attempt to systematize the teachings in the Buddha's discourses in terms of a consistent and complete dharma theory (including a theory of causes and conditions). In Abhidharma dharmas are considered to be ultimate in the sense that they cannot be analyzed further. Also, in Abhidharma, as in later Buddhism, dharmas are impermanent not merely in the sense of being "not permanent" but also in the sense of being strictly momentary. Dharmas last no more than an instant, and the impression of objects that endure through time is a sort of "cinematic" illusion produced by rapid sequences of similar dharmas.

While we cannot go into the intricacies of Abhidharma here, there are some more points that ought to be mentioned. One is that most of the dharmas enumerated in the Abhidharma systems are mental. Because of the overriding Buddhist concern with liberation from suffering, Abhidharma is primarily an analysis of our psychological experience rather than of our experience of the physical world. Within psychology the emphasis is on moral factors, on the psychology of meditation, and on the psychology of the spiritual path. Nevertheless, Abhidharma never questions the reality of the physical world, and a number of dharmas, such as sound and visible form, are physical.

The idea of "two truths" appeared first within Abhidharma. In ordinary language we describe the world in terms of objects like jars, but dharma theory analyzes them into dharmas. Thus there is no jar apart from its constituent dharmas of color, shape, texture, etc. Whatever disappears upon analysis, like a jar, is conventional, or superficial, truth. In contrast, the dharmas themselves cannot be analyzed into more basic constituents; if they could be, they would not be considered dharmas. Whatever withstands analysis (in other words, the dharmas) is considered to be ultimate truth.

Not only objects like jars but also persons like Bill Ames can be analyzed into dharmas and thus are merely conventional truths. In his teachings the Buddha often spoke about persons and objects, though at other times he spoke in terms of dharmas. According to the Abhidharmists, when the Buddha spoke in terms of conventional truth, he did so that his audience could understand him. Abhidharma, on the other hand, always speaks in terms of ultimate truth, that is, dharma theory. We will see the idea of the two truths reappearing later in the development of Buddhism, but with a quite different content.

CLASSICAL PHYSICS

Before going on to later developments in Buddhism, I would like to turn to physics and discuss classical physics because, as we shall see, it has some interesting parallels with Abhidharma. First of all, what do we mean by "classical" in the context of physics? Essentially, it refers to the body of experiment and theory that began in the seventeenth century with the work of Galileo and others and continued through the end of the nineteenth century. The publication of Albert Einstein's paper on the special theory of relativity in 1905 marks the beginning of modern, postclassical physics, though a more radical break with classical physics came somewhat later in the twentieth century with the development of quantum theory.[2]

Compared to ancient and medieval physics, classical physics makes much greater use of mathematics in its description of how the physical world works. For example, Newton's second law of motion is expressed as an equation: $F = ma$. This equation states that the force, F, acting on a body is equal to its mass, m, multiplied by its acceleration, a. Acceleration is the rate at which velocity changes. Thus the equation means that if the mass of a body is constant, then the greater the force applied to a body, the more rapidly its velocity changes.

One interesting point to note is that if the force is zero the acceleration will also be zero, and the body's velocity remains constant. (This is, in fact, Newton's first law of motion.) This principle stands in contrast to the sort of "gut-level" physics we use in daily life, where we generally assume that an object will slow down and eventually stop unless some force is applied to keep it moving. This happens because friction is pervasive in our environment. But Newton, generalizing an idea of Galileo's, realized that a better theory of matter and motion could be constructed by making a different assumption, namely, that a body will keep moving in a straight line at a constant speed unless some force is applied to it. Friction is considered to be an applied force. If friction were always present, this way of looking at things might be somewhat artificial. Newton's laws of motion, though, are vastly superior in accounting for the motion of celestial bodies in space, where friction is virtually nonexistent. Even where friction cannot be ignored, Newtonian physics provides a consistent, quantitative way of taking it into account.

WILLIAM L. AMES   291

Thus classical physics represents a step away from our everyday view of the world, though it is a small enough step that "common sense" can be made to accommodate it with relatively modest revisions. Another feature of classical physics that is already apparent from "F = ma" is that its mathematical description of the world permits quantitative calculations of, for example, the motion of a body under an applied force. In principle, the equations used in classical physics permit one to calculate exactly the future evolution of any collection of material bodies; in other words, classical physics is deterministic.

In practice, though, exact calculations are usually possible only in the simplest cases. Thus the predictions of scientific theories usually have some fuzziness because of the approximations made in doing calculations. Likewise, experimental measurements always contain some experimental error, and it is important to estimate the range of the error accurately. One is normally comparing, not only in classical physics but also in science generally, approximate measurements with the results of approximate calculations. Thus the predictions of scientific theories can be confirmed only as being "within experimental error." An experiment will be unable to distinguish between two competing theories if the predictions of both lie within the range of experimental error.

Newton's laws belong to the branch of classical physics known as mechanics, the study of matter moving under the influence of forces. Other areas developed later. In the nineteenth century it was realized that electricity and magnetism are closely related, and James Clerk Maxwell developed a mathematical theory of electromagnetism that gave a unified description of electricity and magnetism. Like Newton's mechanics, Maxwell's theory was deterministic, and, like Newton, Maxwell was able to use his theory to account for a wide range of phenomena. The greatest triumph of Maxwell's theory was his insight that light is a form of electromagnetic radiation. Thus his theory was able not only to unify electricity and magnetism but also to unify both with optics, the study of light.

Besides classical mechanics and electromagnetic theory, other important areas of classical physics were thermodynamics (the study of heat) and statistical mechanics. Thermodynamics developed important concepts of energy and entropy, while statistical mechanics accounted for thermodynamic phenomena such as heat and temperature in terms of the motion of large numbers of particles.

Thus, by the end of the nineteenth century, the successes of classical physics were impressive indeed. Not only had new theoretical understanding of the physical world been achieved and new phenomena predicted and discovered, but scientific knowledge had also been applied to produce powerful technological achievements. The idea became widespread that the physical universe and perhaps all of reality could be explained as unchanging particles of matter interacting by means of forces described by deterministic mathematical laws.

CLASSICAL PHYSICS AND ABHIDHARMA

Despite some significant differences, Abhidharma and classical physics can be seen as broadly similar. Both reduce the world to impersonal, relatively simple units of analysis which are causally related to each other. Whether the units of analysis are particles and forces, on the one hand, or dharmas, on the other, physical objects and living organisms are seen as just complicated combinations of these simple units.[3] In classical physics particles and forces are related by physical laws that are usually expressed as equations. In Abhidharma dharmas are related by the various kinds of causes and conditions summed up under the heading of dependent origination.

For example, in daily life we may say, "There is a book on the table." For both Buddhism and physics this statement is only conventionally true. In classical physics the book and the table are fundamentally a collection of atoms interacting by means of forces. For Abhidharma the book and the table are made up of dharmas that influence each other according to various causes and conditions.

The units of analysis in both classical physics and Abhidharma are taken to be ultimately real. The particles and forces of classical physics are held to be really, objectively there; the same holds for Abhidharma's view of dharmas. Both dharmas, on the one hand, and particles and forces, on the other hand, have definite, knowable properties. Though there may be some practical problems, such as experimental error, the properties of both dharmas and particles/forces are in principle well defined, with no inherent fuzziness.

Of course, there are some important differences as well. The dharmas are closer to our immediate experience than the mathematically described particles and forces of classical physics. Dharmas are known through examining our own experience. Each dharma is said to "bear its own mark" by which it is known. Particles and fields are known through being part of a theory that is found to be consistent with experiment. The fact that the theory is formulated in mathematical terms makes it possible to derive quantitative predictions that can be compared with quantitative experimental results.

Another important point is that most of the dharmas are mental rather than physical, whereas physics deals exclusively with the physical world. (Whether life and consciousness can ultimately be explained in physical terms is a separate question.) Another difference between Abhidharma and classical physics is that dharmas are momentary, while the atoms of classical physics are unchanging. In classical mechanics change is due to the motion of atoms and to the forces that produce changes in motion. Finally, Abhidharma is part of the soteriological project of Buddhism. It helps us to gain insight into how things really are and is thus part of the path to liberation. While many people have had the idea that science contributes to human betterment, strictly speaking, such an idea is not part of physics.

QUANTUM THEORY

Classical physics saw the world as composed of particles with definite properties interacting according to deterministic laws, but by the early twentieth century it was clear that classical physics had difficulty in accounting for some phenomena. Some of these shortcomings of classical physics led Einstein to devise the special and general theories of relativity. Quantum theory was developed by a number of physicists, including Niels Bohr and Werner Heisenberg, in an effort to explain other phenomena, especially phenomena on the atomic and molecular level, which could not be accounted for classically.

Some of the predictions of quantum theory have been verified by the most accurate experiments ever performed in physics, and, as a mathematical framework for calculating the results of experiments, this theory is universally accepted by physicists. On the other hand, there is considerable disagreement about the interpretation of quantum theory, that is, what the theory is telling us about the way the world is. Probably most physicists simply use the theory while ignoring questions of interpretation, and here I will try to discuss the features of the theory that are relatively independent of one's choice of interpretation.[4]

What are the most salient features of quantum theory? I think that any list would have to include the following: quantization, wave-particle duality, complementarity, uncertainty or indeterminacy, probabilistic prediction, the quantum measurement problem, and nonlocality.

Quantization means that, in many circumstances, physical quantities like energy, momentum, and so on can have only certain discrete or discontinuous values. For example, in the case of an electron bound in an atom, the electron in its orbit about the nucleus can have only certain discrete energies. This means that only certain orbits are possible and not others. In contrast, in classical mechanics, a planet can orbit at any distance from the sun.

In classical physics one has waves and particles, but there is no way that something can be both. In quantum theory particles such as electrons behave like waves under some circumstances, and electromagnetic waves, for example, sometimes manifest particle properties. One way of looking at the quantization of the energy of electrons bound in atoms is to say that the electrons are behaving as standing waves in a closed space. In such a case only certain wavelengths are possible; in quantum theory there is a one-to-one relationship between the wavelength of the wave aspect of a "particle" and the momentum of the particle. Thus because the bound electron as wave can have only certain wavelengths, the same electron as particle can have only certain orbits. Similarly, electromagnetic waves behave in some circumstances like particles called photons, with the momentum of the photon corresponding to the wavelength of the electromagnetic wave.

Even in quantum theory one never sees something behaving as a wave and a particle at the same time. This fact is known as complementarity. One can choose to do experiments that bring out either the wave nature or the particle nature of the object one is studying, but one will never observe both in a single measurement. For example, in some types of experiments an electron will act like a wave; in others it will act like a particle, but it will never act like a wave and a particle at once. The difficulty for common sense comes in trying to reconcile the wave behavior at one time with the particle behavior at another.

Another way of looking at wave-particle duality is to say that an electron in itself cannot be defined as either a wave or a particle. It can be said to have a wave nature or a particle nature only in relation to a given experimental situation. Thus at least some of the electron's properties belong to the electron's context as much as to the electron itself.

In classical physics all the properties of a particle are well defined. An electron's position and momentum, for example, are influenced by external forces, but there is no doubt that the electron does have a definite position and a definite momentum at each moment. According to quantum theory, it is impossible to determine the position and the momentum of a particle simultaneously with absolute accuracy. (This is an example of Heisenberg's famous uncertainty principle.) One often hears this fact explained by the idea that we unavoidably disturb something when we measure it. A closer analysis shows that quantum theory implies that the position and momentum of an electron, for instance, are objectively indeterminate.[5] That is, it is not that the electron really has a definite position and momentum but we cannot know what they are; rather, the electron's position and momentum have a certain irreducible "fuzziness."

The uncertainty principle is closely related to wave-particle duality. To the extent that the electron can be considered as a wave, it is not surprising that it usually does not have a definite position, since waves are spread out in space.

In classical physics it is possible, at least in principle, to measure the present state of a physical system exactly and then, knowing the forces acting, to calculate exactly the future evolution of the system. (Here we overlook the practical difficulties of measurement and calculation.) Thus one would know precisely what the outcome of any future measurement performed on the system would be. Quantum theory, on the other hand, generally gives only probabilities for the outcome of a measurement. It is impossible to predict, for example, when a particular unstable atomic nucleus in a sample of a radioactive element will decay.

In other words, classical physics is deterministic, while quantum theory is probabilistic. Nevertheless, in the case of many repeated measurements or many individual quantum events of the same type, probabilities become exact percentages and yield exact patterns of events, even if individual events cannot be exactly predicted. In the case of a radioactive substance, one can predict with great certainty that a certain fraction of the nuclei will decay in a given amount of time, even though one does not know which nuclei those will be. Thus, despite the probabilistic nature of quantum theory, there is still room for certain kinds of definite predictions.

The probabilistic nature of quantum theory is reflected in its mathematical formulation. In quantum theory the state of a physical system is described mathematically by a "wave function." For example, the wave function associated with an electron specifies the probabilities for different possible values of the position, momentum, etc., of the electron; but in most cases it does not give a single definite value. Thus the wave function does not so much represent the wave nature of the electron as it represents something more abstract, a sort of wave of probability associated with the electron.

As long as no measurement is made, the wave function changes in time according to an equation known as the Schrödinger equation.[6] (Erwin Schrödinger was another of the founders of quantum theory.) Interestingly, the Schrödinger equation itself is a perfectly deterministic equation, just like the equations of classical physics. The probabilistic element comes in when a measurement is made. Just as an electron, for example, is never observed to be a wave and a particle at the same instant, so its position or momentum is never measured to have more than one value simultaneously. Rather, it will be measured to have one of the possible values allowed by the wave function; but if more than one value is possible it is impossible to say in advance with certainty which value the measurement will give. What one can predict are the probabilities of the different possible values, as given by the wave function. If one repeats the same measurement on many identical quantum systems, the probabilities give the percentage of the measurements that will show each of the possible values.

When a measurement is made on a quantum system, we get some definite value for a physical quantity such as energy or momentum. In between measurements the wave function for the system generally shows multiple possible values for a particular quantity. According to standard quantum theory, the wave function tells us all that there is to know about a quantum system. Thus, in between measurements, the physical quantities characterizing a system have to be considered to be indeterminate in most cases. This brings us to the quantum measurement problem. When a measurement is made, how and why does an indeterminate state turn into a definite measured value? (We should note that this is a problem in the interpretation of quantum theory. If we ignore the quantum measurement problem, we can still calculate probabilities for the outcomes of measurements.)

A somewhat more formal way of looking at the quantum measurement problem is the following: in between measurements, the wave function is usually one that corresponds to multiple possible values for any physical quantity. Immediately following a measurement, the wave function has changed to one in which the measured quantity has only the measured value. The wave function then continues to evolve in time according to the Schrödinger equation until the next measurement, usually returning to a state in which the quantity that was measured has more than one possible value.

The abrupt change in the wave function when a measurement is made is called the "collapse" or the "reduction" of the wave function. The wave function collapses from a state in which multiple values are possible to one in which only one value is possible. Thus the quantum measurement problem becomes, How and why does the wave function collapse when a measurement is made? In one sense it is the problem of reconciling the deterministic evolution of the wave function in the time interval between measurements with the probabilistic nature of wave function collapse.

The final aspect of quantum theory that I want to discuss is nonlocality. Classical physics conceives of the physical world as composed of separate and distinct physical objects that interact with each other. When one adds special and general relativity to classical physics, it becomes clear that none of these interactions can travel faster than the speed of light. This fact is called "locality," meaning that a distant object cannot influence a physical system instantaneously or in less time than it takes light to travel the distance between them.

The situation is more complicated in quantum theory. Suppose two objects interact with each other, move off in different directions, and later become widely separated. According to quantum theory, the objects remain in a strange way intertwined, subtly influencing each other instantaneously. Due to the probabilistic nature of quantum theory, these "influences" cannot be controlled and used to transmit a message faster than light. Moreover, even though the "influences" seem to travel faster than light, no matter or energy is transported, so the theory of relativity's ban on faster-than-light speeds is not violated. In fact, rather than something traveling faster than light between two distant objects, it may be that the objects are somehow not separate, somehow fundamentally connected even though separate in space.

This nonlocality, whether it is thought of as instantaneous influences between distant objects or as inseparability of distant objects, seems to be not only a property of quantum theory. In 1964 a physicist named John Bell proved a theorem that showed that, in certain types of experiments, *any* local theory would have to predict results that obeyed a certain restriction. Such experiments were eventually done, and it was found that the results violated the restriction Bell's theorem placed on predictions of local theories. Thus the experimental results could not be accounted for by local theories; a nonlocal theory is required. The results agreed with the predictions of quantum theory (a particular nonlocal theory), but this is less significant than the fact that, even if quantum theory eventually turns out to be wrong, it seems any physical theory that hopes to agree with experiment will have to include nonlocality.[7]

MAHĀYĀNA

In discussing Abhidharma, I alluded to the existence of more than one early Buddhist school. Traditionally, there were said to be eighteen schools, which differed on various points of doctrine and monastic discipline. (For example, the Abhidharma of the Sarvāstivāda school recognized seventy-five dharmas, while the Theravāda school recognized eighty-two.) The only one of these early schools to have survived to the present day is the Theravāda school of Sri Lanka and Southeast Asia.

Thus there was diversity within Buddhist thought from an early stage, though the differences between the various early schools could be considered relatively minor. This diversity increased with the rise of Mahāyāna Buddhism. (The earliest Mahāyāna *sūtras* that still exist may date from the second or first century B.C.E.) Though firmly rooted in early Buddhism, Mahāyāna has its own characteristic emphasis on universal compassion, which aims to liberate all sentient beings from suffering, and on wisdom (*prajñā*), which comprehends the emptiness of all phenomena. The Mahāyāna ideal is exemplified by the bodhisattva, who, motivated by compassion, seeks to perfect wisdom and skillful means in order to attain complete enlightenment for the benefit of all beings. Here, because of its emphasis on the idea of emptiness, we will focus on the Mahāyāna philosophical school called Madhyamaka.

MADHYAMAKA

We have seen how classical physics was replaced by a theory that deviates much more sharply from commonsense ideas. In Buddhism, too, Abhidharma's picture of the world was challenged by a more radical understanding. The Madhyamaka school was founded by Nāgārjuna, who lived around 150 or 200 C.E. Madhyamaka can be seen as a philosophy based on the perfection of wisdom (*prajñāpāramitā*) *sūtras,* some of which are among the earliest Mahāyāna *sūtras.* The *sūtras* expound emptiness in a discursive way, while the Mādhyamikas[8] use systematic argument.

The Mādhyamikas agree with the Abhidharmists that living beings and material objects have only a conventional existence. But they go further and argue that even the dharmas themselves exist only conventionally. The Mādhyamikas base their argument on the fundamental Buddhist idea of dependent origination. All conditioned dharmas arise in dependence on causes and conditions. Thus, according to the Mādhyamikas, dharmas have no independent, self-contained existence and no intrinsic nature of their own. They are said to be "empty," meaning "empty of intrinsic nature."

Here it is important to realize that emptiness does not mean that nothing exists. This would amount to nihilism, a position that all Buddhists reject along with the existence of permanent conditioned entities. It is undeniable we have experiences as well as thoughts about whether the experiences are real or not. The question is how do phenomena exist, conventionally or absolutely? By using the term *emptiness* the Mādhyamikas deny any absolute or ultimate existence of phenomena, but they do not deny that phenomena exist conventionally.

Like Abhidharma, Madhyamaka speaks of two truths, but the content is not the same. For the Mādhyamikas the conventional, or superficial, truth is the existence of dharmas in dependence on causes and conditions. The ultimate truth is their emptiness or lack of intrinsic nature.

If emptiness means being empty of intrinsic nature, then a good way of understanding the meaning of emptiness is to look at what the Mādhyamikas mean by intrinsic nature. ("Intrinsic nature" translates *svabhāva,* literally, "own-nature" or "own-being.") In his major work, the *Mūlamadhyamaka-kārikā,* Nāgārjuna says, "Intrinsic nature is not contingent and not dependent on another" (MMK 15–2ab) and "The alteration of intrinsic nature is never possible" (MMK 15–8cd). Thus the intrinsic nature of a thing is what that thing is inherently, independent of any causes and conditions. Intrinsic nature is unalterable because it is independent of all external circumstances.

Why then do the Mādhyamikas say that things have no intrinsic nature? A good example is the heat of fire. Conventionally, and in the Abhidharma, heat is the intrinsic nature of fire because fire is always hot. Heat is invariably present in fire, independent of any other causes and conditions. In contrast, water may or may not be hot, depending on causes extraneous to the water itself. Thus heat is not the intrinsic nature of water.

The Mādhyamikas do not deny this conventional usage of the term *intrinsic nature,* but they deny that heat is the intrinsic nature of fire in any ultimate sense. The heat of fire depends for its existence on precisely those causes and conditions responsible for the existence of the fire itself. The heat of fire is thus contingent and dependent; and so, for the Mādhyamikas, it cannot be the intrinsic nature of fire or anything else. Likewise, no other properties of fire qualify as an ultimately real, intrinsic nature. Fire ultimately has no intrinsic nature, no independent and unchanging essence that makes it what is.

The Mādhyamikas hold that, as with fire, so all phenomena, all dharmas have no intrinsic nature. Conditioned dharmas and all their properties occur in dependence on causes and conditions. Thus dharmas and their characteristics are dependent and contingent. The existence of each dharma is sustained by dharmas other than itself; no dharma is self-sufficient.

Another way of saying this is the following: because a dharma depends for its existence on dharmas other than itself, it is nothing in itself, that is, when it is considered in isolation from everything else. If we focus on a particular dharma in an effort to distinguish its own intrinsic nature from that of other dharmas, we find that it disappears. The process of excluding from consideration everything but the dharma in question removes the very conditions on which its existence depends. Thus we do not find any inherent identity in it, any intrinsic nature that makes it what it is and that is independent of everything else.

We might also say that a dharma's identity is not self-contained but relational. And since the other dharmas to which it is related also exist only relationally, there is no "fixed point," no self-established entity anywhere. Even dependent origination, even emptiness, the absence of intrinsic nature, are conventional, relational facts and not ultimate entities. Emptiness is itself empty of intrinsic nature.

As we have seen, the Mādhyamikas deny that things exist by an intrinsic nature, but they do not deny that things exist in any sense. Nāgārjuna compares the way in which things do exist to the mode of existence of mirages and magical illusions. Like such illusions, things appear in dependence on causes and conditions, but they are not appearances of intrinsically existing entities. This is not to say that there is no distinction on the conventional level between, say, physical objects and optical illusions. The point is that both occur dependently and have no independent essence.

Thus emptiness, lack of intrinsic nature, in no way excludes causal regularity, expressed in Buddhism by the principle of dependent origination. In fact, most of the arguments that Mādhyamikas give to support the idea of emptiness rest on the fact of causality. Looking at it another way, one can say that if things had intrinsic nature causal relations would be impossible because everything would be independent of everything else. In this sense, it is emptiness that makes causality possible.

COMPARISON WITH QUANTUM THEORY

There is much more to say about Madhyamaka, but this should be enough for the purpose of comparing it with quantum theory. We recall that in quantum theory many of the properties of, for instance, an electron are not intrinsic to the electron itself. They depend not only on the electron but also on the type of experiment that is being performed.

In Madhyamaka, too, attributes are relational and not intrinsic. A dharma by itself has no nature, any more than an electron can in itself be said to be either a wave or a particle. The major difference is that Madhyamaka is more complete in its negation of intrinsic nature. In quantum theory some of the properties of an electron, such as its rest mass, are intrinsic to it; and, of course, physics deals only with the physical world. For the Mādhyamikas all phenomena without exception are empty of intrinsic nature.

There are other aspects of quantum theory that can be compared with Madhyamaka. In quantum theory the observer does not play a purely passive role. Whether an electron behaves as a wave or a particle depends on the type of experiment being done, and it is the observer who decides what sort of experiment to do. Thus quantum theory seems to be describing what the physicist John Wheeler calls a "participatory universe."[9] The observer does not simply record an objectively existing electron. Instead, he or she is partially responsible for determining what the electron is. As Wheeler puts it, "No elementary phenomenon is a phenomenon until it is a recorded phenomenon."[10]

Madhyamaka has its own version of the "participatory universe." In line with the general principle of dependent origination, subject and object, knower and known, observer and observed exist only in relation to each other. Neither has an independent, "objective" existence. They are all empty of any self-contained, intrinsic nature. Again, Madhyamaka is more thoroughgoing than quantum theory. Not all the properties of an electron are affected by the conditions of observation, whereas, for the Mādhyamikas, subject and object are fully relative.

Finally, we ought to remember that Madhyamaka, like all of Buddhism, is intended as a means to liberation,[11] whereas physics has more modest aims. Buddhism and Western physics have come out of different cultures, and they have different starting points, methods, and goals. This makes it all the more remarkable that they have produced some very similar ideas.

Notes

1. Some Abhidharma schools recognized another category of dharmas, "conditioned factors dissociated from matter and mind," for phenomena that seemed to be neither material nor part of the conscious operation of mind. Also, all schools recognized at least one unconditioned dharma: *nirvāṇa*.
2. The beginnings of quantum theory can be traced to a paper by Max Planck published in 1900. It took time, however, for the theory to be developed; and it was not until the late 1920s that (nonrelativistic) quantum theory was reasonably complete. It took even longer for its radical implications to be understood.
3. Strictly speaking, physics is concerned only with nonliving matter; but the view is widespread, among scientists and others, that living organisms can be explained in purely physical terms.
4. For a more detailed presentation of the crucial discoveries and unresolved problems in quantum mechanics, see George Greenstein and Arthur G. Zajonc, *The Quantum Challenge: Modern Research on the Foundations of Quantum Mechanics* (Boston: Jones and Bartlett, 1997).
5. The position can be exact if the momentum is totally indeterminate and vice versa.
6. The Schrödinger equation applies under circumstances where the effects of special relativity are negligible, that is, when relative velocities are small compared to the speed of light.
7. Some physicists have attempted to preserve locality by abandoning other cherished assumptions, such as "counterfactual definiteness."
8. Generally, *Madhyamaka* is the name of the school and its philosophy; a follower of the school is called a *Mādhyamika*.
9. John Archibald Wheeler, "The 'Past' and the 'Delayed-Choice' Double-Slit Experiment," in A. R. Marlow, ed., *Mathematical Foundations of Quantum Theory* (New York: Academic, 1978), p. 41.
10. John Archibald Wheeler, "Beyond the Black Hole," in Harry Woolf, ed., *Some Strangeness in the Proportion: A Centennial Symposium to Celebrate the Achievements of Albert Einstein* (Reading, Mass.: Addison-Wesley, 1980), p. 356.
11. For this reason, it is considered necessary in Buddhism to experience ultimate truth personally rather than simply to understand it intellectually. Intellectual understanding is very helpful on the path to liberation, but it is not able to take one the whole distance.

In the interface between Buddhism and psychology, it can be very useful to compare specific theories and methods in both disciplines, for they are deeply concerned with many common issues. Moreover, the first-person introspective methodologies of Buddhism may well complement the third-person modes of empirical inquiry of the cognitive sciences, thereby enhancing insights in both fields. But, as Victor Mansfield points out in the following essay, when bringing Buddhism and physics into dialogue, it is more fruitful to focus on philosophical issues. These are not at all irrelevant to physics itself, for theoretical and empirical research in physics always takes place within a philosophical context, which has a strong influence on the type of questions that are posed.

In this paper Mansfield draws out the deeply human elements of the Buddhist Madhyamaka view, explaining how the predilection to grasp onto the seemingly real, inherent nature of phenomena, including oneself and others, lies at the root of desire, aversion, and their resultant suffering. In rejecting the notion of "essences," which Western philosophy inherited from Plato and later Descartes, the Prāsaṅgika Madhyamaka vision of reality emphasizes the relative nature of all phenomena. Einstein's theory of special relativity is explained here, providing empirical, scientific evidence demonstrating the noninherent nature of time (as well as mass and spatial dimension). But such relativity is confined to objective, inertial frames of reference, whereas the Madhyamaka theory asserts that the nature of all phenomena is also relative to the conceptual identification or designation or phenomena. In other words, all imaginable phenomena arise into existence relative to the conceptual framework in which they are conceived. That is what is meant by "conventional existence," but this does not imply the existence of phenomena is merely a matter of personal or cultural whimsy. For all conditioned phenomena arise as dependently related events, dependent upon their own causal factors. And a central concern of Buddhism and science is to discover the regular patterns of those causal interactions, often called in science the "laws of nature."

One of the meanings of the Sanskrit term dharma *is "law," and Buddhism strongly emphasizes the exploration of the laws of causality as they pertain to human conduct and experience. Here is where Mansfield draws out the practical applications of the Madhyamaka view for the cultivation of compassion and altruistic service. For if self and others are not inherently different, if we all live in interdependence, then compassionate concern for all beings is the only authentic way to relate to others and lead a meaningful life. Thus, Mansfield points out in his concluding comments, "as we stand on the threshold of ever more powerful theories in science, it is more urgent then ever that we find a coherent worldview that can guide our science as well as our moral actions."*

Time is the substance I am made of. Time is a river which sweeps me along, but I am the river; it is a tiger that devours me, but I am the tiger; it is a fire that consumes me, but I am the fire.

—*Jorge Luis Borges*

Victor Mansfield

Time and Impermanence in Middle Way Buddhism and Modern Physics

In the midst of working on this paper, I learned that a friend of ours, an extraordinarily beautiful woman in all senses of the word, found that her equally beautiful nine-month old boy has a virulent strain of muscular dystrophy. For that bright-eyed and laughing little boy with a genetic time bomb, the future points to progressive wasting, immobility, and death before adulthood.

It is easy to see in this little boy the transformations already affecting his body and to feel the sharp sting of how things will unroll in time. There is a clear sense of inevitability, of time being "a river which sweeps him along." Although just as true for ourselves, we easily see in him that time is a devouring tiger and a consuming fire.

I'll show that understanding something about time in Buddhism and modern physics deepens our sense of how "time is the substance I am made of." Such understanding also helps us appreciate that we are the devouring tiger and the consuming fire. Beyond its inevitability and destruction, time has other crucial features.

We can reflect on past events and learn from them, but we cannot influence them. The past has a fixity that contrasts sharply with the more malleable future, where we make choices and influence events. Therefore, we experience a directionality to time, expressed by a metaphorical arrow pointing from the past, through the present, and into the indefinite future.

In contrast, the fundamental equations of physics are all time symmetric, meaning that they have no directionality in time. All the fundamental interactions can proceed in the reverse direction without violating any laws of physics. For a simple example, bounce a ball off the floor and take a movie of it. If you run the movie backward, nothing looks strange because the time-reversed motion violates no laws of physics. Or, take a movie of our solar system from a distant star and play it backward. All the rotations and revolutions of the sun and planets will be reversed, but no laws of physics are violated and nothing looks strange. The same is true for quantum mechanical examples. Let an excited atom decay and emit a photon. Run the process backward and you have an atom absorbing light and ending in an excited state.

Yet many complex processes do display clear temporal directionality. The ruptured balloon dangling from the tearful child's hand never spontaneously reassembles itself back into its inflated condition. Such irreversible processes like the rotting of food and the decay of teeth are in sharp contrast to the time reversible laws of physics. Our little sick friend's inevitable ride down the river of time, along with our own, is full of irreversible transformations, leading to death, the one we most fear. Therefore, despite the symmetry of the fundamental interactions, nature clearly has many asymmetric and irreversible processes. As we will see below, the physicist's explanation for this asymmetry, within symmetric underlying laws, can help us understand some of the deepest lessons from Middle Way Buddhism.

The two decades that this little boy can look forward to seem criminally short from here, yet time may seem to crawl unendurably in his final days. However, in this digital age most believe that, despite such subjective experiences, time is absolute. Two decades is a well-defined interval that all observers can agree on, despite their subjective biases. Again, appreciating how physics destroys this apparent absoluteness can also deepen our understanding of Middle Way Buddhism.

I hope to show that understanding a little about time in modern physics helps us more deeply appreciate some of the most profound ideas in Buddhism. Furthermore, I will also suggest that some appreciation of Middle Way Buddhist ideas could aid in the development of physics. Thus a nontrivial synergy between these two very different disciplines is possible, one that results in deeper understanding and more compassionate action. While time may be a devouring tiger, appreciating these ideas might help us attain equanimity and encourage us to act more compassionately toward each other and the planet.

CARROTS AND EMPTINESS IN THE MIDDLE WAY

I'll review the principle of emptiness within the Middle Way Consequence School (Prāsaṅgika Madhyamaka, which I abbreviate by Middle Way) through a little story. Nearly thirty years ago a very holy man gave me some fresh carrot juice to drink. What a tasty elixir! I returned home determined to grow some fresh carrots of my own on our little farm. However, the soil in my part of the world is heavy and stony, and the carrots that first year were stubby and misshapen. I thought, "If only I had a garden tiller, I could whip that heavy soil into the most beautiful carrot bed." I could not afford one of those fancy tillers that a delicate ten-year-old girl can operate with one hand. My rototiller is a test of my manhood, a bucking bronco requiring strength and stamina. Of course, time destroys both people and equipment, and my tiller soon suffered from a long list of woes. It requires the patience of an advanced Bodhisattva to start, it only works at the deepest setting, it no longer has a reverse, and it cannot run in place and so bolts ahead . . . when you can manage to start it. However, I only use it a few hours a year, so I suffer with it and consider it a perverse sort of challenge.

One beautiful spring day a few years ago, the rototiller was taking me for my annual ride while it bathed me in the blue smoke of burning oil. I was musing on carrots and rototillers and suddenly had a tiny enlightenment. The second of Buddha's Four Noble Truths tells us that suffering is caused by desire. My desire for that delicious carrot juice had chained me to this rototiller for a quarter of a century! A desire for fresh, sweet carrot juice initially seemed innocent and "spiritually correct," in that good health is an aid to practicing dharma, but look where it led. Desire does generate suffering. However, those blue clouds bellowing from the burned out muffler along with that shattering noise and vibration urged me to deeper reflection. Upon what is that carrot-desire based?

The Middle Way clearly answers that desires and aversions are based upon the false belief in independent existence, the idea that beyond my personal associations, relationship, and names for carrots there is a real, substantial, inherently existent entity. This substantially existent object, that entity that "exists from its own side," is the basis upon which we project all our desires and aversions, all our craving for and fleeing from objects.

This innate and unreflective belief in inherent existence divides into two pieces. The first is that phenomena exist independently of mind or knowing. We unreflectively believe that "underneath" or "behind" the psychological associations, names, and linguistic conventions we apply to objects like carrot or rototiller, something objective and substantial exists fully and independently from its own side. Such independent objects appear to provide the objective basis for our shared world. Second, we falsely believe these objects to be self-contained and independent of each other. Each object being fundamentally nonrelational, it exists on its own right without essential dependence upon other objects or phenomena. In other words, the essential nature of these objects is their nonrelational unity and completeness in themselves.

Since it is so critical to identify inherent existence carefully, let me say it in other words. Consider the carrot stripped of its sense qualities, history, location, and relation to its surroundings. All but an advanced practitioner of the Middle Way believes that this denuded carrot has some unique essence, some concrete existence that provides the foundation for all its other qualities. This core of its being, this independent or inherent existence, is what the Middle Way denies. The carrot surely has conventional existence; it attracts rodents and makes great juice. It functions as a food. However, it totally lacks independent or inherent existence, what we falsely believe is the core of its being. In other words, the object or subject we falsely believe independently exists is not actually "findable upon analysis." When we search diligently for that entity we believe inherently exists, we cannot actually find it. Its independent being does not become clearer and more definite upon searching. Instead, phenomena exist in the middle way because they lack inherent existence but do have conventional existence.

While reifying carrots, I simultaneously reify the one who desires carrots and consider him as inherently existent too. Out of the seamless flux of experience, I falsely impute or attribute inherent existence to both the subject and its object of desire and thereby spin the wheel of *saṃsāra*. In this way perception is a double act that simultaneously generates a false belief in inherently existent subjects and objects, gentleman farmers, and their carrots. Then our time is occupied with cherishing our personal ego, putting its desires before all else, pushing others aside to satisfy those desires, and running after objects we falsely believe inherently exist. We think those objects will make us happy, but in fact they can never satisfy us. Perhaps time "is a fire that consumes me, but I am the fire." Was this not the point of the Buddha's fire sermon?[1]

According to the Middle Way, we can put out the fire by deeply appreciating the doctrine of emptiness, the lack of inherent existence in all subjects and objects, in all phenomena. This requires not only an intellectual formulation as given here but a profound transformation of our whole being at many levels—a process that usually takes many lifetimes.

The description of emptiness given so far is negative, a thoroughgoing denial of what we wrongly believe is the core of existence. Next, let me turn to a more positive description of phenomena, including carrots. If phenomena don't independently exist, then how do they exist? The Middle Way tells us that they dependently exist in three fundamental ways. First, phenomena exist dependent upon causes and conditions. For example, carrots depend upon soil, sunlight, moisture, freedom from rodents, and so forth. Second, phenomena depend upon their parts and attributes. A carrot depends upon its greens, stem, root hairs, and so on. Third, and most profoundly, phenomena depend upon mental imputation, attribution, or designation. From the rich panoply of experience, I collect the sense qualities, personal associations, and psychological reactions to carrots together and name them or designate them as "carrot." The mind's proper functioning is to construct its world, the only world we can know. The error enters because along with naming comes the false attribution of inherent existence, that foundation for desire and aversion.

For the Middle Way dependent arising is a complementary way of describing emptiness. We can understand them as two different views of the same truth. Therefore, contrary to our untutored beliefs, the ultimate nature of phenomena is their dependency and relatedness, not isolated existence and independence.

One of the difficulties in understanding emptiness is that we can easily assent to the importance of relatedness while falling prey to the unconscious assumption that relations are superimposed upon independently existent terms in the relation. In fact, it is the relationships, the interdependencies that are the reality, since objects or subjects are nothing but their connections to other objects and subjects.

We might ask what phenomena would be like if they did in fact inherently or independently exist. The Middle Way explains that inherently existent objects would be immutable, since in their essence they would be independent of other phenomena and so uninfluenced by any interactions. Conversely, independently existent objects would also be unable to influence other phenomena, since they would have to be complete and self-contained. In short, independently existent objects would be immutable and impotent. Of course, experience denies this since our world is made up of continuously interacting phenomena, from the growth of carrots nourished by sun, rain, and soil to their destruction by rodents. From the subjective side, that we do not independently exist implies it is possible to transform ourselves into Buddhas, exemplars of infinite wisdom and compassion.

Critics of the Middle Way often say that if objects did not inherently exist, they could not function to produce help and harm. Carrots lacking independent existence could not give sweet juice or make soup. The Middle Way turns this around 180 degrees and answers that it is precisely because objects and subjects lack independent existence that they are capable of functioning. So the very attribute that we falsely believe is at the core of phenomena would, if present, actually prevent them from functioning.

Now how does all this relate to the Middle Way notion of time? As I mentioned above, if phenomena inherently existed, then they would, of necessity, be immutable and impotent, unable to act on us or we on them. Since, in truth, phenomena are fundamentally a shifting set of dependent relations, impermanence and change are built into them at the most fundamental level. That the carrot exists in dependence upon causes and conditions, its parts and attributes, and on our attribution or naming is what makes it edible, allows me to experience it and be nourished by it. More important for impermanence, these defining relations and codependencies and their continuously shifting connections with each other guarantee that all objects and subjects are impermanent, ceaselessly evolving, maturing, and decaying. In short, emptiness and impermanence are two sides of the coin of existence and therefore transformation and change are built into the core of all entities, both subjective and objective. In this way the doctrine of impermanence is a direct expression of emptiness/dependent arising. Because I lack inherent existence and am most fundamentally a kinetic set of shifting experiences, with no eternal soul, as we normally understand it, then "time is the substance I am made of." Borges's compact sentence seems like a Middle Way aphorism. Being substantially of time guarantees my continuous transformation and death. Indeed, time "is a fire that consumes me, but I am the fire." These philosophic truths of emptiness and impermanence are central to Buddhist practice, and I return to them later. Now let us turn to physics and its view of time.

TIME IN MODERN PHYSICS

RELATIVITY OF TIME

As mentioned in the introduction, we all have a natural belief in the absoluteness of time, meaning that, for example, one minute is the same for all observers. Let me again proceed by way of example.

My carrots take 70 days to harvest time. Our belief in the absoluteness of time or its independent existence appears in the view that this time is something intrinsic to the carrot. As long as the growing conditions are normal, it does not matter how this time is measured or who measures it. It has an independent or absolute nature. However, let an astronaut take the same seeds and grow them in a space ship traveling at 90 percent of the speed of light relative to the Earth. Then relativity theory tells us that the days to harvest (as measured by an Earth-based observer) would be 161 days.[2] Figure 1 shows the days to harvest, as observed on Earth, plotted against the velocity of the space ship, relative to Earth, divided by the speed of light, c. So, for example, when $v/c = .9$ then we move straight upward from that point on the horizontal axis and intersect the curve at 161 days. Only in a reference frame at rest with respect to the observer (the rest frame) is the days to harvest 70 days.

Relativity emphatically states that no value of the days to harvest time is any more real or intrinsic than any other. For example, if the astronaut looked back at my garden she would correctly measure my time to harvest as 161 days. Since time intervals depend directly upon the relationship between the object and the observer, they are essentially relational. *We cannot consider time independent of a particular reference frame.* In Middle Way language, it lacks independent existence. If the seed manufacturers were devotees of relativity, they would state on the package, "The

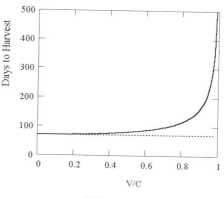

FIGURE 1

time to harvest is 70 days only in the rest frame. For other reference frames consult figure 1." We can attempt to evade this relational nature of time by saying that humans never travel at any significant fraction of the speed of light, and so this is just an academic consideration. This move denies the conceptual import of relativity's view of time and the thousands of experiments done all over the planet every day that rely on it.

If we clarify the idea of the present moment, the essentially relational nature of time intervals, whether decades or microseconds, is complemented by a thoroughgoing relativity of the present. Take the reasonable definition that all the simultaneous events taking place for an observer at one time define the present moment. Let's say I plant my carrots at exactly 9:00 A.M. on a given day and at that moment a friend in New Delhi boards a plane, while my son enters a classroom in a distant city. Relativity teaches that those simultaneous events defining the moment of carrot planting are only simultaneous in my garden's reference frame. If our farmer-astronaut, moving at 90 percent the speed of light, passes directly over my garden at 9:00 A.M., he observes a different set of simultaneous events and thus his present moment differs from mine; while a second astronaut, traveling at a different velocity over my garden at 9:00 A.M., finds yet a third set of simultaneous events and thus a different present from mine or the first astronaut.

Therefore, relativity makes both time intervals and individual moments relative to a given reference frame, leaving our old absolute view of time far behind. There are similar things to say about other primary qualities of objects, but these points about the relativity of time are enough for the present. A more interesting and profound quality of time comes from understanding how it has an arrow.

ARROW OF TIME

We store our carrots in the cellar where there is a cool, even temperature. However, even there, they rot after four to six months. We have never seen rotten food return to its fresh state. Rotting, whether of vegetables, teeth, or our entire bodies, is an irreversible process. Given that the quantum mechanical laws that govern the chemical changes of rotting are time symmetric, this is mysterious. The great Austrian physicist Ludwig Boltzmann made the first significant progress in understanding this mystery. He realized that irreversibility comes from reversible underlying laws only when you have large numbers of particles in the system.

Boltzmann started by considering a simple box containing many gas particles governed by Newton's laws. In analyzing this system he assumed that it was totally isolated from the rest of the universe. There were no influences of the universe on the box and its contents or vice versa. Now this should give anybody influenced by the Middle Way philosophy real discomfort, since he is assuming that the system independently exists. More about that later.

Boltzmann then imagined a partition in the middle of the box with all the particles in just one half of the box. The other half is totally empty. To proceed further we need to understand the concept of entropy or measure of disorder. The more disorder, the less knowledge we have about the details of the system, the higher the entropy. When the partition is removed, the overwhelmingly most probable configurations of the new equilibrium condition involve the gas spreading evenly throughout the box. In principle it is possible for the gas to bunch up in only one quarter of the box. However, it is overwhelmingly more probable that it will attain a new equilibrium configuration diffused throughout the box. Such equilibrium states have maximum entropy. Through this reasoning Boltzmann proved the famous Second Law of Thermodynamics, which says that any isolated system's entropy must either stay the same or increase. Therefore, when the partition ruptures, the gas is overwhelmingly likely to go to a state of greater entropy. What is more, the increase in entropy defines the direction of the arrow of time. Time advances in the same direction in which entropy increases— what we call the future. This does not deny that there are local decreases in entropy, like the growth of a child, but the global entropy relentlessly increases with time.

For several years, I taught our junior-senior level course on statistical physics. We used the standard textbook and followed Boltzmann's derivation of the Second Law of Thermodynamics, with the appropriate level of mathematical sophistication. In the last few years I found that there were arguments as far back as 1877 that showed Boltzmann was deeply wrong. I review some of these problems elsewhere in nontechnical language.[3] Here, I take a different approach and follow an elegant and simple argument by P. C. W. Davies.[4] As we will shortly see, entropy increases, but not the way Boltzmann thought. Why several revisions of this famous text persist in the error is a mystery.

The basic difficulty, which can be seen in several independent ways, is that completely isolated systems, like the box of gas, can generate no direc-

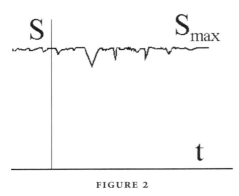

FIGURE 2

tionality to time because of the time-symmetric laws governing the system. Figure 2 displays the entropy, S, of an isolated box of gas plotted versus time, t. We see that the random gas motions give occasional deviations below the maximum. Although it is unlikely, the random motions spontaneously generate states of greater order or lower entropy, which are then brought back to maximum disorder by the same randomization. This is like the shuffling of playing cards that, on rare occasions, puts them into states of greater order, with continued shuffling returning them to disorder.

Now imagine the following experiment illustrated in Figure 3. We just patiently monitor the system until its entropy spontaneously drops to the value S_1 or below at a time t_1. If we choose S_1 low enough, this could take a long time. The virtue of choosing a small value of S_1 is that, once it occurs, we know we are very likely to be near the bottom of a dip in the entropy curve, rather then part way down a larger dip. This is simply because the even larger dips are so much less likely. At t_1, when the low entropy, S_1, occurs, since we are very likely at the minimum of a dip, an increase in entropy with time happens in either direction. At time $t_1 + \varepsilon$, where ε is some small time interval, the entropy increases. We consider this the future. However, the entropy also increases in the past at $t_1 - \varepsilon$. Therefore, the symmetry of the underlying laws of physics gives no directionality to entropy increase or time.

Even before I began getting instruction from my rototiller twenty-five years ago, the problem of the arrow of time had largely been resolved, although there are still technical subtleties. Much to the delight of the Middle Way, the main problem lies in assuming we have a totally isolated system independent of interaction with its environment.

We now understand that we must account for how Boltzmann's box got

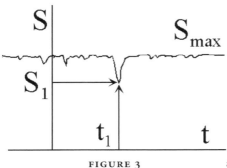

FIGURE 3

into the low entropy state of all particles in just one half. This did not result from just waiting a long time for random motions to throw the gas all to one side but from Boltzmann evacuating one half and placing gas in the other. Preparing the box in a low entropy state must generate more entropy elsewhere in the universe. For example, Boltzmann consumed calories from lunch and radiated energy from himself and his equipment that eventually went into deep space. In other words, the box had its entropy put into a low condition by processes outside itself, but at the expense of a much greater entropy increase elsewhere in the universe.

Let me give an example closer to the garden. I walk in the garden to check on whether the mice have eaten the carrots. My footprint in the soft soil gives it more order and structure, thus lowering its entropy. However, this lower entropy comes from a much greater generation of entropy from my metabolic processes, which eventually degrade to heat radiated to the universe.

As we have long known, the energy emitted into deep space from our activities can only radiate into space because the universe is expanding. If the universe were not expanding, then it is so large that any line of sight from the Earth, when extended far enough, would land on a hot star surface. Then the effective temperature of deep space would be that of the surface of stars, which is typically 6,000 K, rather than the 3 K it actually has. However, since the universe is expanding, it shifts the radiation from distant stars to very low temperatures. Since entropy can increase only when energy moves from high to low temperature regions, the simple process of radiating our body's energy into space would be blocked in a static universe. Thus there would be neither a Boltzmann nor the ability to reduce entropy locally in the box by generating more entropy elsewhere in the universe.

All systems organizing themselves or decreasing their entropy, whether the growing of a carrot, a snowflake, or a child, are decreasing entropy in one location that must be accompanied by a greater entropy generation in another. Not only is the energy from Boltzmann's food and his equipment eventually traced back to our sun, but the sun's low entropy is critical. Energy generation processes, whether the digestion of our food or the workings of a nuclear power plant, are totally dependent upon our solar system being in a low entropy condition. What causes the sun and other stars to be in a low entropy condition? This occurs because the expansion of the universe was faster than the nuclear generation rates in the first three minutes of the Big Bang. Then, when nearly all the helium (about 25 percent of the total mass of the universe) was formed, the universe expanded so quickly that after three minutes it was too cool for nuclear reactions to occur. If the expansion and associated cooling were much slower, then all the matter in the universe would have formed into iron-56, a very stable isotope of iron, an inert and high entropy condition. Then the stars would not shine, there would be no great entropy gradients in the universe, no time asymmetry, and, of course, no life.

Local time asymmetry, such as the decay of any biological system, from carrots to our own bodies, must be accounted for by connecting it to the expansion of the universe and its earliest evolution. This extraordinary beautiful result has many technical twists and turns, but the central idea is clear: increasing entropy and time asymmetry owe their existence to the largest and earliest processes in the universe and its continued expansion. This is a long way from the notion of an isolated and noninteracting system, so abhorrent to the Middle Way. In this way, when you put cold milk into your coffee and the mixture comes to the same temperature and a higher entropy than when the fluids were separated, you are profiting from the universe's expanding and cooling before iron-56 could form. Similarly, that we must all face the irreversible process of death, with its massive entropy increase, is traceable to the earliest and largest processes in the universe. In other words, the impermanence and decay found all around us is due to the earliest and most distance process in the universe and its continued expansion.

On a more positive note, irreversible processes are essential to life. If metabolic processes did not irreversibly transform my lunch, not only would I get indigestion, I would not live. That which sustains me also destroys me. Indeed, time "is a fire that consumes me, but I am the fire."

COMPARISONS AND CONNECTIONS

As I have said in my recent ruminations about the relationship between physics and Buddhism,[5] it is a mistake to connect any Buddhist principle too closely with any particular phenomena from physics. Physical theories are prime examples of impermanence. What happens if I make an argument that some physical effect verifies some great principle of Buddhism and then the physics is replaced by a new theory? Does that damage Buddhism? Are the foundations of Buddhism to tremble at every scientific revolution?

A more fruitful dialogue between Buddhism and science can occur when comparisons and connections are done at a more philosophic level. For example, here I have tried to focus on emptiness, the philosophic heart of Buddhism, and make connections with questions of comparable philosophic significance in physics. If the connections mutually illuminate both the physics and the Buddhism, without trying to reduce one to the other, then our understanding of both disciplines deepens. In the present example the erroneous assumption of a thermodynamic system being completely isolated from any form of external interaction was a critical error. This error could have been avoided if the philosophic principle of emptiness were more widely understood and appreciated in the scientific community.

Physics is always done in a philosophic context. In the case of classical statistical physics and thermodynamics, it was done within Cartesian dualism. Although Descartes's vision helped both physics and Western philosophy, it has also hindered us in more ways than we can count. I suggest that the principle of emptiness, if more fully appreciated within science, could actually further the scientific enterprise.

What does Buddhism gain from such connections and comparisons as attempted here? I see at least two benefits. First, understanding such things as the relativity of time (the seventy days to harvest example) and the relativity of the present moment helps us appreciate the closely parallel arguments made in the Middle Way about time's lack of inherent existence. There is a well-known and difficult section in Nāgārjuna's *Mūlamadhyamakakārikā* that analyzes time and leads to the modern interpretation, "Time is thus merely a dependent set of relations, not an entity in its own right, and certainly not the inherently existent vessel of existence it might appear to be."[6] Such critical but difficult points are illuminated by understanding Einstein's relativity of time. In short, science can help us understand ancient, pivotal philosophic aspects of Buddhism.

Second, Buddhism is a portable religion that has wandered far from the home of the original Prince. In each movement, whether to China, Japan, or Cambodia, it takes on the hues of the local culture without losing its original spiritual impulse. Science is clearly a cultural dominant in the West. Therefore, if Buddhism is to come to the West, in the best and fullest sense of the term, then interaction with science is both inevitable and necessary for a real transplant to take place. The present effort at understanding some common ground and even synergy between Buddhism and science can be part of the effort to translate Buddhism into terms that are easier for a Westerner to assimilate.

REFLECTING ON THE RELATIVITY of time, and how the irreversible nature of my little friend's disease connects to the first few minutes of the universe and its continued expansion, gives me little comfort. Yes, intellectually these ideas strongly support the principle of emptiness, that both the mother and the little boy along with the one who writes these words lack independent existence. Yes, we are all a system of interdependent relations and thereby subject to the law of impermanence. Nevertheless, the heartache remains. That little boy will be consumed by the "fire of time" before he reaches the age of my two sons.

According to the Middle Way, my inveterate projection of that false quality of independent existence is the foundation for my attachment and consequent suffering. It all comes back to my inability to put these ideas fully into practice. This is often the plight of those who can articulate ideas but not fully live them. Or, being kinder to myself, perhaps I have assimilated just enough of the principle of emptiness to give me a deep appreciation of the mother's sorrow, but not enough to dispassionately see it all as an embodiment of the First Noble Truth, that all experience is suffused with suffering. What then do we do?

The Middle Way advises us to take refuge in the Three Jewels: the Buddha or fully enlightened One, the Buddha's teaching, and the community of those seeking enlightenment. The Buddha shows that we can do it. We suffering humans, nurtured and destroyed by time, can become full embodiments of wisdom and compassion. We too can break free from the fetters of time, from the suffering of *saṃsāra*, the endless torment of repeated death and rebirth. The Buddha's teaching, which includes emptiness and much more, is the work at hand among those who support our efforts at realizing these great truths—including the mother and her sick child.

If I could reflect deeply enough on the relativity of the twenty years as the maximum allotted to this child and that the very irreversibility of his condition, and my own, is due to deep cosmological connections, then perhaps my sense of connectedness to others and the cosmos could increase. Could I realize more deeply that my ego and yours are dependent, not inherently existent, but fundamentally codependent systems of relationships? Could I profoundly appreciate that there is no speaker without a listener, no griever without a dependently related object of grief? If I could, then the centrality of my own ego and my self-cherishing would surely diminish. Such a realization of my ego's emptiness and our mutual codependency must result in compassion, not just for this little boy and his mother, but for all sentient beings. Assimilating these great truths and shifting my ego off center stage is surely not easy, but the promised increase in understanding and compassion keeps me trying.

If I could deeply appreciate that any irreversible process, whether the rotting of carrots or my body, is due to its connection with the earliest and largest-scale structure of the cosmos, then how much easier it would be to appreciate that my neighbor's loss or gain is not separate from mine. Then the suffering in one cell of the body of humanity is truly the suffering of all. Perhaps we could even realize that compassion is actually in our own enlightened self-interest and that the survival of our very planet requires a profound understanding of our codependence.

In contrast, we could ask what happens when our philosophic view embraces the false notion of independent existence. The late David Bohm, known for both the depths of his physics and philosophy, said it directly when he wrote:

> It is proposed that the widespread and pervasive distinctions between people (race, nation, family, profession, etc., etc.), which are now preventing mankind from working together for the common good, and indeed, even for survival, have one of the key factors of their origin in a kind of thought that treats things as inherently divided, disconnected, and "broken up" into yet smaller constituent parts. Each part is considered to be essentially independent and self-existent.[7]

According to Bohm, many of the evils of our modern world are traceable to a view where "each part is considered to be essentially independent and self-existent." In other words, one in which things inherently exist. I tried to show above that, although we commonly assume for simplicity that a system, such as Boltzmann's box, is independent from its surroundings, such a view misleads us. This is bad enough in physics, but when a race, nation, or person views themselves as fundamentally independent, then the stage is set for calamity—the stuff of our daily headlines.

As we stand on the threshold of ever more powerful theories in science, it is more urgent then ever that we find a coherent worldview that can guide our science as well as our moral actions. Consider how the advent of quantum mechanics and relativity brought about the wonders of the information age along with our horrendous weapons of mass destruction. Then imagine what wonders and horrors might be released by a grand unified theory or "theory of everything" that today occupies some of the best minds in physics. What benefits and horrors can we expect from the revolution already underway to understand the complete genetic code?

I'll conclude with one small example. Despite it not being "spiritually correct," I enjoy watching professional football on TV. I usually hope for a close game with plenty of action. Occasionally, I find myself rooting for one team. I urge them on to victory—and even try to exert mental influence through my TV set. I catch myself and wonder what I am doing. "Hey, these guys are getting millions of dollars to beat each other up, what do I care who wins?" After a little reflection, I realize that "my teams" are those I have some connection with, even it if is only because they are from the state of New York or I go through the Pittsburgh airport on most of my flights. These flimsiest of connections give me affection and concern for those gladiators.

What would happen if I could more deeply appreciate the profound interdependence implied by the Middle Way? What would happen if I could more deeply appreciate, as more than interesting physics, how the irreversible processes that sustain and destroy our lives occur because of our connection to the first few minutes of the Big Bang and the continuing expansion of the universe? Then how much would my loyalties expand? If I could appreciate that the relativity of time is logically extended to all my subjectivity, then how could I rationally support my selfishness and self-cherishing?

It is overwhelming to think about extending my loyalties beyond a small circle of family and friends to the cosmos. Now that we know of more planets outside our solar system than within, does the Bodhisattva vow of working for the liberation of all sentient beings embrace even those beyond our solar systems? Surely, experiencing the sadness of more parents and their mortally sick children would crush me. How then can I possibly cultivate compassion on a cosmological scale?

Perhaps the ecological activists can offer guidance. In the face of daunting global ecological problems, they advise us to "think globally and act locally." Following their counsel, I try to keep the cosmological picture in mind and simultaneously act in the present with the person in front of me. Then it seems small ripples of compassionate action gradually flood beyond my little circle of family and close friends. The ideal is to extend our concern out in ever widening radii, until it encompasses more and more of the great suffering body of humanity. If, in fact, I lack inherent existence, then my present limitations are not fixed, in place for eternity, and I can work toward this ideal. Let us begin to widen the circle of concern beyond the narrow confines of *our* team and *our* friends. How else can we live with that devouring tiger of time that inexorably includes our final irreversible process?

Notes

It is a pleasure to thank Professor B. Alan Wallace for inviting me to present these ideas. As always, I offer special thanks to my consort, wife, and best friend, Elaine Mansfield, for her careful reading and suggestions for improvement on an earlier version of this manuscript. I warmly thank Devon Cottrell and Andrew Holmes of Carmel, California for several useful comments and encouragement on an earlier version of this paper. I offer my deep gratitude to His Holiness the Dalai Lama for encouraging the dialogue between Buddhism and science and showing the power of wisdom and compassion in action. Finally, I offer my deepest gratitude to the late Anthony Damiani, founder of Wisdom's Goldenrod and great exponent of dharma in many forms, who ignited our desire for some personal realization of wisdom and compassion.

The epigraph to this essay is from Jorge Luis Borges, "A New Refutation of Time," in D. A. Yates and J. E. Irby, eds., Labyrinths: Selected Stories and Other Writings, p. 234 (New York: New Directions, 1964).

1. See Walpola Rahula, *What the Buddha Taught* (New York: Grove Weidenfeld, 1974), pp. 95–97.
2. The time interval $\Delta t = \Delta t_0 / \sqrt{1 - (v/c)^2}$, where Δt_0 is the rest frame value (seventy days in our example) and v/c is the relative velocity between the system and the observer divided by the speed of light, c.
3. Victor Mansfield, "Time in Madhyamika Buddhism and Modern Physics," *Pacific World Journal of the Institute of Buddhist Studies* 11/12 (1995, 1996): 10–27. Available at http://lightlink.com/vic/time.html.
4. P. C. W. Davies, "Stirring Up Trouble," *Physical Origins of Time Asymmetry* (Cambridge: Cambridge University Press, 1994), pp. 119–130.
5. Victor Mansfield, *Synchronicity, Science, and Soul-Making* (Chicago: Open Court, 1995); Mansfield, "Time in Madhyamika Buddhism and Modern Physics."
6. Jay Garfield, *The Fundamental Wisdom of the Middle Way* (New York: Oxford University Press, 1995), p. 257.
7. David Bohm, *Wholeness and the Implicate Order* (London: Routledge and Kegan Paul, 1983), p. xi.

In the preceding essay Victor Mansfield emphasizes the importance of developing a coherent worldview that can guide our science as well as our moral actions in the modern world, and this quest has been rigorously pursued in the following essay by Michel Bitbol. Drawing on the philosophy of Kant and Nāgārjuna, and the discoveries of quantum mechanics, he presents here an alternative to the nihilism that is explicit in many versions of cultural relativism and implicit in scientific materialism.

In comparing the Kantian notion of the noumenon with the Madhyamaka theory of emptiness, Bitbol, following Jacques May, rightly argues that emptiness can in no way be construed as an underlying ground of phenomena, for this would entail a reification of emptiness. And that move is systematically avoided in Madhyamaka writings. Another important difference is that, according to Kant, the noumenal ground of phenomena is forever unknowable, whereas the experiential realization of emptiness is a central goal of Mahāyāna Buddhist practice. Indeed, such contemplative insight is crucial for healing the mind of all its afflictions, such as craving, hostility, and delusion, which arise as the result of grasping onto the true, inherent existence of phenomena.

When one gains nonconceptual insight into emptiness, what appears to the mind? Does one, as Jay Garfield argues, see the very same world experienced by everyone else, but perceiving it now as empty, dependent, impermanent, and nonsubstantial? If all phenomena come into existence by the power of conceptual designation, as advocated by Candrakīrti and other proponents of the Prāsaṅgika Madhyamaka school, then what exists for a mind that has ceased (at least temporarily) to conceptually designate anything? According to some Mādhyamika contemplatives, the phenomenal world vanishes.

Is the Madhyamaka radically inconsistent with any notion of a transcendent ground of being? According to Dzogchen, alluded to in the concluding essay of this volume by Piet Hut, there is an ultimate ground of both saṃsāra and nirvāṇa, and many proponents of this view, including H. H. the Dalai Lama, claim that it is perfectly consistent with the Madhyamaka. This ultimate ground is variously called "primordial awareness" and the absolute space of phenomena (dharmadhātu) in Dzogchen literature. Perhaps the relation between emptiness and primordial awareness can be understood by way of the following analogy. While *dreaming, you may encounter a Mādhyamika teacher who leads you to realize the noninherent nature of all the subjective and objective phenomena appearing in the dream. You recognize their empty nature and their "dreamlike" existence, as these phenomena appear to be real but in fact are not. Thereafter you encounter a Dzogchen teacher who leads you to the realization that the world you experience is not merely dreamlike but is actually a dream. This insight brings you to a state of complete lucidity: while dreaming you now recognize, with your waking intelligence, that you are dreaming. And you now realize that all the phenomena in the dream are actually manifestations of your waking consciousness, which is analogous to primordial awareness.*

In the following essay Bitbol comments on the resistance of many scientists to question their realist interpretations of scientific theories, despite so much evidence against this position, especially from quantum mechanics. One crucial reason for this is that whoever speaks with greatest authority about the real nature of the universe commands enormous power, influence, and financial resources in society. In medieval Europe this was the Roman Catholic Church, which insisted on a realist interpretation of scripture but an instrumentalist interpretation of scientific theories, such as Copernicus's heliocentric view. Now the holders of such power are the institutions of science, which are also the main bastions of the creed of scientific materialism. In this paper Bitbol attempts to weaken the intellectual ground of this dogmatic position and provide us with a hint as to how to integrate the many strata of our scientific, philosophical, and spiritual understanding in today's world.

Michel Bitbol

A Cure for Metaphysical Illusions: Kant, Quantum Mechanics, and the Madhyamaka

My purpose in this paper is to show that the transcendental approach, first formulated by Kant, and then elaborated by generations of neo-Kantian thinkers and phenomenologists, provides Buddhism in its highest intellectual achievement with a natural philosophy of science. I take this highest achievement to be the Madhyamaka dialectic and soteriology,[1] which was developed in India from the second century C.E. to the seventh century C.E. by masters such as Nāgārjuna, Āryadeva, and Candrakīrti.

Yet I am aware that we are likely to meet obstacles in the course of this attempt at establishing a threefold relation between science (especially modern physics), transcendental philosophy, and Madhyamaka Buddhism. Every possible mutual relation between these three terms has been studied in the recent past, and each one of them has raised serious doubts. My preliminary task, in sections 1, 2, and 3, is therefore to locate the obstacles. Then, in section 4, I suggest a promising way to overcome these obstacles. In sections 5, 6, and 7 I use the ideas developed in section 4 to give three examples of a possible synergy between a neo-Kantian philosophy of science and the Madhyamaka. One example concerns reifications in particle physics, another one develops the dialectic of determinism and indeterminism in various readings of quantum theory, and the last one deals explicitly with the concept of *relation* in quantum mechanics.

1. KANT, MODERN PHYSICS, AND THE MADHYAMAKA: THREE DIFFICULTIES FOR A COMPARISON

To begin with, what are the obstacles?

First, the relevance of Kant's philosophy for modern physics has repeatedly been challenged during the first half of the twentieth century by the very creators of the new theories. According to Kant, space, time, causality, etc. are "norms" imposed in advance by our sensibility and understanding onto the "matter" of sensations. These forms are supposed to hold true "for all times and for all rational beings."[2] But modern physics has undermined this invariability clause. According to Einstein, general relativity has jeopardized an important aspect of Kant's a priori forms of sensibility (space and time) and, according to Heisenberg, quantum mechanics has shown the lack of universality of Kant's a priori forms of thought (the categories of substance and causality). A large majority of philosophers of science currently accept these claims. Following the pioneering work of Moritz Schlick, Emile Meyerson, and Hans Reichenbach in the 1920s, they thus agree that most features of Kant's original a priori forms are outmoded, or at least that their validity is restricted to the cognitive ordering of the local mesoscopic[3] environment of humanity. Moreover, these philosophers implicitly consider that this failure of Kant's original philosophy of physics condemns any renewed transcendental approach of modern physics.

Second, a huge amount of work has been done in order to draw parallels between the most striking features of modern physics and several trends of Eastern thought, including the Madhyamaka. But a high proportion of this work (e.g., Fritjof Capra's, Michael Talbot's, or even David Bohm's) was disparaged by the academic world at the very time it was arousing a large popular interest. Undoubtedly, part of this academic discredit was due to an overestimation of science as the only acceptable source of truth. Another part of it expressed a misapprehension of the high rational standards of many schools of Eastern philosophy (especially the Madhyamaka), which triggered a spurious fear of "obscurantism." But there were also good reasons to distrust the most popular parallels between science and Eastern philosophies. One of these reasons was the poor methodological background of the attempted comparisons. No systematic assessment of the difference of status between the two terms to be compared was made, no discrimination of the points on which the confrontation between physics and Eastern spiritualities do or do not make sense was undertaken, and no clear idea of what can or cannot be expected from the comparison emerged. With a few recent and remarkable exceptions,[4] this type of reflection thus resulted in little more than mere analogy at an ill-defined level of the two discourses, with obvious apologetic purposes.

Third, T. R. V. Murti already proposed, years ago, a Kantian reading of Madhyamaka thought.[5] But this reading raised a series of sound objections which were remarkably expressed by Jacques May,[6] and by other authors.[7] I will elaborate on this problem in the two following sections, for it has been less documented than the former ones. But readers who are more interested in solutions than in problems can perfectly well skip sections 2 and 3. After all, the aim of this essay is to undo some conceptual knots of our current belief system; it is not to indulge in philosophical technicalities.

2. KANT AND THE MADHYAMAKA: SOME SIMILARITIES

The manifest resemblance between Kant's critical philosophy and the Madhyamaka system bears on at least four points:

1. The Madhyamaka is intended, even etymologically, as a middle way between absolutism and nihilism, that is, between the view of an absolute self-subsisting reality and the view of no reality. Due to its insistence on holding no view about reality,[8] it was wrongly accused of holding a no-reality view.

Similarly, Kant's transcendental philosophy was construed from the outset as a middle way between dogmatic rationalism (which tends to identify the ideas of reason with absolute realities) and sceptical empiricism (which radically challenges the claim of reason in regard to the possibility of gaining anything like objective knowledge). Due to his strong criticism of dogmatic transcendent realism, Kant was wrongly accused of defending a form of subjective idealism.

2. The Madhyamaka and the philosophy of Kant both involve an analysis of the dialectic of reason. On the one side, Nāgārjuna undertakes a systematic rejection of all opposing metaphysical views either by a form of logical reductio ad absurdum or by pointing out an absence of empirical proof.[9] On the other side, Kant develops an analysis of the internal conflicts of pure reason, which culminate in the so-called antinomies.

At a more detailed level, one may notice a striking equivalence between (i) Kant's first cosmological antinomy and (ii) Nāgārjuna's symmetric rejection of the view that the world is limited and of the view that the world is

unlimited.[10] Even the ways in which Kant and Nāgārjuna explain the antinomic character of any assertion about the world taken as a whole are surprisingly close to one another. Kant insists that since cosmological questions bear on an ideal absolute totality, namely, on a closed and static entity
called "the universe," they transgress the bounds of human experience. Indeed, for human beings, significant questions can concern only the open series of phenomena and the unended progress of knowledge tending toward
a synthesis.[11] As for Nāgārjuna, he similarly suggests that the reason neither
the finitude nor the infinitude of the world as a whole makes sense is that
the world should not be construed as a single absolute entity of which
something can be significantly predicated. If anything, the world should
rather be construed as an indefinite "series of flickering events"[12] compared
to the flame of a butterlamp.

    3. Kant restricts the validity of the concepts of our pure understanding,
such as the category of substance or causality, to the formal ordering of the
empirical contents; he also restricts the validity of the ideas of our reason to
a "regulative" use, namely, to providing us with an inaccessible goal (a *focus
imaginarius*) that motivates the unended process of the ordering of phenomena. If we do not recognize these restrictions, we can easily take the
form given by our intellectual faculties to phenomena for the form of the
things in themselves. We take the risk of projecting the a priori structure of
the knowing subject onto the world, thus mistaking it for a pregiven worldly structure. This confusion defines what Kant calls the "transcendental illusion," which, unlike the ordinary empirical illusions, is all-pervasive and extremely difficult to recognize and to compensate for.

    On the other hand, unlike subjective idealists, Kant accepts that our
senses are affected by an "external" thing in itself taken as an absolute reality. He often explains that the *ground* of phenomena has to be found beyond
the immanence of the phenomena in a "transcendental object."[13] But this
affecting thing in itself, this ground of phenomena, is, by definition, beyond
any possibility of knowledge; it can but be for us a *noumenon,* a purely intelligible reality whose epistemological function is to be formally opposed
to phenomena.

    T. R. V. Murti then displays some analog features of the Madhyamaka
system. He insists that, in the Madhyamaka as in Kant's philosophy, "causality is of empirical validity only,"[14] that causality is not (and cannot be) a
process of substantial production giving rise to an intrinsinc being out of
another intrinsic being. Moreover, whereas the Madhyamaka invites us to

accept bodily forms as part of an empirical reality, it rejects them at the same time as not ultimate, not absolute.[15]

The basic illusion, in the Madhyamaka as in Kant's philosophy, thus amounts to taking the empirical reality as it is molded by our perceptive automatisms, basic presuppositions, concepts, and conventions for some intrinsic reality. Disclosing this illusion, Murti says, would mean "disabusing the mind of its presuppositions" and, to begin with, recognizing these presuppositions as such. Yet, criticizing the absolutization of the elements of the empirical world does not mean denying the existence of any absolute reality. In fact, according to Murti, the Madhyamaka system is a variety of absolutism.[16] But the absolute it sketches is "utter indeterminateness and non-accessibility to reason. . . . Even existence, unity, selfhood and goodness cannot be affirmed of it."[17] Murti concludes by equating boldly Nāgārjuna's distinction of two truths, namely, *saṃvṛti* (usually translated as "conventional")[18] and *paramārtha* (translated as "ultimate" or "absolute"), with Kant's distinction between *phenomenon* and *noumenon*.[19]

4. Denunciation of false absolutes is associated, in both Kant's philosophy and the Madhyamaka system, with a strong emphasis on relations, constitutive relativities, or relative existence.

Kant describes two classes of relations, which one may call "transversal" and "lateral." First, there is a "transversal" relation between the thing in itself and a knowing subject.[20] Second, there are direct "lateral" relations between consecutive perceptions, even though these perceptions admittedly arise from a human subject's being "transversally" affected by the thing in itself.

According to Kant, in the empirical world we only know relations between phenomena.[21] Matter itself is construed by him as a bundle of relations, since the only characteristics by which it manifests itself are (attractive or repulsive) forces. Developing this conception systematically in the *Transcendental Analytic* of his *Critique of Pure Reason*, Kant replaces any statement of inherence (say about substance or productive causality) by a corresponding a priori law of succession of phenomena. These laws, imposed onto phenomena by our understanding, are constitutive of objectivity. Indeed, objectivity is understood by Kant as universal validity, for any subject, of a certain mode of relational organization of phenomena rather than as intrinsic existence.

Not surprisingly, the most important categories among those that impose laws onto phenomena are those falling under the rubric *relation*. They are derived from the class of judgments that state relations either between a

predicate and its subject, or between a premise and its consequence, or between the terms of a disjunction. In the empirical network of interrelations
the only elements that can be taken as absolute are the very principles that
govern the relations among phenomena, since they are the conditions of
possibility of there being an experience of phenomena at all. To summarize,
one could say that, according to Kant, we have access only to phenomenal
relations that are themselves in turn constituted by a basic epistemological
relation.

Now, we find a fairly similar pattern in Madhyamaka thought. Despite
some obvious differences with Kant, to be discussed in the next section, the
idea of a constitutive epistemological relation can, for example, be recognized in Nāgārjuna's following remark: "Someone is disclosed by something. Something is disclosed by someone. Without something how can
someone exist? Without someone how can something exist?"[22] This sentence, and other ones in the same chapter or in other texts,[23] can easily be
understood as a way of emphasizing "the corelativity and interdependence
of subject and object."[24] More generally, as it is well known, Nāgārjuna considers that emptiness, understood as universal reciprocal relativity (or "dependent coarising"), is the very condition of existence of empirical (or
"conventional") reality. Disclosing the true nature of this reality here means
only perceiving it as empty, or as constituted by reciprocal relations of dependent coarising. Both the absolutists who think that existence can only be
intrinsic and the nihilists who think accordingly that their denial of intrinsic existence amounts to a denial of existence *tout court* are thereby re-
butted. Their construal of causality as a process of metaphysical production
having been extensively criticized, it is replaced, in the Madhyamaka system
as in early Buddhism, by lawlike codependence of consecutive forms.[25]

3. KANT AND THE MADHYAMAKA: THE DIFFERENCES

Jacques May and other authors have given some reasons to regard virtually every point of the former parallel between Kant's philosophy and the
Madhyamaka as approximative. They have thus reached the conclusion that
there could be only superficial analogies between these two systems of
thought whose status and aims are utterly different.

There are, to begin with, many noticeable differences between Nāgārjuna's and Kant's dialectic.

To be sure, Nāgārjuna's dialectic is much more radical than Kant's.

Whereas Kant carefully analyzes the antinomies, and considers them to be unavoidable (but unwelcome) consequences of an otherwise valuable functioning of reason, Nāgārjuna treats them, according to J. Garfield, as nothing other than pairs of "nonsensical verbal formulations."[26]

Even the structure of the dialectic is much stronger in the Madhyamaka system than in Kant's philosophy.

In his studies of the (cosmological) antinomies Kant shows that we are able to derive two mutually contradictory conclusions from principles that are selected according to conflicting interests of reason. He then divides such couples of contradictory conclusions into two classes. In the first class the two conclusions (i.e., that the world is limited and that the world is unlimited) are both necessarily false, because they both apply to an ideal totality whose concept goes beyond any possible experience.[27] In the second class the two contradictory conclusions (i.e., that there is free will and that everything is ruled by the law of nature) are both true, because each one of them expresses a partial but significant aspect of the situation. To summarize, Kant displays either negative or positive *dilemma*.

By contrast, Nāgārjuna most often uses a negative *tetralemma*. In a tetralemma he denies (by challenging their logical coherence [*na yujyate, nopapadyate*] or their factual relevance [*na vidyate*]) the four following forms of a thesis:[28] P, ¬P, P& ¬P, ¬P& ¬ ¬P. Moreover, unlike Kant in his second type of antinomy, Nāgārjuna is careful not to endorse any one of the available theses. Indeed, even if a thesis were to express a significant aspect of a situation, this would mean that its truth is relative to a certain point of view and that it is thus, once again, merely "conventional."

Another difference concerns the situation of dialectic in the respective systems. In Nāgārjuna's *Mūlamadhyamakakārikā*, dialectic is all-pervasive; it is already there in the first verses of the first chapter, which state a basic tetralemma about causality. But in *Kant's Critique of Pure Reason* the *Transcendental Dialectic* comes quite late, after the *Aesthetic* and the *Analytic*. Even if one agrees with Murti that the dialectic could have been Kant's starting point,[29] this factual difference of order and emphasis cannot be ignored. It manifests very clearly that Kant's priorities are diametrically opposed to Nāgārjuna's. When Kant marks the bounds of reason, his purpose is to secure the mathematics and the (Newtonian) science of nature within these bounds. It is to drive back the illusion outside the bounds of a proper application of reason in order to provide science with a new illusion-free foundation. Here, again, a typical Mādhyamika thinker such as Nāgārjuna is much more radical. For him illusions arise not only from an extension of concepts beyond their empirical domain but also from their application to this domain. Indeed, the very fact that these concepts are successfully used within their range favors forgetfulness of their having a merely pragmatic-conventional value. As May points out, the Madhyamaka does not try to document the empirical validity of concepts, but rather to convey their nonvalidity at the ultimate level.

To recapitulate, Kant's dominant intention was to provide objective scientific knowledge with firm (though not ontological) ground. But Nāgārjuna's exclusive mission was to free everyone from the spell of a reified conventional truth, including science understood as an exceptionally efficient (but thereby also exceptionally liable to reification) part of pragmatic-conventional truth.

Last but not least, the divergences about the ultimate and its status should not be minimized. In Kant's *Transcendental Aesthetic* there appears to be a relation of transcendence between the affecting thing in itself (the ultimate or the absolute) and the affected human subject. This dual relation is underpinned by a presupposed duality of (sensory) matter and (intellectual) form of knowledge. Indeed, in the wake of Kant's "Copernican revolution" the form of the experienced world is specifically ascribed to the subject in general, not to the object(s), whereas the so-called matter of knowledge is considered as the byproduct of a subject's senses being affected by the thing in itself. A strongly dualist structure thus persists in the *Critique of Pure Reason*, despite many opposite tendencies such as the criticism of the idea of a substantial self (the soul) in the paralogisms of pure reason. In that respect Kant's philosophy is predominantly an epistemology. Ontology creeps in only when ethics is at stake. In the latter context Kant ascribes free will to the subject as a thing in itself, and determination by the law of nature to the subject as a phenomenon.[30]

By contrast, says May, the Madhyamaka can by no means be construed as an epistemology. It is, so to speak, ontological from the outset, despite the fact (i) that its exposition of ontology is apophatic[31] rather than dogmatic and (ii) that this exposition has the status of a factor of transformation in being, not of a discourse on being. In the Madhyamaka a modification in attitude and in knowledge is also an internal mutation of being (or rather, if one wishes to impute some slightly less inappropriate words on it, a change in the direction of becoming). Thus one cannot say, as Murti does in order to strengthen the parallel with Kant, that the function of *prajñā* is to induce "an epistemic (subjective), not an ontological (objective)"[32] change, even less that *prajñā* prompts a transformation of our attitude, not of the real. In the Madhyamaka the duality of subject and object is empty; it is part and parcel of the "conventional" truth (it is even one of its most crucial departure points). It is not asymmetric and hierarchical, as Kant's dualities of form and matter or of subject and thing in itself; it is rather symmetric and reciprocally coconstitutive. One central function of *prajñā* is appeasement of the latter duality in the immanent flux of becoming of which we are participants; it is therefore flatly mistaken to see it as operating at a purely subjective level.

This being granted, it is clear there can be no relation of transcendence in the Madhyamaka system, as some expressions of Kant suggest there is between phenomenon and noumenon. Not even between *saṃvṛti* and *paramārtha* or between *saṃsāra* and *nirvāṇa*. In no way can one say, for instance, that the ultimate truth points toward an underlying ground of the conventional truth, by analogy with Kant's referring to the thing in itself as the ground of empirical reality. The lack of any such relation of transcendence in the Madhyamaka thought is made as clear as possible by Nāgārjuna, in the following celebrated sentence: "There is not the slightest difference between *saṃsāra* and *nirvāṇa*."[33] Everything, including *nirvāṇa*, is embedded in the same immanent plane, in the same network of relative coarising. To be in *nirvāṇa*, according to J. Garfield, means seeing the very same things that appear to the deluded consciousness of *saṃsāra*, but seeing them "as they are—as merely empty, dependent, impermanent, and nonsubstantial." It does not mean "to be somewhere else, seeing something else."[34] An even less inappropriate expression could be found by avoiding the expression "seeing as," which still conveys an epistemological connotation, and replacing it by "living as" or "being as." This would help to dissolve any residual picture of a transcendence.

4. FUNCTION RATHER THAN ANALOGY: A METHODOLOGICAL TURN

At this point one can see that part of the difficulties that hinder these comparisons arises from a static and reified conception of discourses and doctrines. No serious use of the evolution of the doctrines can be made if, from the outset, they are considered to be immutable claims of truth. Further, analogies between two closed systems of thought fail to be convincing if one does not display extensive elements of isomorphism bearing on their contents, presuppositions, and scope. No other relation than similarity and dissimilarity is conceivable between them.

Another part of the difficulties comes from a dominant representational conception of knowledge. It was implicitly accepted in some of the previous analogies that comparing two theories means showing that they provide the same picture of the world. It therefore proved easy to criticize these analogies by showing that the pictures (i.e., that of modern physics and that of the Madhyamaka) are only superficially similar and that they are based on very different bodies of evidence.

But giving up the static and representational outlook is likely to allow a thorough renewal of our conception of the threefold relation between modern physics, transcendental philosophy, and the Madhyamaka. Let us then consider a scientific theory or a system of thought as an operator within an open network of practices, rather than as a closed set of truths or as a (more or less) faithful representation of a pregiven reality. Let us construe scientific theories as operators of structuring our actions within the world and of anticipating their outcomes. Let us construe philosophical doctrines as operators of mutual adjustment between our possibilities of action (stated by scientific theories) and the set of values, scopes, and representations that define our culture. And let us construe the Madhyamaka dialectic: (i) as a patient reminder of the all-pervasive impermanence and emptiness of appearances and, accordingly, (ii) as a universal operator of self-transformation.

In this case establishing a relation between modern physics, Kant's philosophy, and the Madhyamaka does not amount to displaying their strict isomorphism; it means showing that, as operators, they fit well enough to be articulated into a higher-order, broad-range operator. Here, the analogies have no value by themselves; they are only signs indicating the most appropriate locus of articulation between the operators. Moreover, insofar as they are nothing but tools (operators) the three terms to be related must be taken as plastic and evolutive; each term has to be seen in the context of its history, of its potential developments, and of the dynamics of its possible coadaptation to the other terms rather than treated as a closed doctrinal system.

True, the widespread trend toward strict separation of domains between science, philosophy, and religion, which culminated at the end of the nineteenth century, may make this idea of a higher-order integrated operator look quite odd. But, actually, partial integrations work daily in the making of science, philosophy, and broader outlooks (or "forms") of life. Science is driven by extrinsic values, aims, motivations, epistemological conceptions, or metaphysical pictures, and it modifies them retroactively.[35] So much so that saying that scientific theories are nothing but guiding operators of action usually seems too narrow a characterization. On the other hand, philosophy is constrained (though underdetermined) by scientific advances at the same time that it provides scientists with general directions of research. As for religious dogmas and forms of life, they have either been shaken by changes of values, behavior, and representations related to science or forced to protect themselves by community closure and explicit denial of some scientific theories.

Furthermore, nowadays it is widely accepted that, from the end of the middle ages to the first half of the eighteenth century, Western science was given its impetus by Christian theologies and, more indirectly, by simplified versions of Jewish and Moslem metaphysical speculations.[36] Disclosing the fabric of God was no small motivation for the dawn of science. Concepts such as the laws of nature or absolute space were directly derived from belief in an omnipotent and omnipresent God. And the dominant realist-representationalist philosophy of science was clearly favored by creationism associated with theological foundationalism. Such a genealogical link between representationalism and theological foundationalism holds true notwithstanding the fact that the first reaction of the Catholic Church at the time of Galileo was to confiscate the benefit of realism for its own dogma and to impose a purely instrumentalist status on science.

This historical many-level organization having been recognized, the call for separation that has prevailed since the second half of the nineteenth century can be read retrospectively as an expression of felt failure. It reveals the breakdown of the original compromise between science, a predominantly representationalist philosophy of science, and Christian theology. True, the separation, and the correlative feeling of failure, had some positive consequences: an increased concentration on specialized tasks and a better definition of the respective domains. But it also had very negative consequences: (i) a schizophrenic appraisal of indissociable aspects of human life and (ii) a variety of nihilism, as Francisco Varela defines it,[37] namely, a state of mind where we are perfectly aware that our system of values and beliefs is incoherent, but where we cannot do without it.

The components of this contemporary nihilism are well documented:

scientists who look toward religion for an ethical guarantee, even though they are deeply sceptical, philosophers of science who try to save realism at any cost in spite of the acknowledged resistance of modern physics, or who adopt empiricism with the bitter feeling of having renounced the very meaning of the scientific endeavor, and priests or monks who know deep down that the dogmatic and mythological component of their religion has become untenable (or merely allegorical) but see no other solution than maintaining it because they believe that to be a prerequisite of a truly religious stance (including morals, a contemplative life, and striving toward self-transformation).

Overcoming the failure and moving beyond nihilism is possible only if we identify a new higher-order operator articulating modern science, an alternative philosophy of science, and a nondogmatic soteriology, thus fitting globally with the essential aspects of contemporary human life. The multiple analogies that have been discussed previously can be seen as a few partial steps toward such a higher-order operator. But, as I have already pointed out, most of them are definitely clumsy because they rely on the very (static and representationalist) assumptions about doctrines and knowledge they purport to challenge. So our task now is to show in some detail how the many-leveled articulation can be secured: (i) by relying on the dynamic potentialities of doctrines and theories rather than on their canonical text, (ii) by fully recognizing their functional-operational status, and (iii) by disentangling, in the available unself-conscious presentations of scientific theories and philosophical doctrines, components coming from various layers of a half-forgotten but still efficient past higher-order operator.

The difficulties that hindered the attempts at establishing relations between modern physics, Kant's philosophy, and the Madhyamaka can be thus overcome. Let us take them in the same order as in section 2.

First, the obvious discrepancy between Kant's original a priori forms and some prominent aspects of modern physics does not mean that the very idea of a transcendental reading of science has failed. To see this one has only to come down to the central idea of the transcendental philosophy (below the particular shape that was given to it by Kant) and take into account its aptitudes to development as they have been displayed by the neo-Kantian philosophers of the nineteenth and twentieth century.

What is then the central idea of the transcendental philosophy? It is to construe each object of science as the focus of a synthesis of phenomena rather than as a thing in itself. And it is to accept accordingly that the very possibility of such objects depends on the connecting structures provided in advance by the procedures used in our research activities. Thus something is objective if it results from a universal and necessary mode of connection of phenomena. In other terms, something is objective if it holds true for any (human) active subject, not if it concerns intrinsic properties of autonomous entities.

Here science is not supposed to reveal anything of a preexistent underlying absolute reality, nor is it a more or less random aggregate of efficient recipes. Science is rather the stabilized byproduct of a dynamic reciprocal relation between reality as a whole and a special fraction of it.[38] Defining this special fraction of reality qua subject is the reverse side of its actively extracting objectlike invariant clusters of phenomena.

Somebody who shares this philosophical attitude is metaphysically as agnostic as empiricists, but as convinced as realists that the structure of scientific theories is highly significant. For, from a transcendental standpoint, the structure of a scientific theory is nothing less than the frame of procedural rationalities that underpin a certain research practice (and that, conversely, were constrained by the resistances arising from the enaction of this practice).

A conception of science based on this central idea is perfectly capable of developing nowadays, provided it drops the residual static and foundationalist aspects of Kant's system. Instead of accepting Kant's uniqueness and invariability claim about his forms of intuition and thought, one should acknowledge, as Hermann Cohen[39] and Ernst Cassirer[40] did, the possibility of change of the so-called a priori forms and their plurality as well. Recent flexible and pluralist conceptions of transcendental philosophy include Putnam's and Hintikka's transcendental pragmatism. According to Hilary Putnam, for instance, each a priori form has to be considered as purely functional (he also calls it a quasi a priori). Each quasi a priori is relative to a certain mode of activity, it consists of the basic presuppositions of this mode of activity, and it has therefore to be changed as soon as the activity is abandoned or redefined.[41] As for Jaakko Hintikka, he characterizes the transcendental philosophy, in a neopragmatist style, as a process of redirecting attention from the objects to our game of seeking and finding.[42] We shall see in section 8 that a neotranscendental philosophy of science developed along these lines is able to account for quantum mechanics to a much larger extent than either scientific realism or empiricism.

Second, the gap that separates science and the Madhyamaka, due to obvious differences of methods and scope, could be filled in only by a third intermediate term. This is the bridging function I ascribe to a neotranscendental philosophy of science (see my third point below).

But even before any precise assessment of this threefold articulation is attempted one should identify the level at which an articulation, be it indirect, between a scientific theory and a dialectical-soteriological system is acceptable at all. To begin with, one must avoid the temptation of drawing from modern science a sort of monolithic official mythology, in order to display its superficial analogies with a popular Eastern mythology. Instead, one should insist on the manifest underdetermination of scientific theories and models by experiment, and on the fact that, in the history of science, this underdetermination was de facto removed by additional, extra-empirical, constraints. These additional constraints were provided by a demand of coherence between new theories and an older philosophical background[43] whose roots are profoundly embedded in the (partly religious) Western forms of life.

The problem is that these traditional (philosophical) constraints, which have been so easy to cope with in classical physics, have begun to introduce tensions, difficulties, and paradoxes in modern (relativistic and quantum) physics. The traditional conception of a world made of separate material bodies bearing intrinsic properties has not been completely relinquished, but, in order to survive, it has assumed several hardly recognizable forms. The nonlocal hidden variable theories, whose archetype is Bohm's 1952 theory, is the most explicit one. But even the physicists who are most committed to the so-called Copenhagen interpetation still use remnants of the old mechanistic outlook together with fragments of a new nonmechanistic outlook. They use a versatile and flexible language that enables them to speak sometimes as if the particles were individual entities and sometimes as if they were nonindividual quanta of field excitation, sometimes as if objects had monadic properties and sometimes as if one had to think that they are only relational observables, sometimes as if it were possible to ascribe a state to a "physical system" made of a set of particles and sometimes as if the particles themselves reduced to states of the vacuum, and so on (see section 6 for more details).

The quicker solution to eliminate these difficulties and lack of conceptual unity (without resorting to a nonempirical world of hidden processes) would be to jettison both the mechanistic conception of the world and the dualistic epistemology. Unfortunately, there are deep-seated resistances to this seemingly extreme solution. Even our cultural familiarity with the most recent and radical varieties of transcendental philosophy of science (which, as we have seen, are pragmatic, dynamical, relationist, and nondualist) is not strong enough to make us take this step collectively.

But aren't these resistances related to our elementary creeds and forms of life? Aren't they due to our distress about losing ground, if we are left without a belief in a pregiven and prestructured reality? Would we not be deprived of our strongest motivation for making science if we did not have the regulative aim of disclosing a preexistent reality lying, so to speak, in front of us? At this point the Madhyamaka comes in. The Madhyamaka construed not as the purveyor of one more mythology, one more representation of the world, or one more philosophical doctrine but (i) as a patient dialectical deconstruction of the class of substantialist views and dualist epistemologies that we find so difficult to abandon and (ii) as a soteriology, namely, an introduction to a form of life in which losing ground is not a tragedy (it can even promote enlightenment . . .) and in which an alternative (say, pragmatic, integrative, and altruist) strong motivation can be given to science.

To summarize, the meeting point of science and the Madhyamaka is not a common view of the world. It is rather a tension between traditional views of the world and the recent advances of science, which can be formally avoided by transcendental philosophy and relaxed at the deepest level by the Madhyamaka dialectic and soteriology.

Third, some of the discrepancies that were pointed out between Kant's philosophy and the Madhyamaka are not as insurmountable as they appear to be. In order to overcome them one has only to be sensitive to the evolution of Kantian and neo-Kantian thought.

Let us consider for instance the difference between Kant and the Madhyamaka on the status of the ultimate. As we know, Kant's position on this point apparently involves a remnant of substantial dualism (between the thing in itself and the affected subject). As a consequence, a kind of transcendence seems to be ascribed to the thing in itself. By contrast, Nāgārjuna does not consider any other form of epistemological duality than a purely functional-relational one. The duality of subject and object, of perceiving and perception, is not denied, but it is shown to be empty, that is, to arise from a symmetric relation of mutual dependence. Nāgārjuna's critical analysis is thus maintained on a strict level of immanence throughout.

However, Kant's position on this point is much less elementary than what can be inferred from a selective reading of certain texts (such as the *Transcendental Aesthetic* of the *Critique of Pure Reason*). At the end of the *Transcendental Analytic* one finds that the concept of noumenon is only a limitative concept, that it only points obliquely toward the finitude of our sensibility, and that its use is therefore only negative.[44] Some commentators then explain that Kant's thing in itself is nothing beyond the representation, nothing other than the brute fact of this representation (of its givenness, of its not being arbitrarily produced by a deliberate act of our will).[45] And thus the last shadow of dualism disappears.

Later the Marburg school[46] of neo-Kantian philosophy developed an even more explicitly immanentist position. Against substantial dualism, Cassirer recommended that one not construe subject and object as a pair of ontologically closed entities. He rather insisted on a purely methodological distinction between a function of subjectivation and a function of objectivation in the process of cognition.[47] He then stated, after Cohen, the idea of a "reciprocal cobelonging" of the concepts of subject and object. Against transcendence, Paul Natorp also argued that there is no external standpoint from which a relation of causality can be established between a thing in itself and our senses. We can thus see how, in the course of its development, transcendental philosophy has come closer and closer to a crucial feature of the Madhyamaka.

Of course, there remains a momentous difference of scope between them. As we know, neo-Kantian philosophies aim at securing the validity of objective scientific knowledge in its specific domain. But the Madhyamaka has another priority. This priority is to locate science as an integral part of conventional truth and to free us from the temptation of taking any part of conventional truth for an absolute truth. Such a difference clearly invalidates simple analogies or straightforward identifications, but it cannot prevent us from establishing both a relation of complementarity and an operational articulation between the two systems.

COMPLEMENTARITY

Saying, as the Madhyamaka does, that scientific knowledge has only a conventional validity is not tantamount to denying it any validity whatsoever.[48] Exploring the extent and limits of this (admittedly conventional) validity, as transcendental philosophy purports to do, is thus worthy of the effort in a Madhyamaka context. Has not Nāgārjuna pointed out that "without a foundation in the conventional truth, the significance of the ultimate cannot be taught"?[49] In that respect the Madhyamaka system and transcendental philosophy are complementary.

OPERATIONAL ARTICULATION

In exquisite detail transcendental philosophy shows that the credibility of scientific knowledge is in no way based on its correspondence with some immutable absolute reality, but rather on the consistent mutual relation between the processes of defining invariants (objectivation) and setting apart the noninvariant residue (subjectivation). This may well contribute to the effort made by the Mādhyamika masters to dispel reifying illusions; for, in our culture, science is the most powerful source of these illusions. Challenging reification in the domain of science is likely to lower the triggering threshold of the sought after disabusing chain reaction.

Conversely, in the frame of life and thought that is likely to emerge from a self-transformation performed in the direction indicated by the Madhyamaka, an antifoundationalist, immanentist, relationist philosophy of science, such as the neo-Kantian, would be immediately acceptable. The very existential roots of the widespread resistance of those scientists who are afraid to lose their landmarks and their motivation by adopting it would indeed be cut.

In this respect the Madhyamaka system and transcendental philosophy are potentially synergetic, and they are therefore predisposed to operational articulation.

5. ONTOLOGICAL ILLUSIONS IN MODERN PHYSICS

To sum up, whereas neo-Kantian philosophy is concerned with revealing the detailed mechanism of reifying illusions in science, the project of the Madhyamaka Buddhist community is to dispel them from the outset. As we shall see in this section, the purely intellectual stance of neo-Kantianism may have been superficially sufficient in the context of classical physics, but, in the context of quantum physics, the need for a synergetic association with the existential stance of the Madhyamaka becomes manifest.

Let us return to the essential mechanism of the reifying illusions. It consists in projecting upon nature the commitment of human beings to the practices that enable them to relate to their environment and to live in it.

The man in the street is committed to the objects of his action and discourse, and this commitment gives rise to what Arthur Fine[50] called the Natural Ontological Attitude (NOA). As for the scientist, she is committed to the postulated objects of her experimental practice as well as to the heuristic guides of this practice. The latter commitment is not independent of the former one, for it often extrapolates its basic features. It gives rise to a scientific version of the NOA that is so deeply entrenched it tends to resist at any cost. The scientific version of the NOA especially resists the rising tide of tensions and paradoxes induced by its being stubbornly imposed onto modern physics.

Can we do something to overcome this sort of illusion? Kant and his followers did not think so. They believed that nothing can be done beyond mere intellectual recognition of the transcendental illusion. According to them, we can know intellectually that certain subjective rules are mistaken for objective determinations of the things in themselves, but we cannot help seeing the world as if it inherently possessed these determinations[51]—exactly in the same way as an astronomer cannot help seeing the moon bigger when it is close to the horizon than when it is at its zenith, although she knows intellectually the optical mechanism of this illusion.

This rather pessimistic view is clearly expressed in the last part of the *Critique of Pure Reason.* But it is already latent in the first chapters of the *Critique,* where the "constitution" of valid objective knowledge by means of the forms of our sensibility and understanding is at stake. One can see this in the way Kant minimizes the implications of his philosophy for men in the street and scientists. On the one hand, in the *Transcendental Aesthetic,* Kant states that space is not a concept abstracted from our outer experiences but rather the a priori form of all intuitions of the bodily objects that we take as external to us. It is only if this subjective status of space is accepted, he writes, that one can understand how it is possible to acquire knowledge of the necessary propositions of geometry. For an a priori knowledge of necessary truths can only be about something we ourselves produce. But, on the other hand, Kant also explains that, with respect to any possible human experience, everything remains exactly as if (*als ob*) space were an intrinsic feature of the world.[52] The critical attitude thus stems from the meta-standpoint of the philosopher, and it proves mostly irrelevant from the ordinary standpoint of the man in the street or of the scientist. The philosopher is aware of the as if clause, whereas the man in the street and the scientist just make use of it unself-consciously.

This dual, not to say schizophrenic, analysis may have been acceptable as long as the as if procedures worked without too many discrepancies (e.g., in classical physics). Indeed, the internal coherence of the ontological-like discourse of classical physicists made it quite easy for them to forget the as if clause. But, in quantum physics, discrepancies have become so glaring that in order to save something of the Natural Ontological Attitude, especially something of the favorite ontology of material bodies, one needs tortuous (and thus too visible) strategies.

As I have suggested in section 4, these strategies include: (i) flexible use of the substantives and predicates in particle physics, (ii) implementation of new logic or new (quasi) set theories, (iii) call for future theories endowed in advance with the aptitude of solving the paradoxes of quantum mechanics, or (iv) hidden variables theories that carry on some basic features of the classical mechanics of material points.

Let us explore briefly two of these available strategies. One concerns predication, and the other reference. The first one is quantum logic and the other one is the particle label approach. Both of them reveal a strong philosophical and cultural bias in a situation where underdetermination of theories by experiments prevails.

Predication was already perceived as a problem during the period of emergence of quantum mechanics. At first, the formulation of this problem was quite clumsy. In 1927 Heisenberg and Bohr insisted on the fact that, due to the indivisibility of the quantum of action, no phenomenon may be observed without disturbing it appreciably.[53] Therefore, saying that a phenomenon merely reflects a predicate possessed by the (micro-)object is quite dubious. But a few years later (especially from 1935 on, after the celebrated Einstein-Podolsky-Rosen paper), Bohr became increasingly suspicious about the concept of disturbance. As he noticed in 1954, "one sometimes speaks of 'disturbance of phenomena by observation,' or 'creation of physical attributes to atomic objects by measurement.' Such phrases, however, are apt to cause confusion."[54]

Bohr was especially aware of the lack of coherence of the most widespread way of using this concept of disturbance. Indeed, speaking of a disturbance presupposes that something like a property of the micro-object exists in nature, ready to be "disturbed" by the observing agent; it is thus difficult to invoke disturbances in order to prohibit (as some members of the Copenhagen group did) any reference to intrinsic properties of objects. Even worse, supposing that properties preexist but that they cannot be

A Cure for Metaphysical Illusions

known because of disturbances is tantamount to accepting that our knowledge of the hypothetical properties is incomplete and encouraging some physicists in their search for hidden variables. Bohr therefore insisted less and less on the crypto-dualist picture of disturbed (or created) properties and more and more on a holistic definition of the phenomenon in which the hypothetical contribution of the object cannot be dissociated from the contribution of the structure and irreversible functioning of the measuring apparatus.[55]

From the very beginning quantum logic was aimed at restoring realism in quantum physics against Bohr's views. Rather than sticking to "phenomena" or "observation" as Bohr did, quantum logic enabled one to recover the possibility of speaking of "physical qualities,"[56] or of properties of systems, at the cost of changing the algebra (namely, the combination by conjunction and disjunction) of these properties. Instead of a Boolean algebra,[57] one merely had to accept a non-Boolean "orthocomplemented non-distributive lattice."[58] So much so that the whole historical perspective was reversed by later quantum logicians. While history indicates that non-Boolean logic is the realist reply to Bohr's criticism of the ideal of a complete separation between an object and an observing agent, some quantum logicians asserted that "the rejection of the 'ideal of the detached observer' is the Copenhagen response to non-Booleanity."[59] Thus, according to these authors, the world is inherently non-Boolean, and Bohr's holism is a spurious epistemological interpretation of this ontological feature.

Unfortunately for them, however, there is much to be said in favor of Bohr's original standpoint. I personally tend to promote the following argument of simplicity. From the elementary supposition that phenomena are relative to their (sometimes incompatible) experimental contexts of appearance, it is easy to derive: (i) the full non-Boolean structure of quantum logic,[60] (ii) the quantization itself (through the commutation relations between conjugate variables), (iii) the wavelike aspect of certain distributions of discrete phenomena,[61] and (iv) features that concern the hypothetical bearers of properties, beyond the properties themselves.[62] This derivation does not require any well-defined assumption about the structure of the world (with the exception of the nonzero value of the Planck constant).

By contrast, starting from a detailed non-Boolean structure of the algebra of properties of the systems that constitute the world introduces a high amount of arbitrariness in the premises. The derivation of consequences from this kind of premise thus have little explanatory power.[63]

To recapitulate, even though the two starting points, namely, holism-relationism and inherent non-Booleanity, cannot be settled by experiments, there are many good reasons (especially economy, unity and explanatory power) to choose the first one. The only reason that may make the second one more attractive is that a realist interpretation of physical theories seems to be so unquestionably desirable in the framework of our Western view of the world that the best ampliative[64] arguments in favor of another interpretation lose weight. Everything, in the philosophical debate about modern physics, goes as if the following maxim were enforced: "Whenever a realist interpretation of a physical theory is available, you must adopt it, come what may."[65]

Another important case of underdetermination with philosophical bias concerns two views on the traditional bearers of predicates, the particles. According to the first view, the world is made of labeled quasi-individual particles whose momentum exchange mimics the empirical effects of fields, whereas according to the second view the world is made of fields whose nonindividual quanta of excitation mimic the empirical effects of particles. The two views can be made empirically equivalent in virtually every respect,[66] but, here again, they cannot fulfill to the same extent the standards of economy, unity, and explanatory power. Let me state some important differences of this kind between them.

1. In order to account for the quantum (Bose-Einstein and Fermi-Dirac) statistics, the quasi-individual particle view imposes a set of state-accessibility conditions: the restriction of states to their labeled symmetric and antisymmetric forms. But the quantum field view needs neither labels nor imposed restriction of the set of accessible states (only a generalized version of the algebra of commutators that underpins quantum theories). As P. Teller points out, the quasi-individual particle view has the defect of carrying a "surplus formal structure" (the labels) and of accepting a certain arbitrariness (in the choice of the accessible states).[67] Economy thus favors the quantum field view.

2. What the particle view calls "creation" or "annihilation," thus evoking ontological quantum jumps, is construed by the quantum field view as a continuous change of state that reveals itself discontinuously only in experiments.[68] The quantum field conception is thus clearly more in the line of the general rules of quantum theoretical treatment than the particle conception. Conceptual and formal homogeneity thus favors the quantum field view.

3. Both views must accommodate an indetermination in the number of micro-objects. However, they are not explanatorily equivalent. In the particle view this indetermination is imposed, but in the field view it arises quite naturally from the principle of superposition that holds for any quantum state. Furthermore, in the particle view one must cope with the baroque picture of individual objects whose number (and therefore whose being) is not definite, but the quantum field superposition of states may easily be understood as describing a propensity for the manifestation of various numbers of discrete relational events in a given experimental context.[69] Coherence of representations and, here again, economy of thought, thus favor the quantum field view.

4. The explanatory gap between the particle view and the quantum field view becomes even more striking when the problem of the so-called Rindler particles (or quanta) is at stake. The Rindler particles (or quanta) are observed by means of an accelerated detector, in situations (called the "vacuum state") wherein no particle at all is observed with nonaccelerated detectors. It is quite difficult to understand this phenomenon in the frame of the absolutist particle view, for a particle is supposed to exist (or to be devoid of existence) irrespective of the state of motion of the detector. But the Rindler phenomenon raises no problem in the frame of quantum field theory as read by Teller,[70] because each event of detection is here assumed to express a dynamic relation between the environment and the (accelerated or inertial) detector. The quantum field view is thus able to make us dispense with ontological questions, which become almost intractable in certain situations.

Why then should one keep on with the contrived particle view instead of adopting the much more natural relational-propensionist[71] reading of quantum field theory Teller proposes?

Most arguments in favor of the particle view rely on a demand of historical continuity of representations and concepts: historical continuity with classical physics, but also with the Natural Ontological Attitude of everyday life. M. Born already insisted, in his discussions with E. Schrödinger, on the importance of historical continuity between the concept of particle and the concept of material body.[72] As for Bohm's original hidden variable theory of 1952, which develops and transforms the mechanistic picture of a world made of a plurality of material points, it was explicitly motivated by an ideal of historical continuity, not only methodological but also conceptual, between the new theory and classical physics.[73]

True, the majority of realist philosophers of science currently accept that there cannot be an exact ontological similitude between two stages of the development of science. However, they still make a tacit use of what R. Harré[74] calls an ontological type hierarchy. It is usually this choice of developing a single ontological type hierarchy over history, which removes the empirical underdetermination of representations and replaces it by an effective determination. Let me give an example. Elementary particles are not mistaken for material bodies in microphysics, but the historical constitution of their concept, and the standard grammar of the expressions used about them by physicists, show that they belong to a well-characterized type hierarchy whose archetype is the material body of everyday life. The residual affinity of the concept of particle with the material body manifests itself most clearly in popular science, where precautions are dropped and familiar representations dominate again.

The difference between classical physics and quantum physics becomes easily perceptible at this point. In classical physics the type-hierarchical continuity between systems of interacting material points and the "things" of everyday life did not raise any difficulty. As we have seen earlier, awareness of Kant's as if clause was therefore confined to a little circle of philosophers and philosophically minded physicists. The ordinary physicist and the (Western) layman could stick quietly to their reifying and materialist picture of the world. But in quantum physics the distortions imposed by upholding the type hierarchy of material bodies are manifest; the conventional, or normative, aspect of this preservation can hardly be ignored by anyone, and the as if status of the substantive (particle)/predicate (state) mode of expression in the microworld then becomes all the more plausible. Moreover, adopting a radically different conception of physical theories, such as the relational-propensionist reading of quantum field theory, is an increasingly attractive option.

This being granted, the usual attitude, which consists in asserting that the world is made of inherently existent particles yet recognizing that this is not a satisfactory picture and adding lots of qualifications, clearly appears "nihilistic" in Varela's sense. A way out of this sort of nihilism is sorely needed. Now, in view of the previous analysis, the condition for taking (individually and collectively) the way out is nothing less than cutting the favorite ontological type hierarchy at its archetypal root, namely, the material body, recognizing (in the full existential strength of this verb) that the privileged status enjoyed by material bodies in our lives is due only to pragmaticalconventional reasons, seeing or living the all-pervasiveness of the *as if* clause in our material environment. This condition is difficult to fulfill in a Western context (except, perhaps, for a few phenomenologists able to practice the Husserlian "bracketing" of the Natural Attitude), but it becomes almost trivial in a Mahāyāna Buddhist context. Hasn't the Buddha himself "rejected the belief in matter"?[75]

6. A DIALECTIC OF DETERMINISM AND INDETERMINISM

From the standpoint of Western culture both Kant's dialectical critique of metaphysics and Nāgārjuna's symmetric rejection of "views" (*dṛṣṭi*) sound negative. They are felt as a renunciation of the grand project of *Episteme* inherited from the ancient Greeks. In this section I will try to show, on the contrary, that a dialectical reasoning may convey an important positive teaching and may lead one onto the edge of a renewed conception of knowledge.

My example of dialectical reasoning bears on determinism.

It is commonly accepted that the birth of quantum mechanics marks the triumph of indeterminism. But this word, *indeterminism,* is so ambiguous it has generated many misunderstandings about the status of quantum-mechanical laws. It is true that, in quantum mechanics, there is in general no strict predictability of phenomena, that predictions are usually only probabilistic. It is also true that the quantum rules of combination of probabilities[76] are not compatible with the idea that each phenomenon is strictly determined by other phenomena that we just happen to ignore. In other words, the "ignorance interpretation" of quantum probabilities is precluded as long as one holds on to the plane of phenomena. However, this says nothing about the hypothetical "ultimate laws of nature" below the level of phenomena, this says nothing about whether quantum indeterminism is only epistemological or ontological as well. As Bohm pointed out, "The mere uncontrollability and unpredictability of quantum phenomena does not necessarily imply that there can be no quantum world, which would in itself be determinate."[77] Indeed, we now know that there exists a class of processes undergoing chaotic motions, which is both ruled by deterministic laws and remains unpredictable. Microphysics thus does not point toward strict, intrinsic, indeterminism; it rather illustrates the undecidability of ontological propositions by science, be they about the determinist or indeterminist status of the "ultimate laws of nature." As indicated by Jacques

Harthong, this type of undecidability can easily be expressed in a form that mimics Kant's dialectic of pure reason. The antinomy of probabilistic predictions develops thus:

Thesis: "The ultimate law of the world is chance, and any partial determinism that could be found in it results from the law of large numbers."

Antithesis: "The ultimate law of the world is entirely deterministic, and any random phenomenon that could be observed results from deterministic chaos."[78]

This being granted, the strongest argument that can be given in favor of indeterminism in microphysics is that any search for deterministic laws would be sterile, that applying the Leibnizian principle of sufficient reason at any cost would be fruitless, that no experimentally testable consequence would arise from this research.[79] But, as we have seen, this argument is not compelling. Moreover, it does not prevent one from inquiring philosophically into the possible reasons for unpredictability at the level of microphysical phenomena.

What is fascinating at this point is that many results in the literature on the interpretation of quantum mechanics tend to converge toward one explanation of such an indetermination.

Karl Popper,[80] to begin with, noticed that, even in a world ruled by underlying deterministic laws, an observer could not predict a phenomenon if she were herself entangled with the process of its production. Unpredictability here results from a logical limitation in self-prediction. In short, as soon as the observer has predicted what she will do, the very content of the prediction can influence her future behavior. This spurious effect of predicting on the predicted behavior may make the prediction wrong. Then, due to the entanglement of the predictor and the phenomena that have to be predicted, the logical limitation of self-prediction results in a limitation of prediction of phenomena.

Much earlier, G. Hermann,[81] a young philosopher of science who worked with Heisenberg, explained with some detail that one is not bound to assume that quantum phenomena have no cause, only that the causes are not defined in the absolute but rather relative to the very circumstances of the production of the phenomena.

Even more precisely, P. Destouches-Février demonstrated that any predictive theory bearing on phenomena defined relative to possibly incompatible experimental contexts is "essentially indeterminist."[82] Indetermination in the sense of unpredictability is here a direct consequence of the relativity (or context-dependence) of phenomena.

To summarize, a plausible positive teaching of the dialectic of determinism and indeterminism is that microphysical knowledge is contextual, relational, or participatory at the deepest level. At any rate, this is the teaching one is likely to draw from this dialectic if both an absolutist defense of one thesis and a nihilistic reaction to the lack of proof of any thesis are to be avoided.[83]

7. RELATIONAL KNOWLEDGE

It is not precise enough to say that the Madhyamaka and the neo-Kantian philosophies of science are similar in their focus on relations or on relational knowledge. They are also similarly specific about relations. Both of them put relations before (or on the same footing as) the relata, both of them share a nonpolar conception of relations, and for all that they do not reify relations.

When Nāgārjuna equates mutual dependence with emptiness,[84] or lack of inherent existence, this clearly shows that he has not the slightest temptation to think that the relata precede the relation. He even insists something that is "due to a cause and which does not exist in lack of such" is like a "reflection"[85] or like "foam, bubbles, illusion."[86] The relation is what makes the relata emerge (as noninherently existing phenomena), just as much as the other way round. Yet, no ontology of relations is asserted: "Neither connection, nor connected nor connector exist."[87] Indeed, asserting the existence of relations to the detriment of that of the relata would involve the use of an opposition (relation-relata) and the solidification of one of its terms, whereas the two terms of this opposition also arise in dependence.

As for the neo-Kantian philosophers, they are very careful to put function before substance (to paraphrase the title of a book by Cassirer) and relations before their relata. According to Natorp, Plato's most important discovery in the *Sophist,* at the end of his lifelong discussion about being and not being, is that they somehow mutually ground each other.[88] The applications of this discovery came much later, especially in Kant's conception of the synthetic power of thought. But as soon as it was clearly understood it underwent radical developments.

The most prominent neo-Kantian philosophers pointed out that the basic shortcoming of metaphysics consists in "separating correlative standpoints within the field of knowledge itself, and thus transforming what is logically correlative into an opposition of things."[89] They then quickly reinterpreted Kant in this spirit, by showing that his basic method of transcendental deduction[90] was precisely aimed at avoiding such an unwarranted transformation. Thus, notwithstanding Kant's original formulations, Hermann Cohen and Paul Natorp claimed that transcendental deduction should *not* be interpreted as an attempt to return to some absolute foundation of knowledge. According to them, there can be no static relation between an epistemic ground (the forms of thought) and something grounded (objective knowledge), where a dynamical process of mutual accommodation is involved.[91] What is expressed by each special instance of transcendental deduction is only a "constraining reciprocity, in which there is neither *prius* nor *posterius*."[92] In other words, what a transcendental deduction reveals is a perfectly symmetrical relation of coproduction. Yet, here again, no well-rounded ontology of relations emerges: only a tireless study of ever-developing relational cognitive acts.

As previously noted, the special feature of neo-Kantian philosophies, when compared to the Madhyamaka, is that they are explicitly aimed at the justification of science, especially physics. They tend to apply their basic relational insights to the clarification of the nature of scientific knowledge, and they therefore complement the Madhyamaka, where these insights are predominantly used to promote *existential disabuse*.[93]

If applied, say, to the deduction of Newtonian mechanics by Kant (in his *Metaphysical Foundations of Natural Science*), the neo-Kantian construal of transcendental deduction yields an important epistemological teaching. The very fact that part of this physical theory *can* be transcendentally deduced shows that it must not be interpreted as a reflection of some inherent feature of external reality but rather as an expression of the mutual constraints between the two codependent terms of the cognitive relation.[94] More specifically, the extensive use of differential calculus by classical mechanics shows, according to most neo-Kantian thinkers, that only (infinitesimal) relations are accessible and that no monadic foundation of these relations, no absolutized *relata,* can ever be grasped by physics.[95]

In quantum mechanics the relational structure of knowledge is only enhanced.

> (Quantum mechanics) exaggerates the relative character of the description of nature. It abandons the representation according to which the structures of relations are univocally determined by certain connections of things in space and time, and shows their being dependent on the way an observer takes cognizance of the system.[96]

Using the vocabulary of section 3 (point 4), this sentence is to be understood as follows. In quantum mechanics we can no longer content ourselves
with describing "lateral" relations between spatiotemporal objects, thus behaving *as if* the "transversal" cognitive relations did not exist or were irrelevant; we have somehow to take into account the multiple cognitive relations
between the microphysical domain and the measuring apparatuses. Indeed,
due to complementarity (or to the commutation relations), the multiple
microdomain-apparatus relations cannot be reduced to one and then
pushed away in the background. This remark recurs in current neo-Kantian
interpretations of quantum mechanics and in some other interpretations as
well.[97]

Moreover, Jean Petitot insisted that the transversal cognitive relation is
represented by the formalism of quantum mechanics in such a way that its
"subjective" pole (and also, in all likelihood, its "objective" pole) is not
made explicit.[98] Just as in Kant's interpretation of Newtonian mechanics
the spatiotemporal structure implicitly conveyed the relational nature of
macrophysical knowledge, in a neo-Kantian reading of quantum mechanics
the Hilbert space structure implicitly conveys the relational nature of microphysical knowledge while involving no description of the two relata.
From this standpoint the (nonlocal) hidden variable theories are to be understood as desperate attempts at pointing toward an inaccessible world of
relata behind the relational network of standard quantum mechanics.

This feature of quantum mechanics gave rise to interesting developments about the notion of nonsupervenient relations, namely, relations
that do not depend on hypothetical monadic properties of the relata.[99] But
it was already latent in many earlier discussions on the measurement problem and on the entanglement of state vectors. If one recognizes the purely
relational status of the state vector, that is, its being an expression of the
propensity for phenomena under activation conditions, the measurement
problem reduces to a problem of transition from relativities to monadic
properties. It is a problem of breaking the chain of relations expressed by
the entangled state vector of the system (object + apparatus), thus jumping
to nonrelational determinations of both the apparatus and the object. Now
a fascinating proposal for a solution to this problem has been given in the
framework of Everett's original "relative state" interpretation.[100] The solution consists in remarking that, if the experimenter herself partakes of the
network of relations, things may *appear* to her *as if* well-defined nonrelational determinations arose from the measuring interaction. In short, a

state relative to her appears from her standpoint as a well-determined feature of something substantial. This is a good summary of how reifying (or "absolutizing") illusions may arise.

In addition, it can easily be shown that taking into account this deeply relational character of microphysical processes is nearly enough to derive the basic structure of quantum mechanics.[101] In other words, it can be shown that one may formulate a kind of "transcendental deduction" of quantum mechanics. Of course, the type of transcendental deduction that has to be used to derive the overall structure of quantum mechanics is much more general than Kant's. In this general sense a transcendental deduction is not a regression from objective knowledge to its conditions of possibility, as in Kant's *Critique of Pure Reason*. It is a regression from a set of minimal requirements about the scientific process of anticipation of phenomena to a strong anticipative structure as the condition of possibility for these requirements to be satisfied. Now, taking into account two requirements, namely, (i) that the anticipation must bear on *contextual* phenomena and (ii) that the predictive tool must be *unified* under the concept of a *preparation,* the basic anticipative structure of quantum mechanics arises. As Jean-Louis Destouches and Paulette Destouches-Février[102] argued convincingly, the formalism of vectors in a Hilbert space, together with Born's correspondence rule, is the simplest predictive formalism among those that obey the constraint of *unicity* in a situation where decontextualization cannot be carried out. Even the general form of the (Schrödinger or Dirac) equations of evolution can be obtained this way, by a series of direct or bridging transcendental arguments.[103]

This being granted, typical features of microphysical phenomena such as wavelike distributions and quantization, which are predicted by the quantum theory, no longer appear to be contingent aspects of nature. In view of the previous derivation, they rather appear as necessary features of any activity of production of contextual and mutually incompatible phenomena whose level of reproducibility is sufficient for its outcomes to be embedded in a unified system of probabilistic anticipation. Of course, this does not mean that quantum mechanics could have been obtained by mere armchair philosophizing, only that the structure of quantum mechanics has retrospectively revealed its deeply relational nature to the philosophical inquiry. As was the case for Kant's deduction of Newtonian mechanics, the very possibility of a transcendental deduction of quantum mechanics teaches us something important about the status of this theory. It suggests that quantum mechanics should not be construed as a reflection of some (exhaustive or nonexhaustive) aspect of a pregiven nature, but as the structural expression of the coemergence of a new type of experimental activity and the "factual" elements that constrain it.[104]

Here, again, these results and reflections are easily available in the literature. They were strongly promoted by the neo-Kantian trend in philosophy of science. But, in order to become widely accepted, in order to be articulated into a new coherent participatory conception of the world, they will need to overcome the Western urge for foundations, and for reification of the pragmatic categories of everyday life. This can occur only through their integration within a higher-order axiological and existential operator (not to say within an alternative form of life) of which the Madhyamaka dialectic and soteriology is likely to be the central element.

WHAT IS THE USE of this essay, beyond its philosophical content? It does not pretend to be a substitute for the best spiritual writings of the Mahāyāna Buddhist tradition, nor even to add the slightest contribution to them. Even less can it alone induce the self-transformation of other human beings. But it may weaken the intellectual ground of those who still (roughly one century after the alleged fading of scientism) take science as the modern equivalent of late religious dogma. It may also, more important, help those who are already engaged in a process of self-transformation not let themselves be impressed, at an intellectual level, by the substantialist tales of the majority of physicists. It may, above all, give them a hint as to how to integrate the many strata of their life and thought in our modern culture. These effects having, hopefully, been obtained, the present article is to be thrown away as any other step in the Wittgensteinian ladder toward what really matters.

Notes

I would like to thank Rachel Zahn for her careful reading of this paper, and for helping me to adapt it to a broader audience.

1. A soteriology is a doctrine of salvation (*soteria* means salvation in ancient Greek).
2. S. Körner, Introduction to E. Cassirer, *Kant's Life and Thought* (New Haven: Yale University Press, 1981), p. xi.

3. The mesoscopic scale is intermediate between the true (cosmological) macroscopic scale and the (atomic and subatomic) microscopic scale.

4. B. A. Wallace, *Choosing Reality: A Buddhist View of Physics and the Mind* (Ithaca: Snow Lion, 1996).

5. T. R. V. Murti, *The Central Philosophy of Buddhism* (London: Allen and Unwin, 1955); M. Sprung, "The Madhyamaka Doctrine of Two Realities As a Metaphysic," in M. Sprung, ed., *The Problem of Two Truths in Buddhism and Vedanta* (Dordrecht: Reidel, 1973). An early parallel between Kant and Dharmakīrti's logic (with some references to the Madhyamaka) can also be found in T. Stcherbatsky, *Buddhist Logic* (Leningrad: Office of the Academy of Science of the USSR, 1927; repr. Delhi: Motilal Banarsidass, 1994).

6. J. May, "Kant et le Madhyamaka, A propos d'un livre récent," *Indo-Iranian Journal* 3(1959): 102–111.

7. E.g., J. Garfield, *The Fundamental Wisdom of the Middle Way* (New York: Oxford University Press, 1995), p. 305–306, n. 119.

8. See K. Bhattacharya, *The Dialectical Method of Nāgārjuna* (Delhi: Motilal Banarsidass, 1998), a translation and commentary on Nāgārjuna's *Vigrahavyavartani*.

9. See G. Bugault, "Logique et dialectique chez Aristote et chez Nāgārjuna," in G. Bugault, *L'Inde pense-t-elle?* (Paris: Presses Universitaires de France, 1994), p. 260.

10. Nāgārjuna, *Mūlamadhyamakakārikā*, XXVII, 21, in Garfield, *The Fundamental Wisdom of the Middle Way*, p. 350; or see D. J. Kalupahana, *Mūlamadhyamakakārikā of Nāgārjuna* (Delhi: Motilal Banarsidass, 1986), p. 387.

11. I. Kant, *Critique of Pure Reason*, A479–B507.

12. Comments on Nāgārjuna, *Mūlamadhyamakakārikā*, XXVII, 22, Garfield, *The Fundamental Wisdom of the Middle Way*, p. 350–351.

13. Kant, *Critique of Pure Reason*, A379–380, A539–B567.

14. Murti, *The Central Philosophy of Buddhism*, p. 167.

15. Ibid., p. 251.

16. T. R. V. Murti, "Samvṛti and Paramārtha in Madhyamaka and Advaita Vedānta," in Sprung, *The Problem of Two Truths in Buddhism and Vedanta*.

17. Murti, *The Central Philosophy of Buddhism*, p. 229.

18. Literally, *samvṛti* means "covering," or "concealing." *Samvṛti-satya* is, so to speak, the surface truth.

19. Ibid., p. 294

20. Referring to this type of relation, Kant writes that "the absolute must be thought of as external to the empirical world, and the latter only consists of the relation with our senses." I. Kant, Reflections 5968, *Kants Nachlass*, AK XVIII.

21. Kant, *Critique of Pure Reason*, B321, B341.

22. Nāgārjuna, *Mūlamadhyamakakārikā*, IX, 5, Garfield, *The Fundamental Wisdom of the Middle Way*, p. 184–185.

23. Nāgārjuna, *Lokātitastava* 6, 7, 10 (*Hymn to the Buddha transcending the world*), in C. Lindtner, *Nagarjuniana* (Delhi: Motilal Banarsidass, 1987), p. 131, 133. "[An object of knowledge is] no object of knowledge unless it is being known. . . . Therefore you have said that knowledge and the object of knowledge do not exist by own-being."

24. Garfield, *The Fundamental Wisdom of the Middle Way,* p. 185.

25. Bugault, *L'Inde pense-t-elle?* p. 186, 292.

26. Garfield, *The Fundamental Wisdom of the Middle Way,* p. 198.

27. Immanuel Kant, *Prolegomena to Any Future Metaphysics That Will Be Able to Present Itself As Science* (Manchester: Manchester University Press, 1971), 52c.

28. T. J. F. Tillemans, "La logique bouddhique est-elle une logique non-classique ou déviante? Remarques sur le tetralemme," *Les cahiers de philosophie (Lille),* no. 14 (1992): 183–198.

29. Murti, *The Central Philosophy of Buddhism,* p. 295.

30. One can also distinguish these two cases by saying that free will holds from the standpoint of the actor, whereas universal application of the laws of nature holds from the standpoint of the spectator. L. W. Beck, *A Commentary on Kant's Critique of Practical Reason* (Chicago: University of Chicago Press, 1963).

31. "Apophatic" derives from the Greek *apophasis*, which means "denial." Here, the features are denied rather than asserted of Being.

32. Murti, *The Central Philosophy of Buddhism,* p. 273–274.

33. Nāgārjuna, *Mūlamadhyamakakārikā*, XXV, 19, Garfield, *The Fundamental Wisdom of the Middle Way,* p. 331.

34. Ibid., p. 332.

35. L. Laudan, *Science and Values* (Berkeley: University of California Press, 1984); G. Boniolo, *Metodo e rappresentazioni del mondo* (Milan: Bruno Mondadori, 1999).

36. A. Funkenstein, *Theology and the Scientific Imagination* (Princeton: Princeton University Press, 1986).

37. F. Varela, E. Thomson, and E. Rosch, *The Embodied Mind* (Cambridge: MIT Press, 1991), chapter 6.

38. This sounds very similar to F. Varela's theory of cognition. Actually, his autopoietic theory of cognition can easily be interpreted as a naturalized version of the neo-Kantian theory of knowledge. See M. Bitbol, "Physique quantique et cognition," *Revue Internationale de Philosophie* (2000), pp. 212, 299–328.

39. Hermann Cohen (1842–1918) is the founder of the Marburg school of neo-Kantian philosophy. See note 46.

40. Ernst Cassirer (1874–1945) is the most prominent philosopher of the Marburg school. His publications include *Substance and Function* (1910) and the three volumes of the *Philosophy of symbolic forms* (published in the 1920s).

41. H. Putnam, *Pragmatism* (Oxford: Blackwell, 1995).

42. J. Hintikka and I. Kulas, *The Game of Language* (Dordrecht: Reidel, 1983), p. 33.

43. Boniolo, *Metodo e rappresentazioni del mondo*, p. 123.

44. Kant, *Critique of Pure Reason*, A255/B311.

45. See, e.g., L. Ferry, in the preface to Kant, *Critique de la raison pure* (Paris: Garnier-Flammarion, 1987), p. xix.

46. Marburg is a little town in western Germany (land, or province, of Hesse). An important Protestant university was created there in the sixteenth century. H. Cohen taught in this university until 1912, and he had remarkable students, among them E. Cassirer and P. Natorp. The most renowned school of neo-Kantian philosophy in Germany, composed of H. Cohen's followers, was thus referred to as the Marburg school.

47. E. Cassirer, H. Cohen, and P. Natorp, *L'école de Marbourg* (Paris: Cerf, 1998), p. 247.

48. Undoubtedly, this word *conventional,* commonly used to translate *saṃvṛti,* has spurious connotations. A convention is an overt agreement between persons. But the *saṃvṛti-satya* does not arise from an explicit discussion between the members of human societies. Every (human) being is so to speak bound to it by the very fact he/she partakes of a *form of life* that involves efficient practices including the use of language. I thank Christiane Schmitz for having raised this question, and for so many other valuable remarks.

49. Nāgārjuna, *Mūlamadhyamakakārikā*, XXIV, 10, J. Garfield, *The Fundamental Wisdom of the Middle Way*, p. 298. Candrakīrti comments that one must accept (and presumably analyze) the surface truth at first, for it is an instrument to reach *nirvāṇa*. Candrakīrti, *Prasannapadā Madhyamakavṛtti,* trans. [French] J. May (Paris: Adrien Maisonneuve, 1959).

50. A. Fine, *The Shaky Game* (Chicago: University of Chicago Press, 1986).

51. Kant, *Critique of Pure Reason*, A298–B354.

52. Kant, *Prolegomena to Any Future Metaphysics,* 13.

53. See, e.g., N. Bohr, *Atomic Theory and the Description of Nature, The Philosophical Writings of Niels Bohr*, 2 vols. (Woodbridge, Conn.: Ox Bow, 1987), 1:53.

54. N. Bohr, *Essays 1933–1957 on Atomic Physics and Human Knowledge, The Philosophical Writings of Niels Bohr*, 2 vols. (Woodbridge, Conn.: Ox Bow, 1987), 2:73.

55. This being accepted, *complementarity* does not appear any longer as a trick to accommodate contradictory properties. It only expresses (i) the indissociability of object and experimental device in a phenomenon, and (ii) the mutual incompatibility of the devices in the context of which certain classes of phenomena occur.

56. G. Birkhoff and J. Von Neumann, "The Logic of Quantum Mechanics," *Annals of Mathematics* 37(1936): 823–843.

57. This algebra, invented by G. Boole (1815–1864), underpins classical logic. It

involves the following rules of combination of propositions (or properties) by conjunction and disjunction: commutativity, associativity, distributivity, and complementation.

58. R. I. G. Hughes, *The Structure and Interpretation of Quantum Mechanics* (Cambridge: Harvard University Press, 1989), p. 188.

59. J. Bub, *Interpreting the Quantum World* (Cambridge: Cambridge University Press, 1997), p. 12.

60. P. Heelan, "Complementarity, Context-Dependence, and Quantum Logic," *Foundations of Physics* 1(1970): 95–110; P. Heelan, "Quantum and Classical Logic: Their Classical Role," *Synthese* 21(1970): 2–33; see also S. Watanabe, "The Algebra of Observation," *Supplement to Progress of Theoretical Physics* 37 and 38:350–367, 1966.

61. J. L. Destouches, *Corpuscules et systèmes de corpuscules* (Paris: Gauthier-Villars, 1941); P. Destouches-Février, *La structure des théories physiques* (Paris: Presses Universitaires de France, 1951); P. Destouches, *L'interprétation physique de la mécanique ondulatoire et des théories quantiques* (Paris: Gauthier-Villars, 1956). See section 8 of this paper, and, for more details, M. Bitbol, "Some Steps Towards a Transcendental Deduction of Quantum Mechanics," *Philosophia Naturalis* 35(1998): 253–280; M. Bitbol, *Mécanique quantique, une introduction philosophique* (Paris: Flammarion, 1996), chapter 2.

62. See the end of this section.

63. A satisfactory explanation is usually a derivation of a great number of complex and apparently arbitrary observed consequences from a small number of simple and less arbitrary assumptions.

64. According to L. Laudan's definition, ampliative arguments are rational but extra-empirical motives for adopting a scientific theory. They consist in an *amplification* of the purely empirical arguments.

65. It is only if (as it was the case in the period 1927–1952) any such realist interpretation seems out of reach that most scientists allow consideration of an alternative. But they then revert to a purely instrumentalist attitude they feel to be a renunciation, and they are thus usually delighted when realist interpretations become acceptable again, even if these interpretations sound extremely artificial. See Wallace, *Choosing Reality*, for a critique of the couple realism-instrumentalism.

66. W. De Muynck, "Distinguishable and Indistinguishable-Particle Descriptions of Systems of Identical Particles," *International Journal of Theoretical Physics* 14(1975): 327–346. Doubts concerning this empirical equivalence are expressed in J. Butterfield, "Interpretation and Identity in Quantum Theory," *Studies in the History and Philosophy of Science* 24(1993): 443–476.

67. P. Teller, *An Interpretive Introduction to Quantum Field Theory* (Princeton: Princeton University Press, 1995), p. 25.

68. Ibid., p. 138.

69. Ibid., p. 105. The superstring theories bring little change in that respect. But

in order to understand this one has to revert to Feynman path-integral formalism. In standard quantum field theory the propensity structure is described by an integral over (an infinity of) linear paths. Now, the cross-section of one path is a point. Hence the usual talk of point-particles. The problem is that taking seriously this mode of expression is unwarranted since one cannot reduce "what there is" to *one* cross-section of *one* of the multiple paths whose *complete* sum is required to calculate the probability of a final experimental event. In superstring theories the cross-section of each *one* of the tubes that have to be added to give a probability is a string. But, here again, and for the same reason as in standard quantum field theory, taking seriously (i.e., ontologically) the usual talk of string-particles is unwarranted.

70. Ibid., p. 110.
71. Propensions, according to K. Popper, are tendencies to realize a certain state of affairs. A relational-propensionist view refers to tendencies of a given relation (say between an apparatus and the rest of the world) to produce certain phenomena.
72. M. Born, "Physical Reality," *Philosophical Quarterly* 3(1953): 139–149; M. Born, "The Interpretation of Quantum Mechanics," *British Journal for the Philosophy of Science* 4(1953): 95–106. Both articles are reprinted in M. Born, *Physics in My Generation* (Oxford: Pergamon, 1956). See M. Bitbol, *Schrödinger's Philosophy of Quantum Mechanics* (Dordrecht: Kluwer, 1996).
73. D. Bohm and B. Hiley, *The Undivided Universe* (New York: Routlege and Kegan Paul, 1993), pp. 4, 160.
74. R. Harré, *Varieties of Realism* (Oxford: Basil Blackwell, 1986); A. A. Derksen, *The Scientific Realism of Rom Harré* (Tilburg: Tilburg University Press, 1994), pp. 7–8.
75. Nāgārjuna, *Lokātitastava*, Lindtner, *Nagarjuniana*, p. 131.
76. These rules essentially consist in adding the complex *amplitudes* and then squaring the sum.
77. Bohm and Hiley, *The Undivided Universe*, p. 25. These ideas about determinism are developed in chapter 8 of M. Bitbol, *L'aveuglante proximité du réel* (Paris: Flammarion, 1998).
78. J. Harthong, quoted in A. Dahan-Dalmedico, J.-L. Chabert, and K. Chemla, *Chaos et indéterminisme* (Paris: Seuil, 1992); see also J. Harthong, *Probabilités et statistiques* (Paris: Diderot, 1996).
79. This is the case, admittedly, in Bohm's hidden variable theory.
80. K. Popper, *The Postscript to the Logic of Scientific Discovery*, vol. 2: *The Open Universe* (London: Hutchinson, 1982), 22.
81. G. Hermann, "Die naturphilosophischen Grundlagen der Quantenmechanik," *Abhandlungen der Fries'schen Schule*, new series, vol. 6, 2 (1935). French translation and extensive comment in G. Hermann, *Les fondements philosophiques de la mécanique quantique*, ed. L. Soler, trans. A. Schnell and L. Soler (Paris: Vrin, 1996).

82. P. Destouches-Février, *La structure des théories physiques* (Paris: Presses Universitaires de France, 1951), p. 260–280.

83. The idea of a participatory universe, presented by J. A. Wheeler, is defended in a Buddhist context by B. A. Wallace, *Choosing Reality,* op. cit. chapters 14, 15. See also M. Ricard and Trinh Xuan Thuan, *The Quantum and the Lotus* (New York: Crown, 2001).

84. Nāgārjuna, *Mūlamadhyamakakārikā*, XXIV, 18, Garfield, *The Fundamental Wisdom of the Middle Way,* p. 304.

85. Nāgārjuna, *Lokātitastava*, Lindtner, *Nagarjuniana,* op. cit. p. 129.

86. Nāgārjuna, *Acintyastava,* ibid., p. 147.

87. Nāgārjuna, *Mūlamadhyamakakārikā*, XIV, 8, Garfield, *The Fundamental Wisdom of the Middle Way,* p. 219.

88. Cassirer, Cohen, and Natorp, *L'école de Marbourg,* p. 220.

89. E. Cassirer, *Substance and Function* (New York: Dover, 1953), p. 271; my emphasis.

90. According to C. Taylor's simple definition, a transcendental deduction is "a regression from an unquestionable feature" of our knowledge to "a stronger thesis as the condition of its possibility." C. Taylor, *Philosophical Arguments* (Cambridge: Harvard University Press, 1995).

91. See C. Schmitz, "Objectivité et temporalité," in M. Bitbol and S. Laugier, eds., *Physique et réalité, un débat avec Bernard d'Espagnat* (Paris: Frontières, 1997), p. 273.

92. Cassirer, Cohen, and Natorp, *L'école de Marbourg,* p. 55.

93. Moreover, the neo-Kantian philosophers mostly owe their insights to a free play of ideas, rather than to direct stabilized experience of a disabused outlook.

94. Of course, the cognitive relation can change, and its terms as well, in the course of the development of experimental research. This is enough to explain that science is liable to revolutions, notwithstanding the possibility of transcendentally deducing its structure at a given stage of its development. Here, the spurious eternalist connotations of Kant's a priori should not be allowed to impose on us a foundationalist reading of his transcendental deduction of Newtonian mechanics. Once this is recognized, nothing can prevent us from looking for a (similarly nonfoundationalist) transcendental deduction of quantum mechanics. A sketch of this deduction is provided below.

95. Hermann, *Les fondements philosophiques de la mécanique quantique,* p. 116; see also H. Cohen, *Le principe de la méthode infinitésimale et son histoire,* trans. M. de Launay (Paris: Vrin, 1999).

96. Hermann, *Les fondements philosophiques de la mécanique quantique* p. 119. see also E. Cassirer, *Determinism and Indeterminism in Modern Physics* (New Haven: Yale University Press, 1956), p. 131, 182.

97. V. Fock, quoted in M. Jammer, *The Philosophy of Quantum Mechanics* (New York: Wiley, 1974), p. 202; M. Davis, "Relativity Principle in Quantum Me-

chanics," *International Journal of Theoretical Physics* 16(1977): 867–874; M. Mugur-Schächter, "Space-Time Quantum Probabilities II: Relativized Descriptions and Popperian Propensities," *Foundations of Physics* 22(1992): 235–312.

98. J. Petitot, "Objectivité faible et philosophie transcendantale," in Bitbol and Laugier, *Physique et réalité,* p. 207–208.

99. P. Teller, "Relational Holism and Quantum Mechanics," *British Journal for the Philosophy of Science* 37(1986): 71–81.

100. H. Everett, "'Relative State' Formulation of Quantum Mechanics," in B. S. De Witt and N. Graham, *The Many-Worlds Interpretation of Quantum Mechanics* (Princeton: Princeton University Press, 1973). Here it is especially important not to mix up Everett's original *relative state* formulation and its later reading in terms of *many-worlds*. See, e.g., Y. Ben-Dov, "Everett's Theory and the 'Many-Worlds' Interpretation," *American Journal of Physics* 58(1990): 829–832.

101. Bitbol, "Some Steps Towards a Transcendental Deduction"; see section 5 of the present paper.

102. J. L. Destouches, *Corpuscules et systèmes de corpuscules* (Paris: Gauthier-Villars, 1941); P. Destouches, *L'interprétation physique de la mécanique ondulatoire et des théories quantiques* (Paris: Gauthier-Villars, 1956).

103. A bridging transcendental argument establishes a *bridge* between the specific form of transcendental deduction that was used by Kant within the direct spaciotemporal environment of mankind and the generalized sort of transcendental deduction needed in domains of scientific investigation that may go beyond the human *Umwelt.* An example is Bohr's correspondence principle, which ensures a connection between the basic thinglike organization of everyday life and classical mechanics, and the contextual organization of quantum mechanics. See Bitbol, "Some Steps Towards a Transcendental Deduction."

104. See Varela, Thompson, and Rosch, *The Embodied Mind,* for similar remarks in the general framework of the cognitive sciences.

In the preceding essay Michel Bitbol points out that philosophy has always played a role in the formulation of scientific theories, and he suggests that the Madhyamaka might provide the context for new developments in physics. This is precisely what David Ritz Finkelstein does in the following essay, in which he develops a theory of "universal relativity." In this theory the concept of states of being is replaced by the idea of modes of action, events are prioritized over things, and there is a relativization of knowing (fixing a state) and doing (changing a state).

One of the most intriguing elements of Finkelstein's discussion is his explanation of idols, which results from fixing on false absolutes. One sign of an idol is a nonreciprocal coupling, or interaction, between two entities, such as space and time, or matter and space-time. The unaffected partner in such a compound is identified as an idol. To relate this theme to the mind-body question, modern cognitive scientists commonly assume there is no mind separate from brain, so there is absolute brain and mind-brain but no absolute mind. By fixing on the brain (and behavior) as an absolute, the brain takes on the role of an idol. In their essay in this volume, Varela and Depraz challenge the widespread assumption that the brain influences the mind but the mind does not influence the brain. This is the kind of reciprocity failure that Finkelstein discusses. His argument from the history of science suggests that this theory is defective and possesses a false absolute.

In order to discover what absolutes are tacitly assumed within any given theory, we must step outside the theory, be it that of cognitive science or physics, and examine what scientists say, what they do, and the connection between their words and deeds. For both the cognitive sciences and physical sciences, Buddhism provides one such outside vantage point that may be invaluable in revealing such "idols of the tribe" of science. Finkelstein points out that any theory—including the Buddhist and the scientific—is a view from a position, which is then an idol of that theory. This implies that the process of making a theory inevitably introduces idols that can only be corrected by a later theory. And this implies a never ending process.

This raises the crucial question: what is the point of formulating theories? Much of science seems to be aimed at the formulation of an ultimate, all-encompassing, fixed theory of the universe: the final achieve-

ment of a "God's eye view." But Finkelstein's principle of universal relativity, as he says, seems incompatible with a fixed theory of any kind, and in particular with the goal of a final theory, which may be another idol that we must evolve beyond. This perspective parallels the Buddhist theme of not grasping onto any view as being supreme, even the Madhyamaka. It calls into question, in both physics and Buddhism, the optimal role for theories: do they really serve any higher purpose than to lead to more valid and comprehensive experience? And if a theory does not serve that function, what is its purpose?

While the details of this essay will be comprehended only by readers with a solid background in theoretical physics, it is bound to be deeply provocative even for those who are not well versed in physics.

David Ritz Finkelstein
Emptiness and Relativity

RELATIVITY AND INTERACTIVITY

The Buddhist principle that all is empty is understood by some as the principle that all is relative (Thurman 1993). This universal relativity principle is more embracing though less structured than Einstein's general relativity principle, which still admits many absolutes. The major changes in physics in this century have been extensions of relativity at one level or another, and I think a further extension is due, at an even deeper level of physics than the previous. Philosophical inquiry has aided such extensions before, and it could do so again. A philosophical argument for a universal relativity could be a useful guide for future physics.

I consider such a universal relativity principle to be meaningful and perhaps even true in physics.

For the purposes of physics, however, one must be more specific about the nature of the relation that is implied but not specified in the broad term *relativity*. The relation between people that special relativity considers, for example, is that between people in relative motion who communicate with a system and each other by exchanging signals, typically light. The relation underlying quantum theory, however, is that between observers of complementary quantities, and one studies the effect of this relation upon the basic interaction between experimenter and system that holds during a measurement. Interaction is a lower-level concept than communication, in that every communication is made up of interactions. In what follows I usually specify the terms relative to mean interactive, and relativity interactivity.

Quantum mechanics has been characterized as a "nonobjective physics," expressing the idea that it is founded not on objects but on interactions. Is it indeed possible that all is interactive? How far we have already gone in that direction? What absolutes remain?

My main tool in this inquiry is an analysis of Segal (1951) and Inönü and Wigner (1952) that shows us by precept and example how to detect possibly false absolutes and how to relativize them.

The main absolute of physics today that we will discuss is the dynamical law, also called the law of nature, describing how the system develops in time, also called the dynamical process. I revisit the idea that we should relativize dynamical law much as Einstein relativized geometric law in general relativity. Furthermore, I consider taking the dynamical process as the sole variable under study, as Einstein proposed for the geometric law in his unified field theory. Such a more interactive space-time-matter-dynamic unity might embrace general relativity and the standard model and reach beyond them.

IDOLS

Let me indicate how I use three terms basic to this discussion.

Relativity is the part of any physical theory that concerns how appearance—the phenomenon—depends on the observer.

An *absolute* (or *covariant*) property or entity—a noumenon—is one whose presence or absence all experimenters agree on, though they may name it differently.

Reification is imagining an absolute entity where there is none.

An *idol*, in the language of Francis Bacon (1620), is a false absolute resulting from reification. "Idols of the tribe" are those common to a whole community, such as those resulting from innate propensities to reify. "Idols of the theater" are those erected within a particular theory. My usage differs from Bacon's in that I regard idols as inevitable and useful products of the same theory-making process that breaks them.

Relativity came to center stage in the mechanical physical theories of projectiles and planets, where one must relate observers in relative motion.

For example, Johannes Kepler wrote an entire relativistic science fiction novel, *Somnium*, just to relate the views of people on the Moon to those of people on Earth and argue against the commonsense conviction that the Earth was at absolute rest. On the evidence of his relativistic manuscript, his mother was charged with witchcraft and exposed to the instruments of torture. For their relativisms, Bruno had earlier been burned at the stake, Galileo was merely placed under house arrest, and later Einstein was laureated. The intellectual climate is clearly thawing.

Nevertheless physics once again runs into idols that block its development.

We can spot these idols using a detection system that Segal (1951) and later Inönü and Wigner (1952) formulated and applied to classical mechanics and other physical theories. I describe it first and then apply it to present physical theory.

Segal, and Inönü and Wigner, look for partially but incompletely fused structures, which they call nonsemisimple and which will be called *compound* here. I count as an entity whatever everyone in a community can experience, such as an electron, or the Moon, or the time of day, as opposed to optical illusions or hallucinations. An entity is called a *simple* in this context if it properly includes no other entity (except trivial constants). It is called *semisimple* if it is simple or equivalent to a collection of separate simples. It is called *compound* if it is not semisimple.

A compound results when one simple variable has subordinated another without fully integrating it.

A compound looks like a snake that has just swallowed a pig. A strong attachment has been formed but full integration has yet to come. A compound is a distress signal.

Inönü and Wigner applied the Segal criterion only to classical mechanics, where the diagnosis could be checked against the already known outcome. This tested the test more than the theory. The S-I-W (Segal-Inönü-Wigner) test passed its test, by "predicting" the evolution of the relativity of Galileo into special relativity.

The compound of classical mechanics that they studied is space/time.

A solidus as in "space/time" indicates a compound composite, not a quotient, and is followed by an idol. A hyphen indicates a simple fusion like Einsteinian "space-time."

The Galilean compound space/time forms from the Aristotelian simple time and simple space when time "swallows" space. That is, in Galilean thought there is no space separate from time; we cannot recognize the same place at a different time, and to speak of it has no meaning. But there is still time within space/time, and still a unique space at each time, a slice of the tree of history. Galileo has absolute time and absolute space/time but no absolute space.

Space/time is therefore not a composite of two simples, space and time. Yet it contains the simple time. Space/time is, therefore, compound and time is its idol.

To put it differently, Inönü and Wigner look for a one-way coupling between entities. The snake swallows the pig and not the pig the snake. In transforming from one observer to another in relative motion, Galileo couples time into space, but not conversely. Another way to say that space/time is compound, then, is to say that there is this one-way space/time coupling.

This nonreciprocity could have hinted to Galileo or a contemporary that there is likely a missing physical constant coupling space back to time; a speed c, therefore. The speed c would have to be so large that the effect of this coupling from space to time, an effect that must vary as $1/c$, could elude notice in Galileo's day. But if c were not too large, it could become important later, when experimenters attain greater relative velocities or develop more sensitive instruments. The coupling constant c "predicted" by the S-I-W test is the speed of light.

Such one-way coupling is a sure sign of a compound and is circumstantial evidence that the unresponsive partner in the coupling is an idol, a false absolute. The guiding heuristic principle is that actual coupling is always reciprocal. This is not Newton's physical principle of action and reaction, but might be its philosophical grandmother. I find this principle of reciprocity plausible enough to explore its implications here, and elsewhere, with experiment as the court of last appeal.

The more evolved construct, the space-time point of Einstein and Minkowski, is simple, with no nontrivial parts. This is the evolution of Galileo's space/time "predicted" by the S-I-W test in retrospect. Galileo had shown that space was an invalid reification. Einstein's development showed that time was too. Aristotle's two uncoupled absolutes, space and time, had evolved through the compound space/time of Galileo into the one symmetrically coupled absolute space-time of Einstein.

The S-I-W test can show us a possible idol and it can suggest the kind of reverse coupling to look for experimentally, but it gives no indication of how strong this coupling might be, except that it must be weak enough to have been overlooked so far. The actual size of the new coupling coefficient must be learned from experiments that invalidate the theory containing the idol under study.

It sometimes happens that one relativistic evolution compounds previous simples, and then another later evolution simplifies that compound, but creates other idols at the same time. Galilean space/time is the transitional phase between Aristotelian "space & time," a semisimple conjunction of two separate simple space and time entities, and Einstein space-time. Einstein space-time preserves other idols, which I mention below.

It took thousands of years to get from space & time to space/time, and only two centuries more to form space-time. The pace picks up, with each relativization helping the next. To develop skill and confidence with our idol test, I will apply it to three more relativizations that occurred in the first three decades of the twentieth century before tackling one of the *new* millennium. I omit some important relativizations that are not crucial for the story.

GENERAL RELATIVITY

Newtonian physics has a compound absolute that special relativity inherited: not absolute rest, but absolute coasting, nonacceleration. Newton believed that, while there is no standard of absolute rest, there is a standard of absolute nonacceleration. For example, a droplet is spherical if it is not spinning and ellipsoidal if it spins. (Newton used a water bucket for this test, but in free fall droplets work better.) In the spinning droplet each part accelerates toward the center of the droplet. In the nonspinning droplet each part follows a geodesic.

What provides the standard of nonacceleration everywhere in space-time is today a local structure called the metric (field). It is reckoned as part of the structure of space-time.

Through its metric, space-time acts on the rotating droplet or other matter, but in Newtonian physics and special relativity the matter, even if it be the size of the Sun, does not act on the space-time. There is therefore a compound matter/space-time. A simplification of this compound was urged by Mach and carried out by Einstein in his theory of general relativity, a successor to special relativity, bringing us closer to a matter-space-time unity.

In general relativity the dynamical evolution couples matter to metric as well as metric to matter.

The resulting variations in the metric account for gravity, which is locally indistinguishable from the effect of an accelerated observer. The coupling coefficient that corresponds to c for this evolution of physics is usually taken to be G (Newton's constant), henceforth the hallmark of general relativity. It may equivalently be taken to be a small time formed from G, h, and c called the Planck time T_P, whose value is about 10^{-43} seconds. Again the "prediction" of our idol-test agrees with the outcome that we knew in advance.

The relativization from special to general relativity was more dramatic than the previous ones because it introduced a richer new physical entity, the metric field, where before had been a frozen constant, and because both relativizations took place in one mind within one decade. The G relativization has been enormously fruitful. The current standard model of the nuclear forces was modeled on it, with several other local standards playing the role of the standard of coasting.

QUANTUM RELATIVITY

The simplicity of quantum theory emerges from another idol of classical theory by another relativization and idoloclasty. To take this conceptual quantum jump, Bohr emphasized, we must first change epistemologies. One formulation of this change is that we stop defining entities by their states ("ontically") and define them by our actions upon them ("praxically"; Finkelstein 1996). In an action-based (or praxic) semantics, any property of the system is defined by actions of preparation, selection, or registration carried out upon the system by the experimenter. I call this replacing reality by actuality. It helps us empty a concept of essence by making us more aware of how our knowledge of the concept arises from our own actions.

First we point out an idol of the classical epistemology.

In classical physics, since Descartes, the distinction between physical system and mathematical model was intentionally minimized. Some claimed the two were isomorphic and identified them. Transformations between observers were considered to be of a rather shallow kind, mathematical changes of variables amounting to a word for word literal translation from the language of one to the other. They were regarded as relating different but complete views of the same object.

Physicists took for granted that there was a special variable of the system called its *state* (of being, implicitly), independent of the experimenter, and completely describing the system. The state is thus a complete variable by itself, in that it determines all others. In classical particle mechanics the state is the specification of the positions and velocities of all the particles at one instant. Each determination of a classical system by an ideal experimenter simply fixes its state. Each action on the system simply transfers it from one state to another. Classical thought thus builds in an absolute distinction between knowing (fixing the state) and doing (changing the state). A classical relativity theory need merely specify how different experimenters represent these same absolute states in order to determine how they represent the same action.

When an ideal experimenter determines the state, the state couples to the experimenter, who learns something, but the experimenter does not couple to the state, which is fixed. Here the state is the absolute, like the time of Galileo or the space-time of special relativity. As a result, any classical variable—say a pendulum of theoretical mechanics—is compound.

To see this idol most clearly one studies the most elementary actions that define the variable. In classical thought each such action is represented by an arrow from one state to another. The collection of such arrow transformations is "closed." This means that doing two arrow transformations in sequence is again an arrow transformation, if it is defined at all. [We assign the ideal value 0 to the undefined case.]

The key point is that within the collection of these arrow transformations lies another closed collection consisting of those arrows that start and end at the same state, representing acts of selection or knowing. So the collection is not simple.

But the entire collection is not merely the composition of this closed subcollection with another. So the collection is compound.

[Brackets like this are asides to specialists, indicating the mathematics behind the words. In mathematicians' argot: The arrow semigroup of a classical object is not semisimple but a category whose objects are states. The corresponding semigroup of a quantum object is not a category but simple, being the projective semigroup of a vector space. It has no objects, hence no states. To apply Segal's analysis to quantum theory, one must generalize it from Lie algebras to linear algebras.]

By focusing on actions rather than states in this pragmatic way, we can discern the classical compound of doing/knowing. The quantum relativization then fuses the two into one simple concept of operation or action. All classical theories have this absolute, the state, and quantum theory relativizes it.

In quantum physics there is no complete variable. Learning (something about the system) and doing (something to the system) are no longer fantasized as fundamentally different kinds of action. The act of determining a property is an interaction between experimenter and system that now has significant consequences for both. Such reciprocity was expected by some on philosophical grounds long before experimentation at the photon level made it manifest.

The way the game actually played out is surprising, however, and unforeseen. The future value of any one variable may still be prepared long in advance, in principle, but not those of any two. For example, if the system is a particle, I may determine its future position in space at some time, or you may determine its future momentum at that time, but we cannot do both at once. Such complementarity between properties was not imagined before quantum mechanics. It is as alien to wave theory as to particle theory. A quantum acts like neither a wave nor a particle.

Rather than renounce the concept of the absolute state of being, some quantum physicists seize a quantum concept that should be called "mode," as in "mode of action," and call it "state." They thereby violate the correspondence principle, which relates classical and quantum concepts so that the two languages are mutually consistent where they both apply.

[The quantum Hamiltonian corresponds to the classical Hamiltonian, for example. But it is well known that a quantum mode corresponds to a classical construct obeying a Hamilton-Jacobi equation, characterizing a flow of infinitely many possible systems and simply related to the classical concept called action, and not to a classical state at all.]

Sometimes we still talk of "sunrise," "points of time," and "states" as if Copernicus, Einstein, and Heisenberg had never worked here. These locutions still work, if taken relativistically. One really means, "sunrise (or point of time or state) relative to my (or some other specified) frame of reference." In the present, more philosophical context, such implicit agreements cannot be taken for granted, and I avoid them here.

Experimentally fixing a property is now only a special case of an action on the system, and changing the property is another of the same kind. Now we no longer separate them but unite them in Heisenberg's one simple concept of operation without object.

The idea of visualizing anything completely and exactly, a goal of some mental practices, is renounced by Bohr and is alien to quantum mechanics. Since illuminating the system disturbs it unpredictably, completely visualizing anything "as it is" is self-contradictory. "As it is" means without external intervention, in which case the system is sitting alone in the dark, unperceived.

A quantum entity is simple compared to a classical one because among our actions on it there are no longer privileged acts of selection that are not also acts of transformation. There is no "is" here, just a "does."

The coupling coefficient that corresponds to $1/c$ and G for this relativization is h (Planck's constant), the hallmark of quantum theory. Again our idol test works, in that the diagnosis agrees with the outcome that was known in advance.

The quantum theory is so much simpler, more unified, and better-working than its ancestor that I am sure that we shall never go back to classical thought. We must therefore go forward.

In physical theories so far there have always been absolutes, vestiges of being, essences. Indeed, some call Einstein's theory of gravity a theory of invariants, not of relativity. What remains now that is absolute? What must we empty next?

As we have seen, we cannot always detect important absolutes easily from within a theory. By never moving, some idols make themselves invisible. We must step outside the theory and examine both what physicists say and what they do, and especially the connection between these two modes of action—the semantics of the theory—to discover what absolutes are tacitly assumed.

Let us apply the idol test to some parts of present physics and predict their evolutions. Now these are genuine predictions. They may even be wrong.

INTERACTIVE LOGIC

The existing quantum theory still has absolute concepts of predicate and negation. Some Buddhist logicians have paid great attention to the empirical basis of these concepts, and so have some quantum logicians. For Boole and quantum logicians, predicates or classes are defined by selective actions, before or after the fact, as distinguished from more general actions, though the resultant of two selective acts in sequence is generally no longer a selective act in quantum physics as it was in classical physics. And quantum physics still has an absolute negation, relating each predicate P to a unique predicate NOT P.

The absolute NOT of present quantum physics is too conspicuous to omit from our gallery of idols. There is a big difference in the empirical content of the classical and quantum concepts. Classical perception is imagined as ongoing and noninvasive. Our perceptions of quanta are generally limited to the brief moments when we perform input or outtake actions on the quantum, and totally invasive, with a quantum often being created fresh at the beginning of each experiment and annihilated at the end.

In normal macroscopic perception we have time to take cues from the object and readjust our analysis until it fits, usually without conscious effort. This permits us the illusion that we simply see something as it is, without choosing an analysis to which we subject it.

The prototype of all inspections in the quantum case is the polarizer for a photon. We have no time to adjust the polarizer to the photon. We can only say whether the photon does or does not pass the given test. Each test alters the photon uncontrollably, so repeating the test on the same photon is pointless.

In that case, we verify the judgment that IF A THEN NOT B by noting that none of the photons coming from an A filter ever pass a following B filter. An uncalibrated null detector suffices for this measurement. We verify the judgment that IF A THEN B by noting that every photon passing the A filter also passes the B filter. This requires a calibrated measurement or counting rather than mere null detection. For quanta the positive judgment is a bit more complicated than the negative one.

According to present quantum theory, some physical processes respect negation. This means that when predicates evolve under these processes the negation of the evolute of a predicate is the evolute of the negation of that predicate.

[In quantum theory these are the processes called unitary.]

Almost all quantum processes do not respect negation; these may be called negation violating [nonunitary]. For example, quantum interventions such as input, selection, and outtake violate negation.

In classical, prequantum physics, however, all system processes, even interventions, respect negation.

[Otherwise put: in classical theory the intersections of two disjoint sets with any third set are still disjoint, so selection respects negation, but in quantum theory projections of orthogonal mode vectors onto a third vector are generally no longer orthogonal, so selection does not generally respect negation.]

In this sense we may say that negation is inviolate in classical physics but not in quantum physics. Nevertheless quantum logic still has a fixed negation that all dynamical evolutions of isolated systems are supposed to respect between our interventions.

Present quantum theory also mentions processes that are not necessarily interventions, yet do not respect negation, such as the creation and annihilation of an individual quantum. For isolated systems creation and annihilation are not supposed to occur separately, however, but only in pairs that respect negation. The evolution of a closed quantum system between our interventions can be expressed as a sequence of creations and annihilations of that system and is assumed to respect negation.

The negation of quantum theory and the metric of special relativity are both associated with mathematical concepts of orthogonality. This formal analogy, and the fact that the metric of special relativity is relativized in general relativity and becomes interactive, has led some to suggest that the negation of quantum logic also be relativized. This idea has not worked yet.

This story of quantum negation is not finished. When a quantum falls into a black hole, for example, it is possible that a negation-violating process occurs, much as though the quantum were annihilated. We still do not know how to deal with black holes systematically within a quantum theory.

INTERACTIVE SPACE-TIME

General relativity simplifies matter-space-time, but at the same time it creates new idols and new one-way couplings that reactivate the idol alarm. Another relativization is due, and long overdue at that. I expect it to introduce another small physical constant having the dimensions of time, for the following reasons.

The one-way coupling that enters now is between field and space-time. It suffices to consider just the case where the field is defined by a vector at each point, like the flow velocity of the universal fluids of Kelvin, Descartes, or the Stoics. In a transformation from one frame of reference to another with relative acceleration, or to curvilinear coordinates, we must know the point in order to transform the field, but not conversely. The space-time coordinates of a point couple into its field, but a field-value does not couple into its point.

The field/space-time construct violates reciprocity much as the Galilean space/time construct does. In field theory the idol is space-time; for Galileo it was time. [These absolutes are bundle bases. In string theory the idol and base is the string manifold, and the fiber is the space-time coordinate manifold.]

Today space-time is as absolute relative to the field as time was relative to space in the seventeenth century. Space-time subordinates the field now as time then subordinated space. A fiber/base compound occurs in any theory that makes an absolute distinction between base coordinates and fiber coordinates, including general relativity, the standard model, and modern string and membrane theories.

Closer inspection reveals how we created this idol. It is not present in a discrete skeletal or network model of space-time, composed of atoms of space-time, where field vectors are reduced to chords or arrows, pairs of points themselves, representing elementary displacements of the atoms of space-time themselves. Points being simple, classical point pairs or arrows are semisimple. The coupling between two points in a pair is reciprocal, not one-way.

The vector/space-time compound emerges from such a polygonal structure only in the continuum limit of vanishing chord size, where the chord joining two points becomes a tangent vector asymmetrically assigned to one of the points. This is the limit where the differential calculus works. The small physical constant that is neglected in the old physics and will be the insignia of the new, if this prediction comes true, is a cut-off value for the limit of vanishing time-interval and is, therefore, probably a small time.

I infer that the physics of differential equations is a transient phase, and that it will evolve into a purely algebraic physics. Einstein (1936) considered this possibility without committing himself to it.

I call this ultimately small time the *chronon* χ, a word coined by Margenau.

Some have suggested that χ must be the Planck time. But the magnitude of χ probably cannot be set from within the theory where χ is 0 any more than classical mechanics can suggest the values of Planck's constant and lightspeed. To fix χ requires physical data incompatible with the degenerate theory with zero χ. The Planck time is a coupling coefficient from matter to metric, while χ is a cell size and a coupling coefficient from momentum to position. The fact that two coefficients have the same dimensions in the MKS system of units does not mean that they are even approximately equal.

There are only three independent units in the MKS system, and there are more than three couplings going on, so inevitably some coupling coefficients with quite different meanings will have similar dimensions.

Quantum and gravitational theory can be played off against each other to show that field theory has an infrared or long-time limit as well as an ultraviolet or short-time one, the result of black hole formation. The short-time limit χ is greater than the Planck time by as many orders of magnitude as the large-time limit is greater than χ (Finkelstein 1999). The ratio in question might be about 10^{10} or more.

The simple quantum entity replacing the compound tangent vector, the atomic unit of dynamics-space-time that bears the scale-size χ, I also call the chronon. The hunting of the chronon has gone on for some time. For example, Aristotle (1984) discussed and rejected the extended "indivisible lines" evidently proposed by his contemporaries, and the Kālacakra tradition of Buddhism includes space atoms (H. H. the Dalai Lama 1997).

To be sure, the tangent vector might be replaced by a chord, a point pair, in the network model mentioned earlier. A point pair is not simple, but at least it is semisimple. Then the underlying simple entity would correspond to one point of space-time, one end of an arrow.

I do not think that is what happens. In the examples of relativization we considered earlier, an idol merged into a simple, not a semisimple, structure under relativization. Synthesis occurs, not analysis. Points of space-time do not have natural evolutions, but tangent vectors do. Probably therefore the chronon does not replace a space-time point, with four coordinates, but a tangent vector, with eight. Since a tangent vector can be regarded as an operator connecting a point to an infinitesimally nearby point, this chronon also fits better into the historic pattern of replacing states of being by modes of action.

It has taken me some time to take this possibility seriously, because in classical thought a tangent vector is compound. But quantum theory simplifies it by introducing complementarity relations between its two classical components.

Now the separation of the matter-space-time network into field and space-time, inhabitant and habitat, becomes a local and relative one, like the division of space-time into space and time.

Moreover, the locality principle basic to Einstein is alien to quantum theory, though tolerated in our present hodgepodge relativistic quantum physics. From the algebraic point of view that is supposed to dominate quantum theory, there is no absolute difference between position and momentum, for example. One transforms from position modes to momentum modes with a harmonic (or spectral, or Fourier) analysis, an application of the superposition principle.

But locality refers specifically to position, not momentum. For a local interaction to occur between objects, their positions in space must agree at some time, but they can be far apart in their momenta.

Therefore the synthesis of quantum theory and relativity probably will relativize the locality of general relativity, replacing absolute locality by a process of localization carried out by an experimenter, one equal among many.

To indicate how little we have progressed in this direction, I mention that the programs of quantum gravity, supergravity, grand unified theory, string theory, superstring theory, and the standard model all incorporate absolute concepts of locality and space-time. One form of quantum relativity (Finkelstein 1996) that relativizes these concepts is still in a formative phase.

Renouncing such a successful common-sense absolute as the point-event in space-time leaves an emptiness which can be felt either as empowerment and liberation or anomie and nausea, depending perhaps on one's prior practice in coping with emptiness and relativity.

INTERACTIVE LAW

Another persistent absolute element of physical theory is the dynamical law, also briefly called here the *dynamic*. [This is the information usually imparted by giving the action function.] In present physics the dynamic influences the system, but the system does not influence the dynamic. The dynamic thus actuates our idol detector as directly as the metric did.

The inference is that the separation between matter-space-time and dynamic is another transitory one and will dissolve in the evolution of physics. The compound matter-space-time/dynamic will become a semi-simple matter-space-time-dynamic unity.

This process has begun. The relation between dynamic law and geometric is more than an analogy. The geometric law ultimately rests operationally on experiments with light, which is governed by a principle of stationary path time: the path time for the actual path of a light pulse is the same for all infinitesimally nearby paths, to the first order of approximation. The dynamic tells us how light and all other signals actually propagate, by giving us the quantum phase for various histories. The geometric law is merely the dynamical one viewed under coarse resolution and restricted to the special case of a light signal or mass particle.

Therefore the metric and the dynamic cannot be chosen independently of each other. The quantity that is stationary in the geometrical law is actually the quantum phase of the dynamic law, a dimensionless quantity. One converts it to the classical action using Planck's constant h, according to quantum theory, and to the path time using the mass of the signal and the speed of light, according to general relativity. When the dynamic governs the geometric, it governs only an aspect of itself.

This reflexivity is at present a plausible inference from the actual operations of the physicist, but it has not been established or even given a fitting mathematical formulation. It applies to fields other than gravity. In every case the determination of a field in a region is operationally the same as the determination of the dynamical law of a system moving in that region. The difference is our choice of what we take to be the system under study.

As a beginning teacher I would tell beginning students, "Physics is the search for the laws of nature." After I read more of Einstein, this became, "Physics is the search for the Law of Nature." Now I wonder what kind of creature such an absolute Law could be. Where could it exist? How could we perceive it, if we cannot change it? After all, perceiving any entity is operationally inseparable from changing the entity. I found the discussion of physical theory of Bohm (1965, especially the appendix) helpful on this point.

And what is this entity called Nature? Where do I stand to see it all sharply?

Now I am sure that only an atavistic vestige of the commonsense split between space and time inclines us to still think of dynamics as absolute, fixed by Nature. The dynamic represents what goes on inside the isolated system while we wait outside.

In present theories the kinematics—the theory of the descriptions of the system—and the dynamic are separate, and the distinction between them is absolute. But the coupling asymmetry between the two parts of the matter-space-time/dynamic compound implies that a further fusion into matter-space-time-dynamic is in the offing.

BREATHING IN EMPTY SPACE

Our little group at Georgia Tech is attempting a more relativistic theory of the matter-space-time-dynamic unity that I have discussed here, built on connection and complementarity, the cornerstones of general relativity and quantum theory respectively.

The sole variable is now the dynamic connecting past to future. When we describe the dynamic we are also specifying the space-time occupied and its material occupant relative to each frame of reference. The vicinity of an event consists of the other events immediately connected with it by the dynamic.

In a fully quantum theory any sharp kinematic description gives a probability amplitude for any other, and that is what a dynamical law is supposed to do.

The coupling coefficient from matter-space-time to the geometric law, the metric, is already known from general relativity. It is the very small Planck time, which I have suggested above is smaller than χ by many orders of magnitude. There is therefore a large number to account for with no units at all, the ratio of χ to the Planck time, about 10^{10}.

Several absolutes would still remain in such a physics, perhaps to be relativized in some later evolution, if the appropriate couplings ever become accessible to experiment. One is the universe; another is the system. Also, each theory we make today has itself as an absolute. The principle of universal relativity—like that of semisimplicity—seems incompatible with a fixed theory of any kind. The concept of a final theory is another idol that we must break before we can pass beyond the gate it guards.

Leibniz conceived of a theory as having three parts, a combinatoric, a characteristic, and a ratiocinatoric, which today have been called its syntax, semantics, and logistics, respectively. He imagined at least one of these parts, the semantics, as generative and open-ended, able to express new meanings as new experiences required. Only a generative, open-ended theory can incorporate a universal relativity.

One may think of any whole theory as a view, as etymology suggests. A view is a view from a position, which is then an idol of that theory. It seems that the process of making a theory inevitably introduces idols that only a later theory can break, and so the theory process can never be completed.

Extrapolating the evolution of physics has led to some hypotheses that resemble tenets of ancient philosophies, especially where both depart from the Cartesian world system. The Einstein energy-mass equation reminds some people of the *ātman-Brahman* equation of Hinduism, and Bohr complementarity seems to sharpen the Taoist reservations about language expressed as the beginning of the Tao Te Ching, for example.

Given the number of different philosophical positions, some such agreements with contemporary physics must be expected by chance, which can also produce beauty. On the other hand, different people sometimes come up with similar ideas because they have independently learned how to work with what actually goes on, as the squirrel and the squid evolve similar eyes out of different body parts. The counterintuitive physics of today evolved by dint of much physical experiment and mathematical theory, drawing its inferences from reproducible external experimentation, physical induction, and mathematical formulation and deduction. Buddhist conclusions seem to derive from life experience, meditative practice, and scholastic debate. There are some well-known harmonies between their conclusions (Stcherbatsky 1930), especially where both differ from the Cartesian philosophy. For example, they agree on

- the empirical revisability of logic;
- the representation of the world as a pattern of acts of termination and dependent reorigination;
- the atomicity of time;
- the indecomposability of the world;
- the incompleteness of any representation of the world

Rather than coincidences, some of these agreements might be due to the fact that both systems of thought work with entities of extreme sensitivity, for which Cartesian rationalism doesn't work well. When we observe a thought, it disappears; when we observe a photon, it disappears. These facts of experience create similar practical problems for the Cartesian metaphysics, and it is reasonable that they should lead to somewhat similar solutions.

When the S-I-W analysis, and other considerations that I omit only for brevity, led to the extreme relativistic surmise that the matter-space-time-dynamic is one unity, I vacillated for some time without committing myself fully to that hypothesis. I found such a unity frighteningly nonintuitive, but that is no indication it is wrong. Intuition is a lazy, docile ox that has to be trained to carry us where is best for us or it would wander into dead ends and pitfalls. The main problem was that there seemed to be too many possibilities. Exploring one wrong path can devour years. While Newton already proposed that the dynamical law is variable, and many have agreed with him, I have not encountered a development of this theory, let alone one where the law digests the system it swallows, as time has space. The closest precedent is Einstein's suggestion that the metric, which defines the dynamical law of a test planet, be the sole variable of a unified field theory. Even then, Einstein wrote a separate higher-level dynamical law to govern the evolution of the metric, once again splitting the variable governed from the governing law. Lacking a mathematical model of such an autonomous dynamic, I have no reason to suppose that such unification can even be self-consistent, let alone consistent with experiment.

Concepts like atoms of space or time and changing laws are discussed at length in Buddhist treatises of the previous millennium. I have already mentioned their far-reaching relativism. The basic heuristic principle at the root of the S-I-W criterion, reciprocity, is hardly new. Relativistic contemplation could have led to a similar unification of the governed with the governing law long ago. I wondered whether this specific relativization and simplification had already been explored.

For five days in 1997, five physicists (Arthur Greenberg, Piet Hut, Arthur Zajonc, Anton Zeilinger, and I) discussed traditional Buddhist physics and modern physics with the fourteenth Dalai Lama, not as a national or religious leader but as a Buddhist monk versed in the Madhyamaka tradition and interested in science. Two bilingual and bicultural communicators (Thupten Jinpa and B. Alan Wallace) bridged our linguistic and conceptual differences. The discussions are recorded elsewhere.

We found that we held several basic positions in common from the start. For example, there was no recourse to faith or divine revelation. Most of the physicists agreed with the Dalai Lama that knowledge, even of the rules of logic, comes from experience and is revised by experience. Here too was a school of thought as systematically relativistic as I had hoped.

Among many other things, we touched on the questions I have raised here. The Dalai Lama had thought about space atoms and was aware of the modern intuitionistic logics that suspend the law of the excluded middle. He propounded his belief that science must be rooted in compassion. This seems to support Sakharov's equation,

$$\sqrt{\text{Truth}} = \text{Love},$$

the root of truth is love.

When the question arose whether concepts like a variable matter-space-time-law unity had been expressed in the Sanskrit or Tibetan Buddhist literature, verses from Nāgārjuna's Madhyamaka treatise *The Fundamental Wisdom of the Middle Way* were cited. From a recent translation of a translation (Nāgārjuna 1995), it seems that they could indeed be read as saying that space, time, matter, and causation are interactive, with no permanent essence, and that this is inferred from the very fact that we perceive them.

Critical steps in the evolution of physics have required us both to break prior idols and to form appropriate new ones. The Madhyamaka appears to focus on the first part of this process, the emptying of concepts, and not on the formation of new idols, which are perhaps among the concepts called "conventions" in the translation of Garfield (Nāgārjuna 1995).

Laplace and Einstein believed in the existence of an absolute law and took it as the supreme goal of physics. But other Western scientists and philosophers, including Newton, Mach, and Whitehead, like some Buddhist and Hindu philosophers, declared that there is no fixed absolute law of nature, and that it makes sense to speak of a varying law. Bohm's (1965) expression of this philosophy especially influenced me. He views a scientific theory as a specialized extension of normal human discourse. A theory is something that we tell one another. A final all-inclusive theory is as likely as a final all-inclusive story.

Again, Smolin (1997) attempts to account for many details of our present law of nature by a Darwinian evolution of that law.

The simplicity test for idols suggests that the dynamic too, with its one-way coupling to the system, is an idol within a compound system/dynamic. Then the variable dynamic and the variable system are both aspects of one deeper quantum variable, and there must be a reciprocal coupling from system to dynamic through a small physical coefficient that is implicitly treated as 0 in present physics. Combined with the other relativizations we have discussed, this fusion means that what goes on in nature is a simple quantum-space-time-matter-dynamic unity. Perhaps the process that goes on may be represented as law changing. It is moot whether we would describe such an evolution of physics as the end or the true beginning of the dominion of law.

I have argued for the relativity of the dynamic previously (Finkelstein and Rodriguez 1984), but the traditional goal of the one fixed absolute law, which now seems so naive, still disfigured my own effort to marry space-time and quantum theory as late as 1996 (Finkelstein 1996: 16.8.3). Now a promising algebraic setting for the operations of a matter-space-matter-dynamic has presented itself (Baugh et al. 2001), and I have been able to replace the search for an absolute dynamic by the study of interactive ones and their average properties. This work is still too speculative and far from experiment to merit more space here.

When Einstein (1936) considered applying Heisenberg's "purely algebraic method" to space-time, he likened it to "trying to breathe in empty space." Now emptiness has acquired another meaning for me, and his simile seems even more apt. The space-time of Einstein and of physics today is still absolute, full of essence. It seems likely that we must cross at least one more relativistic bridge, marked χ on my map, to reach enough emptiness for the next major evolution of physics to breathe.

References

Aristotle. 1984. "On Indivisible Lines." In J. Barnes, ed., *The Complete Works of Aristotle*. Vol. 2. Princeton: Princeton University Press.

Bacon, F. 1620. *Novum Organum.* Trans. and ed. P. Urbach and J. Gibson. Peru, Ill.: Open Court, 1994.

Baugh, J., D. R. Finkelstein, A. Galiautdinov, and H. Saller. 2001. "Clifford Algebra As Quantum Language." *Journal of Mathematical Physics* 42:1489–1500.

Bohm, D. 1965. *Special Relativity*. New York: Benjamin.

Einstein, A. 1936. "Physics and Reality." *Journal of the Franklin Institute* 221:313–347.

Finkelstein, D. R. 1996. *Quantum Relativity*. Berlin: Springer.

Finkelstein, D. and E. Rodriguez. 1984. "Relativity of Topology and Dynamics." *International Journal of Theoretical Physics* 23:1065–1098.

H. H. the Dalai Lama. 1997. Personal communication.

Inönü, E. and E. P. Wigner. 1952. *Nuovo Cimento* 9:705.

Kepler, J. 1965. *Somnium, sive Astronomia lunaris Joannis Kepleri.* In J. Lear, ed., *Kepler's Dream.* Trans. P. F. Kirkwood. Berkeley: University of California Press.

Nāgārjuna. 1995. *Mūlamadhyamakakārikā.* In J. L. Garfield, ed. and trans., *The Fundamental Wisdom of the Middle Way: Nāgārjuna's Mūlamadhyamakakārikā.*

Segal, I. 1951. "A Class of Operator Algebras Which Are Determined by Groups." *Duke Mathematics Journal* 18:221

Smolin, Lee. 1997. *The Life of the Cosmos.* Oxford: Oxford University Press.

Stcherbatsky, Th. 1930 [1970]. *Buddhist Logic.* St. Petersburg. Osnabrück: Biblio.

Thurman, R. 1993. Personal communication.

In the preceding two essays physicist-philosopher Michel Bitbol and theoretical physicist David Finkelstein present philosophical and scientific theories entailing a good deal of interpretation of the experimental data. In the following essay experimental physicist Anton Zeilinger presents what is tantamount to the "bare data" of a conversation between himself another physicist, H. H. the Dalai Lama, and two other Buddhist scholars. Here he has resisted the powerful temptation to interpret, or even edit, the contents of this encounter. Rather, he presents what was said and leaves readers to make of this conversation what they will.

Following the first Mind and Life conference on physics and Buddhism in October 1997, Zeilinger, who was one of the participants in that meeting, invited the Dalai Lama to a conference in Innsbruck, the International Symposium on Epistemological Questions in Quantum Physics and Eastern Contemplative Sciences. In this lively essay Zeilinger recounts some of the highlights of discussions of epistemological questions in Buddhism and modern physics during that symposium. The participants quoted here are H.H. the Dalai Lama, Anton Zeilinger, fellow experimental physicist Arthur Zajonc, Thupten Jinpa, and myself.

The first issue raised concerns the limits of analysis: by what criteria can you conclude that you have explained as much as can be explained about something and simply say, "That's the way it is"? Or, in other words, at what point do you justifiably stop asking why and simply accept something for what it is with no further explanation? This question is equally pertinent for scientists and Buddhists alike, where the former describe the laws of nature, the latter describe the laws of karma.

In our discussion about atoms, photons, and quanta, Zeilinger points out that scientific theories, such as quantum mechanics, are based on the world of everyday experience. When physicists then probe into the nature of quanta, their features seem to become, in a sense, very unreal. Many physicists have pointed out the incongruity between the world of everyday experience and the world of quantum mechanics, claiming that we can't even imagine what the latter is really like because it is so dissimilar to our experience of the world around us. Such incongruities between exceptional contemplative insights and everyday experience are also very prevalent in Buddhism.

The final topic of discussion cited here is cosmology, addressing the nature of the Big Bang and Big Crunch. In some Buddhist accounts the Buddha simply refused to address issues such as the origins of the universe, commenting that these are not relevant to the pursuit of spiritual awakening. Other classic Buddhist literature, however, claims that the present universe originated billions of years ago in an event comparable to a Big Bang; but this was but one episode in a beginningless sequence of Big Bangs. Thus, a standard model of cosmology in Buddhism is that of not only one oscillating universe but innumerable universes going through successive phases of origination, abiding, and eventual destruction. Nevertheless, many Buddhists are content to leave this as an open question, or even an irrelevant question, for it has little to do with how we lead our lives from day to day, or even lifetime to lifetime.

Like all good scientific discussions, this symposium did not result in any final, definitive answers to the questions raised, but it did grapple with important issues and opened up fresh avenues of inquiry.

Anton Zeilinger

Encounters Between Buddhist and Quantum Epistemologies

When in 1997 my colleague and friend Arthur Zajonc invited me to visit His Holiness the Dalai Lama in his residency in Dharamsala to discuss modern physics, I was very excited. My work on the foundations of quantum mechanics had led me to asking progressively deeper epistemological questions, and I was interested in discussing some of these questions with His Holiness, whom I saw as the representative of one of the large spiritual traditions of the world.

The meeting in Dharamsala itself was organized and arranged by the Mind and Life Institute under its president, Adam Engle. The meeting itself took place from October 27–31, 1997. Every morning, on five consecutive days, one participant would present his views on a question of quantum physics or cosmology. In the afternoons there would be long discussions and debates among all of us. The participants were

His Holiness the Dalai Lama, Tenzin Gyatso
David Finkelstein
astronomer George Greenstein
Piet Hut
Thupten Jinpa
philosopher Tu Weiming

Alan Wallace

Arthur Zajonc

An account of this meeting will be published in due course, edited by Arthur Zajonc.

I should mention that for myself the first encounter with Buddhism was when, on the morning of October 27, 1997, I met His Holiness personally and presented an introduction into fundamental concepts of quantum mechanics. I had never investigated Buddhism in any depth before, nor did I ever have close encounters with Buddhist philosophy. My main surprise during the discussions with His Holiness was to learn that Buddhism is not just a spiritual tradition but represents a very concise and stringent logical and epistemological system. I would also like to mention that the discussions with His Holiness were very rewarding not only on a personal level. He always found the right pertinent questions, like the best student one can imagine having as a teacher. Very impressive as well is His Holiness's openness concerning Buddhist teachings. More than once, he indicated that should we in Western science ever find anything that contradicts Buddhist teaching, then that teaching must be changed.

Since I had been told about His Holiness's interest in technical apparata, I had brought some experiments with me to Dharamsala. One was a setup of the famous double slit experiment and the other one was a demonstration of the polarization of light. I am sure that the double slit experiment was the first quantum experiment ever performed in Dharamsala. As the experiments both apparently succeeded in being very instructive, and as they were able to arouse His Holiness's interest, I dared to invite him to come to Innsbruck, where I was situated at that time, in order to see our laboratory. To my great excitement, he immediately accepted this invitation.

So, in June 1998, I organized a conference entitled the International Symposium on Epistemological Questions in Quantum Physics and Eastern Contemplative Sciences in Innsbruck, again with partial financial support by Mind and Life. The meeting with His Holiness himself took place over two and a half days. Arthur Zajonc, Thupten Jinpa, Alan Wallace, Adam Engle, and I met two days earlier in order to prepare the meeting in depth through intensive prediscussions. We also felt the need for more discussions afterward for a day or two.

During the laboratory visits the discussions focused on various fundamental quantum experiments that I had arranged with my group in our laboratory. Again, it was interesting to see how His Holiness always addressed the deep epistemological issues directly, challenging us again and again to give the evidence for the various statements we made as physicists. If I may dare to say this, His Holiness might have become a great physicist in another world without his duties as spiritual and political leader of the Tibetan people.

It is very difficult to do full justice to the depth and breadth of the debates we had both in Dharamsala and in Innsbruck. It is also very difficult to avoid mistakes in analyzing the exchange of concepts and ideas from one's own perspective. Therefore I feel it best to just recount below some interesting parts of the discussion that focused on issues pertinent to both the Western quantum and the Eastern Buddhist epistemology. I understand that this gives only a very limited impression, but I am sure it will convey some of the spirit of the debates. A somewhat more detailed representation of the debate was published in the German monthly magazine *GEO* in January 1999. I am certain that some time in the future this material will be analyzed carefully and presented in a broader way.

ABOUT THE LIMITS OF ANALYSIS

Zeilinger: What kind of things might we discuss today? We were thinking of talking about some philosophical or general questions that have already come up in the discussion with Your Holiness in Dharamsala. We are interested in discussing the following: there are some instances, or were some instances, in the discussions in Dharamsala where Your Holiness said that a line of reasoning sometimes can only be carried to a point where you finally say, "That's the way it is. That is the nature of the situation." What we really would like to know is, when can we say that? When do we know that we cannot give any further reason? This is important in the discussions about the foundations of quantum mechanics, for there we also have similar situations, and we would really like to know from your tradition when you say, OK, that's it.

. . .

HHDL and AW: [*Tibetan*]

AW: In traditional Buddhist philosophy, if you ask why does the apple fall down, would you not say that that's simply the way things are? In other words, that's it. His Holiness said, no, that would be incorrect. You haven't found an explanation. The mere situation of not having an explanation is not a sufficient criterion for saying that's just the way things are. Then I said, how about the color of the sun? The sun is yellow. Why is it yellow? Would a Buddhist not say, why, that's just the way things are? His Holiness said no. Once again, you don't have an explanation, but that's not a sufficient criterion for saying that's just the way things are. But then you can make another statement, which His Holiness is content with, and that is, if you engage in a positive action, it has a positive result, karmically speaking. Well, why is that? If you engage in generosity, then affluence or prosperity is a karmic result of that. Why is that? Then we'd say, that's the way things are.

HHDL: [*Tibetan*]

AW: And likewise, in terms of the salient characteristics of consciousness. What are they? They are that it has a luminous or clear attribute. One. And that it has a cognitive attribute. Two. Why is it that consciousness has these two salient characteristics? It's just the way it is. Maybe you could also say, why does a photon have these three attributes? Why does it have direction, frequency, and polarization? Would you say, that's just the way it is? Or do you have anything further that even conceivably could be said about it? Is there possibly an underlying explanation to that, or do you have the confidence that's it? That there's no more to the story.

Zajonc: That's a good example.

AW: And Buddhists have a lot of confidence about the nature of consciousness—that that's it. There's nothing obscured here. It's not merely our ignorance. That's just it (Tib. *chos nyid*, Skt. *dharmatā*) the way it is.

Zeilinger: And they have this confidence because of analysis or what?

HHDL, AW, and TJ: [*Tibetan*]

AW: His Holiness is saying that it's not only that we don't *know* any further explanation, there *isn't* any more explanation.

HHDL: [*Tibetan*]

AW: And likewise, you must have situations like that in quantum mechanics. You have confidence not just that you can't *find* an explanation, but that there *isn't* one.

HHDL: [*Tibetan*]

Zajonc: But for us there *are* sometimes some gray areas where *we're* not sure. For example, one is you could say there are certain fundamental units, for example, Planck's constant or the charge on the electron. This is a universal constant. You could ask, Why is Planck's constant exactly the value that it is? Now there's some discussion, and maybe there are cosmological reasons that this is actually not just the way things are, but this is because of the distribution of energy or something or other in the universe. So, there's that. Or why is it, for example, that an object has inertia? Why does it resist motion? Well, you could say that's just the way things are, but there are other people who say, no, that's because it stands in a particular relationship to all the masses in the universe, so it has inertia. *These* are to me some gray areas. It's not sure that you really have a boundary or whether it's still open to some discussion. So, it would be nice to know if there were rigorous criteria that one could apply.

. . .

Zeilinger: Just to come back to consciousness, to analyze that in the spirit of what Arthur said, sometimes it's not completely clear whether this is just the way it is or whether there might be an open inquiry.

AW: [*Tibetan*]. I just asked, How do you distinguish that? What are the criteria?

TJ: [*Tibetan*]

AW: There is no explanation, and I don't have one.

Zeilinger: Right. How do you know that? When do you know that?

AW: You have to take it case by case.

HHDL, TJ, and AW: [*Tibetan*]

TJ: When His Holiness is speaking from the Buddhist point of view, he's taking into account even the existence of Buddha's omniscient mind, which is supposed to know everything. So, even from the Buddhist point of view, to the question "Why does consciousness have these two salient features?" one can respond only with "That's the way it is."

HHDL: [*Tibetan*]

TJ: Similarly, to the question, Why does positive action lead to positive results and negative action lead to negative results.

. . .

Zeilinger: Maybe the best example of something like "that's just the way it is" exists in physics. So far the program is to base everything on certain symmetry principles. Symmetry in a very general sense, not just right/left symmetry but also the following symmetry: for example, that the laws of nature should be the same now and in ten minutes. It need not be the case, but nature seems to be that way. And they are the same here in Innsbruck and in Dharamsala. There doesn't seem to be a difference. These are what we call symmetry principles—that the laws stay the same, even as you change something. Concerning these symmetry principles, we probably have to say, "That's just the way it is." There is no deeper reason.

HHDL: [*Tibetan*]

TJ: So, would you say that many of the natural laws are, That's the way it is?

AW: The laws of nature and so forth? Gravity?

Zajonc: Some of these can be derived from this way of viewing things.

AW: [*Tibetan*]

Zajonc: They are entailed. In one sense you can say, well, yes, that's the way things are because in order for the world to be, yes it has to be this way. But the physicist finds it more elegant to move back to very simple presuppositions, very simple statements, very general ones.

HHDL, AW, and TJ: [*Tibetan*]

AW: As general as possible and as few as possible.

TJ: [*Tibetan*]

Zajonc: Maybe in that sense, the situation would be, "That's the way things are." Those are the laws of nature.

TJ: So the most fundamental ones.

Zajonc: Yes, the most fundamental ones. There are a few very beautiful, very simple principles that then allow you to derive many specific results concerning gravitational theory, electromagnetic theory, quantum theory, and so forth. And then you wonder, well, what's the power of this simple principle? A number of physicists have thought of it in a kind of a metaphysical way—this is so powerful that this must represent a kind of greatest possible order in the universe. This simple law. And there you would say that's just the way things are.

AW: There's a similarity in Buddhism. In Buddhism we speak of four laws of karma. One of them is what His Holiness said: if your action is harmful, you get a negative result back. There are three other ones, and none of these can be derived from the others. So, they are all equal in status. But for each one I think you would say, that's just the way it is. And they are fundamental, and the implications are enormous. So there are very few—they have totally universal application and a massive number of derivatives.

Zajonc: In specific circumstances you may have a law, which if you look at a deeper and deeper level reduces to some combination of more basic laws. You wouldn't say, that's the way things are, necessarily, on the higher, the more incidental level. You'd try to trace it back.

ON ATOMS, PHOTONS, QUANTA

HHDL, AW, TJ: [*Tibetan*]

TJ: We can go up to the minutest particle and the Mādhyamikas would reject the notion of some kind of fundamental, absolute, elementary building block—

some kind of atom, that is, or particle that is indivisible. For the Mādhyamikas, they would argue that that concept is incoherent.

Zeilinger: It's ontological. There is, at least within physics, a breach with a long-standing tradition. The idea of physics until the beginning of the twentieth century was that you can basically explain—at least in principle—you can basically explain why specific things happen. There was this old picture of the universe being a clockwork, which was at some time started by the Christian creator—there was the ultimate clockmaker, God, who built the clockwork—he started it—and now the universe is running deterministically. The idea was simply to explain what the laws are according to how this clockwork runs, and you have to start with—in what we call in physics—the initial conditions. You have to know what the universe looked like in beginning, and then the rest is clear.

Now, in the modern view, we know that such a picture is not possible anymore. In other words, the facts in the universe in five seconds *are* not determined by the facts in the universe now, at least not completely. I think that has consequences for the way we view the world. To me, such an open view of the universe is much nicer than the old view of a closed universe where everything follows its course—it's much more open. It's much more romantic in a sense. It's not so boring.

. . .

Zeilinger: My friend, Abner Shimony, is one of the few people in the world who is both professor of physics and professor of philosophy. He says that most practicing physicists are schizophrenic. They have two parts of their brain. In one part of their brain, when they are in the laboratory and play with things, when they play around and do something, they are realists. They talk about the photon. They talk about this going here and there and so on. But when you tell them, but now let's talk about the foundations of quantum mechanics, they switch to the philosophical side, and they say, oh nothing exists without the apparatus defining it and so on.

. . .

TJ: So the question His Holiness is asking is that from your point of view, from the quantum mechanical point of view, given these ambiguities, would it make sense to talk about reality in general?

. . .

Zeilinger: You know the problem is, if you investigate things in detail, the nature of the things, then you can dissolve everything.

So, the problem looking at quantum mechanics, the way out is that we say we have to start from somewhere. We have to build our worldview on something. This something onto what we build is sometimes called the classical world, or you could also call it the world of everyday experience. We build it. We do not doubt this. And on the basis of the properties of these things that we immediately perceive, we build the rest of our description. Then we observe many things, and the quantum features of objects become, in a sense, very unreal. We should be very cautious then to talk about reality there. From a very purist point of view, you can say that all we really can talk about is these experiences of the everyday world. The rest always has to be taken with a grain of salt.

HHDL, AW, and TJ: [*Tibetan*]

Zeilinger: We should be open about any other statements about the world. *We* should be open to the possibility that this could be completely wrong or changeable, or—well, we should be flexible in that respect. The question then, which is probably one of the most important ones we can ask, is, "If we start from everyday experience, when we build models—could it be that in our way of looking at the world, in Western science, we built just one of many possible models? Maybe a completely different physics will be possible. Could that be? Should we start fresh again? Maybe the turn Galileo and Newton took is not the only possible one. This is to me one of the most fascinating questions. Could we build a science that looks completely different? I would simply like to get an idea how to attack such a question. I don't know.

. . .

AW: Arthur, we're trying to make a relationship with the macroworld, and that is, I was arguing that the photon as the attribute bearer is a cipher. To think of it as having something real, independent of these totally indeterminate, nonlocal probabilities of attributes, is just a cipher. It's just a nominal cipher. But then His Holiness, and also Thupten Jinpa-la, says, well, wait a minute, you get a whole bunch of these so-called ciphers and then you have to squint your eyes. So, how do you move from the discourse about light on the quantum level to the stuff that gives you sunburn? Or, you know, the stuff that makes you squint your eyes? That is, if you're going to make ontological statements that the quantum realm exists only in relation to system measurement and so forth . . .

Zajonc: I think this puts you in a better position, because what you've done, you've said you've taken the measurements, or let's call it generally the observations, really seriously. You don't posit the existence of this thing without attributes on the other side, so what you're squinting at is not this thing that's on the other side. What you're squinting at is brightness. Brightness is given standing, not this attributeless thing. You'd say that what causes sunburn is a particular attribute of light. It's not the thing on the other side that causes sunburn. The materialist feels like you can't give an account of the world that's robust unless you have the thing on the other side.

AW: We're getting into very deep ontological waters here. I love what you said there: we can't posit the photon independent of the measurement system, but we can't say it's simply an artifact. You've set up a classic situation of having to find a middle way. Where somehow subject and object are inextricably, primordially related. This is right in the lap of Buddhism, on the one hand. On the other hand, it seems like the sun is blowing out things in all directions irrespective of any system of measurement. It seems like. How are you relating that which the sun is blasting out.

. . .

Zajonc: It's not anthropomorphic somehow or other. It's not . . .

AW: Waiting for little measurement systems all over the place to say, OK, you photon, you can exist. OK, you can have a polarization. I think we really need to avoid what most scientists have fallen into, and that is localizing all the problems into your lab. Rather, get them out into the macroworld. What implications does this have in the macroworld? And what are the ontological issues—are there ontological repercussions?

Zajonc: It's good.

AW: I think we must do that today. Otherwise this is so [local].

Zajonc: It will be so hard. But it's good.

. . .

HHDL and TJ: [*Tibetan*]

AW: So, we're making very interesting ontological statements. But now, as His Holiness was saying, meantime, light is very bright. The sun is giving light in all directions. So, are these photons coming from the sun, do they also not exist independent of the measurement system? Who's doing the measurement? And should all human beings vanish—with no measurements—who believes that the sun then doesn't give any photons anymore? So, how to relate quantum mechanics with the everyday world, and with the sun, the stars?

. . .

Zajonc: . . . One final example of that is every molecule that binds together by what is called covalent bonding—so, for example, a hydrogen molecule relies on this ambiguity, this quantum mechanical ambiguity, to bind those two atoms together into a single molecule. So without this kind of quantum ambiguity all our chemistry would disappear. Life itself wouldn't be possible. So, it's not only a kind of abstract property of photons in universities with lots of money.

. . .

HHDL: [*Tibetan*]

FROM THE BIG BANG TO THE BIG CRUNCH

TJ: His Holiness was saying that if one has to posit only one Big Bang . . .

HHDL: Then why did it happen?

TJ: Why did it happen? That's really a big question.

HHDL: That's also part of nature. If it continues all the time, then it is much easier to accept.

Zeilinger: Yes, I understand what Your Holiness says, but I would really like to leave it open.

HHDL: Really! [*Laughter*]

AW: It's a safe position. A respectable position.

Zeilinger: You know about the other end—it's still open. It is still open whether there will be a Big Crunch or not. This is undecided.

. . .

Zeilinger: I would not say it does exist . . . it's an empty question. It's the same as—there is supposedly the story that in the middle ages they discussed how many angels can sit on the tip of a needle. How many angels can sit on the tip of a needle? It's a useless question. [*Laughter*]

. . .

HHDL: [*Tibetan*]

TJ: In some Buddhist philosophical writings the claim has been made that in relation to physical objects—matter—there is no beginning in terms of the continuum of the causal chain. But there will be an end to the continuum. But from the point . . .

. . .

Zeilinger: Another question is whether there will be a Big Crunch.

HHDL: Big Crunch . . .

Zeilinger: The two go together. What you need for that is for the universe to have a certain minimum mass, so the gravity is strong enough to pull it together again. Now, it turns out that the mass which we see in the universe, the stars and the galaxies, is, at most, maximally, about 10 percent of the critical mass necessary to get the Big Crunch. So, people have been talking about the missing mass problem. Is there some other mass somewhere in the universe that would make it possible—the Big Crunch? One of the possibilities would be many, many particles that have mass, but that would be very difficult to see, nearly impossible to see.

AW: Because they have no charge?

Zeilinger: Because they have no charge and they have no other properties that make them easy to detect. They only interact very weakly. The candidate for this kind of particle, for a long time, is the neutrino. The neutrino is a very tiny particle. The question was, "Does the neutrino have zero rest mass?" If you stop it, is it like the photon, having no rest mass? Or does it have some rest mass, a little bit of rest mass? It's a long discussion. Because if the neutrino would have even a tiny rest mass and there are so many of them, it could possibly, as we say, close the universe. It could easily make the universe such that it will collapse again. Just two weeks ago, in some experiments in Japan, they claim to have found evidence that the neutrino has a rest mass by studying the neutrinos coming from the sun. And they claimed that they found evidence for a little bit of a rest mass.

HHDL: [*Tibetan*]

Zeilinger: If that is true, then the universe might be closed.

HHDL: We're trapped.

Zeilinger: Not in our lifetime.

HHDL, AW, and TJ: [*Tibetan*]

TJ: The conclusion is that this is pointing towards a Big Crunch?

Zeilinger: Yes. That's right. If that is really true, then it makes a Big Crunch possible again.

AW: So, it's not even on the cusp. It's over the cusp and you can say yes.

Zeilinger: The exact numbers are not there. The exact numbers are not out—what the mass is and so on. But certainly—the question is open again. So, it's easily possible. But it will not happen during our lifetime.

HHDL: I think, billion.

Zeilinger: Billions of years.

Zajonc: It's a good reason for reincarnation. [*Laughter*]

Zeilinger: So, we can watch the Big Crunch. [*Laughter*]

AW: Make your reservations now.

Zajonc: We'll see you there.

Note

I would like to acknowledge the support of all those who made these encounters possible, most significantly Adam Engle, the Mind and Life Institute, and all the people who helped in the organization. In Innsbruck this was perfectly done by Andrea Aglibut.

Piet Hut

Conclusion: Life As a Laboratory

What can be the stage for a dialogue between Buddhism and science? Calling Buddhism a religion is not a very accurate description, and the very notion that science might produce a worldview is not correct, since there is still so much that is left out from a scientific description. At this point it might be more prudent to start talking about mutual respect and inspiration between science and Buddhism, with an eye toward future more detailed discussions. One way of phrasing a possible middle ground between both is to start by viewing life as a laboratory, as an opportunity to examine ourselves and our world, using working hypotheses rather than doctrines.

1. BUDDHISM AND NATURAL SCIENCE

When I was asked to write a paper for a book on Buddhism and Science, I was quite reluctant to agree to do so. I very much appreciate the body of knowledge that science has established, over the last few hundred years. I also very much appreciate the body of knowledge that Buddhism has established, over the last few thousand years. And I definitely see both types of knowledge as pertaining to the same reality, with potentially large areas of overlap that could lead to fruitful dialogues. My reluctance, however, stemmed from the lack of a proper framework within which to carry out such a dialogue.

The last few years have seen an increasing popularity of discussions around the relation between science and religion. After decades during which it was extremely unpopular among scientists to even mention the word *religion,* now the tide seems to be turning. I have seen many colleagues "coming out of the closet," so to speak, as I have done myself by writing papers and attending meetings on the general topic of science and deeply felt human experience, with a nod toward spirituality. At the same time, I feel a deep unease with the way in which many "science and religion" dialogues are framed.

Already the very terms *religion* and *spirituality* I find deeply problematic and, frankly, I wish I could avoid using them altogether. Instead of using those lightning rods, I would prefer to focus on an authentic attention for what it means to live a life from a deep respect for the full human condition, with head and heart and guts and all our faculties, in a fully integrated way. Most any culture has placed the cultivation of a full and all-round form of personhood at the top of their agenda. In China for example, Confucianists and Taoists alike, notwithstanding all their differences, focused on the cultivation of our full humanity, the former starting from our societal embedding, the latter from the way we are still part of nature. Our contemporary Western culture is strangely lacking in this respect, which makes a discussion of the relation between science and religion even more difficult.

First of all, the word *religion* is a European term, a category that has been used to compare Christianity with Judaism, Islam, Hinduism, Buddhism, Taoism, and other ways of life. In practice, talking about "other religions" implies an immediate comparison with Christianity. No longer are these "other religions" labeled as heathen, but by default they are considered to play a role similar to that of Christianity. In academia Christianity has traditionally occupied the slot assigned to theology, and in the last century or so a parallel slot has been assigned to departments of comparative religion. By definition, then, the study of Islam and Buddhism and other ways of life are confined to one academic compartment, as opposed to other compartments such as physics, psychology, law, etc.

In practice, though, this classification is highly problematic. As anthropologists know, importing and imposing one's own culturally bound classification system on a different culture precludes a fair and respectful intercultural comparison. Distorting even the most basic terms at the start of a dialogue, in favor of your own way of using those terms, is not a good way to gain understanding of the other side. By putting a shaman healer and a Buddhist meditator and a Muslim mula all in the box of religion, rather than medicine, psychology, or law, we severely limit the possibilities for meaningful comparisons.

Second, even if there is a willingness to search for terms more fitting than *science* and *religion,* it is hard to find good alternatives within the English language. *Spirituality, mysticism,* and *contemplation* are three terms that I would prefer over *religion,* in a comparison with science. However, each of those has its own problems. The word *spirit* in *spirituality* has no clear place when we discuss Buddhism. I personally like the word *mysticism,* since I feel that at least some forms of medieval mysticism form a better starting point in a dialogue with Buddhism than contemporary Christianity. However, the word *mysticism* is often used to indicate something vague and deliberately unclear, which doesn't help start a dialogue. Finally, the word *contemplation,* perhaps the most innocent of the three, may sound too passive, even though some traditional contemplative movements in Christianity could also provide a good starting point in a dialogue with Buddhism.

How, then, to begin a dialogue between Buddhism and science, especially natural science? Which aspects of Buddhism to start with, and how to classify those aspects? These questions deserve a detailed study all by themselves. I hope that they will be addressed patiently, and in great detail, over the years to come. My guess is that it would be a full-time job for a number of Buddhist scholars in conversations with several scientists, over a period of years, to lay appropriate foundations for a meaningful and ongoing dialogue. The current volume can only give a hint as to a possible direction for the construction of a foundation for such a long-term project.

In this essay I will try to address a much simpler question. By asking myself what type of inspiration I have received from those forms of Buddhism with which I have had contact, over the last thirty years, I hope to provide a case study of how Buddhism and science can come together, at least in the life of an individual. To be precise, let me emphasize that I do not call myself a Buddhist. Although I have great sympathy for many aspects of Buddhism, I have a similar sympathy for aspects of many other ways of life, especially the mystical/contemplative sides of many traditions. And if I were forced to accept a label, I would be hard pressed to choose between Buddhism and, say, Taoism or Hinduism.

402 Conclusion: Life As a Laboratory

For me, personally, I would summarize the overlap between Buddhism and science in my own life through the expression "life as a laboratory." My main theme in the rest of this paper will be to try to convey a sense of what that phrase means for me. In the final section 6 I will explicitly address this notion of a lab life. Leading up to that, I first discuss some methodological issues, in sections 2 and 3, of a more epistemological character, followed by the much harder ontological questions pertaining to the relation between Buddhism and science, in sections 4 and 5.

2. BUDDHIST INSPIRATION

I am not a specialist in Buddhism, and I have never attempted to make an exhaustive study of any particular aspect of any Buddhist tradition. Rather, I have read widely in the available literature in search of practical guidance for personal exploration. Within Buddhism this has led me, first, to Japanese Zen Buddhism and, later, to various forms of Tibetan Buddhism, especially Nyingma Buddhism and within Nyingma to Dzogchen. This exploration started when I was at the end of high school and has been continuing for more than thirty years.

The original inspiration that I drew from Buddhism rested on the fact that I recognized an experiential approach to the structure of reality that seemed to be akin to the experimental approach in science. The meditative techniques in Buddhism that I came across were relatively simple and straightforward, and they were presented as tools for exploration rather than tasks to be undertaken by true believers who had already bought into particular creeds, to be accepted in blind faith. The superficial simplicity of these tools quickly turned out to be deceptive; I realized through my own experience that it took years to get enough sense of these tools to begin to appreciate their deeper effects—not unlike the training required to play a musical instrument well, to learn mathematics, or to really master a sport.

This is not the place to go into a detailed discussion of the various meditative explorations that I have engaged in. Let me just mention two aspects of those journeys that I have found to be particularly useful in all aspects of my life. The first is a type of meditative exercise in which you are expected to just watch whatever arises in your mind stream, without any form of judgment, without holding on to pleasant thoughts or emotions or images or whatever may occur, and without avoiding unpleasant ones. Such a systematic exercise of withholding judgment I found enormously helpful. Over the years my growing ability to abide with this shift to a nonjudgmental attitude has become a form of shelter for me—not so much a protective armor as a form of delicious transparency to life's temptations to pick and choose and grumble.

The second lesson I learned from Buddhist meditative pursuits is to hold something lightly—to hold it in the palm of your hands, so to speak, without latching on to it too tightly. Rather than looking at something with a fixed stare, this light attitude calls for an exercise of a more peripheral form of vision, both literally and figuratively. It is as if one is more concerned with the space around an object or idea than with the usual focus of attention. I'm struggling with words here, to try to convey what I mean, and once again I wish we had a vocabulary. Ah, the joy of physics in which you know what momentum and energy are, and in which you know that other physicists know what they are, and that they are linear and quadratic, respectively, in velocity, and so on. If only we had such clarity of terms in meditative pursuits! Note that the peripheral nature of what I am trying to describe here by itself poses no obstacle for precise terms. In science we have learned to talk about chaos and nondeterministic differential equations and uncertainty relations in very precise terms. One can certainly use clarity to talk about twilight and darkness!

Clearly, what is needed is a patient construction of a framework to allow a contemporary discussion of terms and processes and discoveries in the course of meditative practice. To some extent, this will be a form of reconstruction, since millennia of Buddhist explorations in a great variety of cultures have left us with a legacy of "lab reports" and analyses thereof. However, lack of continuity through many of those disciplines, together with the wide gulf between those cultures and ours, preclude any straightforward attempt at reconstruction. A fresher approach to building up a contemporary reformulation based on rediscovery seems to me to be a more promising path.

One major difference that I felt between the forms of Christianity I had grown up with and the Buddhist practices that I explored later revolved around experience. I relished the Buddhist emphasis on its view of reality as something utterly concrete and accessible, something that could be experienced and realized here and now, by anybody—not something to be stumbled upon only in the afterlife, if one lived a good enough life in blind faith. In short, the whole approach of Buddhism appealed to me more because of its similarity to science, where guidelines are taken as working hypotheses rather than as dogma that one just has to accept.

Having put it this way, I realize that I am being too harsh here in my formulation. For a scientist, to work painstakingly on research based on a single hypothesis, which may or may not turn out to be right, requires more than just curiosity. Unless he or she harbors a strong belief, or at least a very strong intuition, that the hypothesis is right, it is hard to imagine someone putting in years of effort to prove the hypothesis. And for a Christian, in practice faith is rarely, if ever, fully blind. But perhaps I should be more concrete, in order to make comparisons between Christianity, Buddhism, and science more authentically grounded in my own life experience. To do so, in the rest of this section let me focus on the particular brand of Buddhism that most appeals to me. For simplicity, let me call this brand the goal-as-path type. And let me make it clear right away that this brand is far from universal; my guess would be that at most 1 percent of Buddhists worldwide focus their world view and their practice on this type of vision. Traditionally, such views have been considered to be extremely esoteric and not meant for mass consumption, although recently more and more has been written about them.

What I have in mind are forms of Zen Buddhism (to use the Japanese term, from the original Chan Buddhism in China), as well as Dzogchen in Tibet, and perhaps some forms of Advaita Vedānta in India. What sets these goal-as-path views apart from most forms of spiritual training, and from any type of any training in anything, is a subtle yet profound difference in emphasis. Typically, we engage in training in order to gain a certain result. We may or may not have much of an idea what that result might be, but at least we engage in the laborious training program offered in the conviction that something is lacking, and that that something can be obtained through appropriate effort. We lack the ability to speak French, or to ride a bicycle, so we learn to speak French by practicing French, and we learn to ride a bicycle by practicing bicycle riding. At least that much seems to be universal in the learning of any type of skill.

Not so for the goal-as-path forms of Buddhism. The whole idea of starting off ignorant and unskilled and then following an arduous path of training while slowly approaching our goal is uprooted. This whole model of progress simply does not apply. The path one follows is described not as starting at the beginning and leading to a goal but rather as starting at the goal. Training is not seen as a trick to acquire something but rather as a way to celebrate and cultivate in celebration what we already have—and fully have and are.

This way of summarizing may seem paradoxical at best, if not downright silly. Let me try to give at least a flavor of the approach. In Christianity there is a strong emphasis on surrender. Because we, as weak and limited (and sinful) creatures, do not have the ability to reach far beyond ourselves, there is no hope for us to get very far, unless we open ourselves to God's grace. Fortunately, God is always ready to bestow His grace, and the only thing we need do is turn toward Him and accept His grace: Christ is already knocking on our door, so all we have to do is open the door of our heart. So in Christianity, too, we are not far removed from the goal of having direct communion with the divine, and living from that Source—in fact, we are not at all removed.

The challenge of course, in Buddhism as well as Christianity, is the question of how to open the door to our heart (in Christian terms) or how to realize one's true nature (in Buddhist terms). The subtlety in both cases is this: if you don't do anything, accepting the doctrine that everything is already perfect or that we are already close to God, we completely miss the point; while if we try hard to "achieve" what is already there, we also completely miss the point. We are damned if we do and damned if we don't—or so it seems. In Zen Buddhism this quandary has been nicely summarized in thousands of koans, presenting us with seemingly impossible situations and asking us for an answer where clearly no answer seems to be in sight.

My Christian upbringing thus prepared me for a way of thinking that was at least somewhat familiar to the Zen and Dzogchen views of practice as starting at the end of the path, at the goal rather than at the beginning. Not only that, after diving deeper into various Buddhist practices, I did find to my delight various Christian sources that seemed to have a similar approach, at least in spirit. Various medieval mystics, like Ruusbroek, or the anonymous author of the *Cloud of Unknowing*, or Meister Eckhart, seemed to live and breathe this paradoxical view of cultivating what is already there, rather than searching for anything. And there are other examples, less directly of a goal-as-path type perhaps, but for me equally inspiring. Saint Francis for one, who lived his Christianity in many ways as a Taoist in tune with nature in the widest sense of the word, is someone whom I found deeply inspiring. He, too, seemed to breathe at least the spirit of an "already there" view.

I plan to elaborate elsewhere the various points I have touched upon here only in passing. Within the framework of a single chapter I cannot provide more than a glimpse of what it is in Buddhism that has inspired me for so many years. Let me move on, to present a corresponding glimpse of what it is in science that has also inspired me tremendously, for an equal number of years.

3. SCIENTIFIC INSPIRATION

As a practicing scientist, a theoretical astrophysicist by profession, I cannot help but let my daily work color my view of the world. I have learned particular approaches to problem solving in my job, and it is natural that I will fall back on some of those approaches when exploring reality in its more contemplative (mystical, spiritual) dimensions.

One such approach, which I find myself practicing in my work, is what I would call the "on-the-table" method. When I am confronted with a problem that I cannot solve, even after using the standard methods for a while, I back off a bit and stop trying to solve anything, for the time being. Rather, I try to put all the pieces of the puzzle on the table, trying to be careful not to overlook anything and at the same time trying to be careful not to smuggle in pieces that are unrelated or that are based on premature guesses of what the solution might be. In order to arrive at "the truth, nothing but the truth, and the whole truth" one has to have all the pieces of the truth at hand, no more pieces and no fewer pieces.

Once I feel reasonably comfortable that I have collected all relevant pieces, I let my eyes stroll over the table, so to speak, lingering on each piece to study it as if for the first time. While I need to take time in an unhurried way to complete the first phase, putting everything on the table, I really need to take a leisurely approach in the second phase of "staring" at each piece. Any attempt to hurry would break the fragile sense of letting an overall understanding crystallize out, starting from each local piece and hopefully growing into a more global understanding. The third phase then simply involves a sitting back and looking at the whole table, with all its pieces, letting everything "sink in." More often than not, at the third stage new insights have emerged for me. In some cases I can see the solution of the problem in a flash, in other cases I suddenly get an idea for a new and fruitful approach, in yet other cases I realize, suddenly or gradually, that the problem as it is posed has no solution.

In daily life, too, I find myself using a similar approach in many situations that at first seem too complex or confusing. Almost always I realize, after taking my time to lay out a situation and "stare" at it, that my confusion stemmed from either dragging in too much or overlooking a crucial part of the situation. And also the same method seems to be fruitful in experiential attempts to understand more of myself and my relationship with others and with the world. The term *contemplation* is a particularly apt way to describe this aspect of various forms of "spiritual" practice. And, of course, there are obvious connections with what I described in the previous section as nonjudgmental forms of Buddhist meditative training.

Did my exploration of Buddhist (and other) forms of practice shape my approach to scientific problem solving? Or was it my scientific curiosity that drove me to explore the fabric of reality through the lens of Buddhist investigations? Or does it simply reflect aspects of my temperament and personal history, which somehow drove me to study both science and Buddhism? I find it impossible to give clear answers to these questions. I certainly don't see the connections as a one-way street, neither from science to Buddhism, nor the other way around. If I were to guess, I'd bet on a mutual process of sensitizing, of learning to make more and more subtle distinctions and thus inviting more clarity, examining my own life and the world around me. I can't very well test such a guess, since I cannot go back to my childhood to try out a life of only doing science and another life of only focusing on Buddhism.

Another approach that I stumbled upon in my work, over the years, also seems to carry over in other aspects of life. This one I would call the "space around" method. It relates to the way I have learned a piece of mathematics or physics, something that always seems to happen to me in three stages. At first, I study the individual parts of the method or theorem and I familiarize myself more and more with each part, until I can clearly see how it works, or in the case of mathematics, why it has to be correct, based on the axioms. After some time I then naturally reach a point where I begin to see the whole structure of the method or theorem.

This second stage feels as if one explores an object by studying each side separately at first and only later obtains a full three-dimensional sense of the object. The emergence of the gestalt, this full presence of the object, can happen gradually or suddenly, and, even if it happens suddenly, it can then be refined and deepened further in a gradual way. Another example is the learning of a language: after having learned words and grammar, there comes a delightful moment in which one realizes that, for the first time, one just spontaneously spoke a whole sentence or could understand a whole sentence being spoken.

Depending on the complexity of the concepts and the intricacy of the method or derivation, this first phase may last anywhere from hours to days to weeks or months. The second phase typically lasts much longer and often is the end point for me of any given engagement with a problem, because I will have moved on to a study of another aspect of the field in question or even a different field. However, there are occasions where I keep coming back to the same problem, for whatever reason. Perhaps I am teaching a specific concept or method a number of times, or I happen to use the same method in a variety of different applications in my research, or my curiosity drives me to return to a particular method with no obvious motivation other than that I feel a form of intellectual attraction.

In such cases, often many years after I learned the method or derivation for the first time, it may happen that I begin to get a sense for the "space around" the situation at hand. In the second phase the problem had become fully embodied and more or less transparent, with all its parts blending harmoniously and functionally together, but the whole problem was still hanging in the air, so to speak, without a clear context of neighboring problems and approaches. Only in the third phase can I begin to see some of the richness of the variety of all possible approaches to this problem—using a mathematical metaphor, I begin to see more of the space of all possible approaches. Richer dimensions unfold in which I "see" the problem as being embedded.

One way in which such an insight translates itself is that I can now clearly imagine, for the first time, how someone might have proved this theorem or found this solution for the very first time. Seeing a solution fully and clearly is one thing, but seeing how one could have derived it without any prior knowledge is quite something different. And another way in which this third phase announces itself is more practical: often I can see wholly new approaches, either to the same or to different problems, now that I have tapped into the type of creativity needed to solve the problem for the first time. Far from this being a useless exercise, to "deeply feel" how one could discover something that had already been discovered anyway, this lifting of the fog "around" a problem has always been profitable for me in other applications.

As with my earlier example, I have found this space around method to be extremely useful, both in daily life and in experiential contemplative quests. In both cases a long-term familiarity, over many years, coupled with a stubborn return to feeling out the problem and going over the same terrain again has often resulted in an unexpected dividend in terms of opening new horizons. And yes, there are clearly connections with what I have described in the previous section as a form of peripheral attention. I have even used the same term of *space around* to try to describe both ways of deepening and ripening insight. Again, it is hard for me to answer questions concerning the chicken and the egg, and I really don't know in detail how these two approaches, meditative and scientific, evolved in my own life. But I do know that for me they live together in a comfortable complementarity; so much so that attempts to view science and religion as pertaining to wholly different domains strike me as patently incorrect.

In itself, one might object, a similarity in method does not prove that we are dealing with an overlap in area of applications. Agreed. A similar mathematical formula may describe the frantic dance of electrons and ions in a glowing plasma in the laboratory and the majestic dance of stars moving through a galaxy (electrostatic and gravitational forces both fall off according to the inverse square of the separation between two particles). This does not imply that electrons and stars are players on the same stage (well, they are, but on such vastly different scales that effectively they seem to live in a different world).

In contrast, science and Buddhism are not talking about vastly different objects, such as a star that is a trillion trillion trillion trillion trillion times more massive than an electron. Rather, both science and Buddhism are dealing with the same world we live in, with our human bodies and minds as they appear in the world we live in. If in addition the methods of investigation show significant similarities, it is hard to avoid the question "What contains what?" Is science a precise form of exploring and describing a specific facet of a wider reality that Buddhism claims to cover? Or is Buddhism a way in which humans have learned to cope with the complexity of mind and world and, as such, subordinate to applications of neuroscience and physics, for example? Granted that there are epistemological similarities, which of the two can legitimately claim ontological pride of place?

In the remainder of this chapter I will try to address these questions, albeit with considerable trepidation. I feel that currently neither our scientific knowledge nor our understanding of the relation between Buddhism and Western academic knowledge is deep enough for a comprehensive attempt to come to terms with these questions. So in what follows I will limit myself to make only a first attempt at an attempt to grapple with these issues.

4. THE VIEW FROM SCIENCE

Let us return to the question of the relation between science and Buddhism or, more generally, science and other ways of knowing. Buddhism seems to have remarkably little trouble accepting the results of scientific investigations and incorporating scientific insights into a Buddhist world-view. In contrast to Christianity, there seems to be little concern with Darwinian evolution, for example, and the emphasis in physics on the central role of cause and effect is familiar to a Buddhist as well: it plays a major role in Buddhist descriptions of the world, both on physical and the psychological levels.

Whether a scientific worldview leaves room for a Buddhist way of looking at the world is a more difficult question. In fact, the very question is already wrongly posed, since there is as yet no scientific worldview, and I don't expect there to be one for at least a century. For an approach to reality to be comprehensive enough to be called a worldview, at the very least such a view should have room for human life, meaning, dignity, responsibility, and other aspects of what it means to be human. Mythologies can provide a worldview. Religions can provide worldviews. But the current scientific description of reality leaves out far too much to deserve the name *worldview*. As for future developments in science, let us postpone that question to the next section.

Scientists do have views about the world, though, and they sometimes express their views emphatically. When you listen to the voices of spokespersons for science, you hear a wide variety of opinions about the extent to which science covers all reality, in practice and/or in principle. And they hold a similarly wide range of opinions concerning the amount of room that is left for other ways of knowing, such as those advocated in Buddhism. As to whether scientific and Buddhist views can coexist, some scientists will answer in a widely affirmative way, whereas others are adamantly in opposition.

When you probe coffee table conversations among practicing scientists, the picture is yet different: by and large questions about the relation between science and Buddhism, or any other worldview, are simply not asked. Until recently it was implicitly understood that such questions were not proper. It just was one of the many unwritten rules of the guild of scientists: to keep a clean separation between your personal views and the objectivistic atmosphere in which science was performed.

Fortunately, this climate of denial is beginning to change now, for a variety of reasons. Whether it is the end of the cold war, the decrease of science funding, the generational turnover of scientists, the shift of emphasis from physics to biology, the rapid progress in neuroscience and the questions triggered thereby—whatever the reasons are, it is certainly a positive and liberating experience for those of us who have been suppressing our real interests for so long.

An ironic aspect of this tacit suppression of spiritual and contemplative values is the fact that science never has been able to stand on its own, as far as worldviews go. There simply never has been anything close to "a scientific worldview." Whether there ever will be is a different question, but for now, at least, science is still far too young to weave a full story about the world we live in, about what it is to be human, about values and meaning and beauty and responsibility. The recent attempts of sociobiology and evolutionary psychology to put together narratives that aim at analysing and "explaining" values, etc., are not much more than "storytelling." Sure, there are elements of truth in what they put on the table, and there are many interesting threads worth following. However, to suggest that such stories could replace a full worldview in the near future is simply preposterous.

Of course, many traditional worldviews are now considered hopelessly naive and outdated, as seen through our modern eyes. Whether such a critical attitude says more about limitations in the older views or about limitations in our current attitude is an interesting question in itself, but there is no need even to go into that question. Already, it is perfectly clear that older worldviews did provide a place for humankind, a role in the cosmos, a world that had order and meaning. This is something science has never aspired to and, not surprisingly therefore, has never delivered.

This is another ironic aspect of modern science. Ask a scientist whether science has anything to say about meaning and values, and chances are high that the scientist will make a point of explaining how science avoids that question, excluding that from its terrain of investigation right from the start. And then, moments later, in a full and contradictory reversal, this same scientist may be heard to speak in a denigrating way about all kinds of nonscientific views as being forms of wishful thinking and superstition, not up to the same standard as scientific views. Talk about having your cake and eating it too! To exclude whole areas of human life, and especially the most important ones, from scientific analysis, and then to use the self-assigned nonscientific character of those areas as a reason not to take them seriously is nothing but a logical fallacy.

It is not so hard to guess what is behind such irrational behavior. The implication is the belief that, first, some day science will grow far and wide enough to cover all aspects of human life and that, second, that future science will show anything nonscientific to have been a form of an unsubstantial, unreliable, and more or less superstitious way of looking at the world. Perhaps I'm putting it somewhat too sharply, but I have met enough colleagues who have reacted that way so I may testify that this kind of attitude is quite prevalent among scientists. And barring the possibility of peeking into the far future, such an attitude is simply pitting belief against belief: the belief in a clean, clear, reductionistic form of future science telling us all that is worth knowing versus the belief in prescientific ways of knowing. Note the often implied denigration in the very use of the term *prescientific*, as if that by itself disqualifies such ways of knowing as somehow having outlived their shelf life!

A truly rational attitude would not fall back on such promissory stories about how a future science will prove the current expectations of a full scientific worldview to be redeemed. But given that such an attitude is quite widespread, among scientists and lay persons alike, I would like to present an alternative. Let me sketch how I expect science to develop.

5. THE FUTURE OF SCIENCE

For the last four hundred years, since Galileo introduced the scientific method of a combined, observational, experimental, and theoretical investigation, science has focused on the object pole of experience. In any scientific description the describing scientist has taken care to step out of the picture and stay hidden behind the camera. In every observation the observing scientist has systematically hidden both the subject of the observation and the lived act of observation, leaving only the object of observation as a residue. When the act and the actor are described, those two are made into objects as well, in a third-person form of description of generic actors and generic acts. Anything resembling first-person experience, let alone the presence of first-person subjects, has been filtered out.

Initiating this process was not at all a bad move. In many ways it was a brilliant way to get an inroad in the complexities of the structure and behavior of matter. Trying to analyze and understand the even greater complexities of subject, object, and experiential acts altogether might well have proven to be too difficult to have a chance to succeed. And, indeed, the objectivistic turn has been enormously successful. However, success tends to lead to arrogance and narrow-mindedness, and now we have reached a point where we have to face up to the limitations of this clever move of neglecting two-thirds of the fiber of every experiential situation we normally find ourselves in.

Fortunately, there is no need to put pressure on science to change itself. Science, unlike scientists, does not get easily stuck in one mold. The wonderful thing about science is that it has enormous resilience. Scientists may, and often do, get stuck in their ways, but there have always been young rebels who have shown novel and better ways to make progress, given the impasses that a previous generation found itself in. The fact that science hands out browny points for new ideas that can explain the observed data in more parsimonious ways, no matter how unconventional the theoretical structures proposed are, is the condition of possibility for its progress.

It is this single fact that separates science from most types of worldviews, from ideologies to religions to various other ways of knowing. Very few human organized activities have this emphasis on honoring successful innovation. Note that I am taking here a long view of science, averaged over at least half a century: many individual innovators were ignored if not abused for decades, if not longer, until they were finally exonerated. However, in the long run, science, more than any other way of knowing that I am familiar with, has managed over and over again to shed its earlier skin, in order to make room for fresh growth into a larger skin.

So what can we expect to happen in the future? Will science continue to be locked into its one-sided objectivistic development? I don't believe that for a moment. The signs to the contrary are all too obvious. In physics, quantum mechanics has taught us that a straightforward description of the world in terms of objects, independent of how they are being observed, is untenable. The observing subject, be it a person or a machine, plays an essential role in defining even how an object can appear. In biology and medicine, neuroscience is getting closer to being able to provide us with a translation table between objective, third-person descriptions of electrochemical events in our brain and subjective, first-person experience reported by the person whose brain is being studied. And in computer science and artificial intelligence, building robots provides us with the challenge of figuring out how to construct artificial subjects rather than traditional tools in the form of objects.

These three inroads into a study of the subject are examples of how science naturally grows and transforms itself. While pushing the envelope of what it means to limit oneself to a study of the object pole of experience, naturally science will expand further and move by its own momentum into a study of the subject pole of experience. At first this will happen at those points of intersection where the extension into the world of the subject is simply unavoidable, as in the examples I have given above. But, after a while, the study of the subject will undoubtedly become a regular part of science. How long will this take? My guess is that a full-blown study of the subject will emerge on a timescale comparable to that needed to build up a detailed study of objects. If we start with Galileo, the era of the object has spanned four hundred years. Perhaps we will proceed faster now, but it may well be that a study of the subject is more difficult intrinsically. If I were to venture a guess, I would bet it would take another three hundred years before we have a well-balanced science, equally focused on subject as object.

6. LIFE AS A LABORATORY

Returning to the question of a dialogue between science and Buddhism, I see various possibilities. In the long run, such dialogues will become normal and natural, the more so when science gets deeper into the study of the subject. However, this may not happen in our life time. On a shorter timescale, individual scientists can of course find inspiration in Buddhism, while individual Buddhist scholars, monks, and lay persons can find inspiration in science. As I mentioned earlier, I have been deeply inspired by the laboratory-type approach that I find in Buddhism, with respect to the study of mind, self, and world. Personally, I expect to continue to view my life as a laboratory, as a stage on which to examine myself and others, trying out various approaches in gradual attempts to find better ways to live my life.

Socrates' judgment that the unexamined life is not worth living is pertinent here. Now that Socrates has inspired a hundred generations of seekers who were born after him, we can ask ourselves how we can formulate his injunction in modern terms. Socrates started his rational inquiry through a search for definitions, in what I see as an attempt to use rationality to go beyond rationality, showing in a rational way the intrinsic limits of a rational approach. Following the thread of rationality, we have embarked on a detailed study of nature, which now begins to lead to a detailed study of ourselves as the human subject studying nature and living a full life as a human being. Know thyself, study thyself—in modern terms: consider your life as a laboratory, as an opportunity to refine your understanding of all that comes your way. Viewing all our ideas and all that we have learned so far as working hypotheses, we can avoid getting glued to our own prejudices.

In addition to a future public role for Buddhism and science, and a contemporary private role, we can of course begin to discuss possible connections in a tentative, exploratory fashion. This volume is an example of such an approach. I hope that the next generation of scientists and Buddhists, as well as Buddhist scientists, will find ways to build up a structure and vocabulary to extend these tentative beginnings into a more firmly grounded exchange. One project that aims toward this goal uses a series of summer schools, gathering graduate students from the sciences and from science studies such as philosophy, sociology, or history of science, to openly discuss issues concerning science and experience. More information about these summer schools can be found on the home page of the Kira Institute that sponsors these events (see the Kira website: http://www.kira.org).

Note

I thank Roger Shepard, Steven Tainer, Bas van Fraassen, B. Alan Wallace, and Arthur Zajonc for their comments on the manuscript.

Appendix: A History of the Mind and Life Institute

The Mind and Life dialogues between His Holiness the Dalai Lama and Western scientists were brought to life through a collaboration between R. Adam Engle, a North American businessman, and Dr. Francisco J. Varela, a Chilean-born neuroscientist living and working in Paris. In 1983, both men independently had the initiative to create a series of cross-cultural meetings between His Holiness and Western scientists.

Engle, a Buddhist practitioner since 1974, had become aware of His Holiness's long-standing and keen interest in science as well as his desire to both deepen his understanding of Western science and share his understanding of Eastern contemplative science with Westerners. In 1983 Engle began work on this project, and in the autumn of 1984 Engle and Michael Sautman met with His Holiness's youngest brother, Tendzin Choegyal (Ngari Rinpoche), in Los Angeles and presented their plan to create a week-long cross-cultural scientific meeting. Rinpoche graciously offered to take the matter up with His Holiness. Within days, Rinpoche reported that His Holiness would very much like to participate in such a discussion and authorized plans for the first meeting.

Varela, also a Buddhist practitioner since 1974, had met His Holiness at the 1983 Alpbach Symposia on Consciousness. Their communication was immediate. His Holiness was keenly interested in science but had little opportunity for discussion with brain scientists who had some understanding of Tibetan Buddhism. This encounter led to a series of informal discussions over the next few years; through these conversations His Holiness expressed the desire to have more extensive, planned time for mutual discussion and inquiry.

In the spring of 1985, Dr. Joan Halifax, then the director of the Ojai Foundation and a friend of Varela, became aware that Engle and Sautman were moving forward with their meeting plans. She contacted them on Varela's behalf and suggested that they all work together to organize the first meeting collaboratively. The four gathered at the Ojai Foundation in October of 1985 and agreed to go forward jointly. They decided to focus on the scientific disciplines that address mind and life, since these disciplines might provide the most fruitful interface with the Buddhist tradition. That insight provided the name of the project, and, in time, of the Mind and Life Institute itself.

It took two more years of work and communication with the Private Office of His Holiness before the first meeting was held in Dharamsala in October 1987. During this time the organizers collaborated closely to find a useful structure for the meeting. Varela, acting as scientific coordinator, was primarily responsible for the scientific content of the meeting, issuing invitations to scientists and editing a volume from transcripts of the meeting. Engle, acting as general coordinator, was responsible for fund-raising, relations with His Holiness and his office, and all other aspects of the project. This division of responsibility between general and scientific coordinators has been part of the organizational strategy for all subsequent meetings. While Dr. Varela has not been the scientific coordinator of all of the meetings, he has remained a guiding force in the Mind and Life Institute, which was formally incorporated in 1990 with Engle as its chairman.

A word is in order concerning these conferences' unique character. The bridges that can mutually enrich traditional Buddhist thought and modern life science are notoriously difficult to build. Varela had a first taste of these difficulties while helping to establish a science program at Naropa Institute, a liberal arts institution created by Tibetan meditation master Chogyam Trungpa as a meeting ground between Western traditions and contemplative studies. In 1979 the program received a grant from the Sloan Foundation to organize what was probably the very first conference of its kind: "Comparative Approaches to Cognition: Western and Buddhist." Some twenty-five academics from prominent North American institutions convened. Their disciplines included mainstream philosophy, cognitive science (neurosciences, experimental psychology, linguistics, artificial intelligence), and, of course, Buddhist studies. The gathering's difficulties served as a hard lesson on the organizational care and finesse that a successful cross-cultural dialogue requires.

Thus in 1987, wishing to avoid some of the pitfalls encountered during the Naropa experience, several operating principles were adopted that have contributed significantly to the success of the Mind and Life series. These include

choosing open-minded and competent scientists who ideally have some familiarity with Buddhism;

creating fully participatory meetings where His Holiness is briefed on general scientific background from a nonpartisan perspective before discussion is opened;

employing gifted translators like Dr. Thupten Jinpa, Dr. Alan Wallace, and Dr. José Cabezón, who are comfortable with scientific vocabulary in both Tibetan and English; and, finally,

creating a private, protected space where relaxed and spontaneous discussion can proceed away from the Western media's watchful eye.

The first Mind and Life conference took place in October of 1987 in Dharamsala, later published as *Gentle Bridges: Conversations with the Dalai Lama on the Sciences of Mind*. The conference focused on the basic groundwork of modern cognitive science, the most natural starting point for a dialogue between the Buddhist tradition and modern science. The curriculum for the first conference introduced broad themes from cognitive science, including scientific method, neurobiology, cognitive psychology, artificial intelligence, brain development, and evolution. The Dalai Lama, at our concluding session, asked us to continue the dialogue with biennial conferences.

Mind and Life 2 took place in October 1989 in Newport Beach, California, with Robert Livingston as the scientific coordinator. The conference focused on neuroscience and the mind/body relationship. Coinciding fortuitously with the announcement of the award of the Nobel Peace Prize to His Holiness, the two-day meeting was atypical for the Mind and Life conferences both in its brevity and its Western venue. The dialogue was published as *Consciousness at the Crossroads: Conversations with the Dalai Lama on Brain Science and Buddhism*.

Mind and Life 3 was again held in Dharamsala in 1990. Daniel Goleman served as the scientific coordinator for the meeting, which focused on the relationship between emotions and health, and has been published as *Healing Emotions: Conversations with the Dalai Lama on Mindfulness, Emotions, and Health*.

During Mind and Life 3 a new mode of exploration emerged: participants initiated a research project to investigate the neurobiological effects of meditation on long-term mediators. To facilitate such research, the Mind and Life network was created to connect other scientists interested in both Eastern contemplative experience and Western science. With seed money from the Hershey Family Foundation, the Mind and Life Institute was born. The Fetzer Institute funded two years of network expenses and the initial stages of the research project. Research continues on various topics such as attention and emotional response.

We met for the fourth Mind and Life Conference in Dharamsala in October 1992, with Francisco Varela again acting as scientific coordinator. The dialogue focused on the areas of sleep, dreams, and the process of dying and was published as *Sleeping, Dreaming, and Dying: An Exploration of Consciousness with the Dalai Lama.*

Mind and Life 5 was held in Dharamsala in October 1995. The topic was altruism, ethics, and compassion, with Richard Davidson the scientific coordinator. The dialogue was published by Oxford University Press in November 2001 as *Visions of Compassion: Western Scientists and Tibetan Buddhists Examine Human Nature*, edited by Richard Davidson and Anne Harrington.

Mind and Life 6 opened a new area of exploration beyond the previous focus on life science, moving into the new physics and cosmology. The meeting took place in Dharamsala in October 1997, with Arthur Zajonc as the scientific coordinator. The volume covering this meeting is in preparation.

At the invitation of Anton Zeilinger, who was a participant in Mind and Life 6, the dialogue on quantum physics that had begun in Dharamsala was continued at a smaller meeting, Mind and Life 7, held at the Institut für Experimentalphysik in Innsbruck, Austria, in June 1998. That meeting has been described in the cover story of the January 1999 issue of *GEO* magazine of Germany.

In March 2000 we met again with His Holiness in Dharamsala for Mind and Life 8, with Daniel Goleman acting again as scientific coordinator and Alan Wallace as philosophical coordinator. The subject of this meeting was destructive emotions, and the book covering this meeting is being written by Daniel Goleman with the assistance of Zara Houshmand and will be published by Bantam Books in January 2003.

Mind and Life 9 was held at the University of Wisconsin at Madison in cooperation with the HealthEmotions Research Institute and the Center for Research on Mind-Body Interactions. Participants were His Holiness, Richard Davidson, Francisco Varela, Matthieu Ricard, Paul Ekman, and

Michael Merzenich. This two-day meeting focused on how to most effectively use the technologies of fMRI and EEG/MEG in the research of meditation, perception, emotion, and on the relations between human neural plasticity and meditation practices.

The Mind and Life Institute was created in 1990 as a 501(c) 3 public charity to support the Mind and Life dialogues and to promote cross-cultural scientific research and understanding.

Mailing address: 2805 Lafayette Drive, Boulder, CO 80305
Website: www.mindandlife.org
E-mail: info@mindandlife.org

Index

Abhidharma, 77, 179*n*31, 283, 298–99, 302*n*1; and neuroscience, 54; and physics, 288–93; and Theravādin Abhidhamma, 174*n*11

absolutes, 70, 73, 83, 127, 133, 329, 355*n*20; false absolutes, 363, 365–69; in quantum theory, 373, 376–80; *see also* essence; idols

absolutism, 132, 327, 329, 350; absolutists, 25, 330; *see also* eternalism

acceleration, 290, 375; nonacceleration, 369–70; accelerated detector or observer, 346, 370; *see also* momentum

aesthetics, 22–24, 26; in Kantian literature, 331–332, 340, 342; aesthetic joy, 259

afflictions; *see* mental afflictions

agency, 123, 134–135, 162–64, 183*n*42, 213; agent, 114, 122–25, 132, 200, 288, 343–44; freely acting or rational agent, 106, 179–80*n*36, 180*n*36, 181*n*38

aggregates, five psychophysical, 26, 101, 136, 147, 156, 174*n*11; synopses of, 92–93, 103*n*3; *see also* conditioned phenomena; feelings; form; perception; recognition

aggression; *see* aversion

agnosticism, 133; agnostics, 337

ālayavijñāna; *see* substrate consciousness

alienation, 56, 129, 137, 170, 285

Alpbach Symposia on Consciousness, 417

altruism, 56, 259, 262, 273, 278, 420; altruistic motives or actions, 269, 303, 339

Ames, W. L., ix, 283–84, 285–302

amplitude, 359*n*76, 380

anātman; *see* no-self

Anderson, B., 187–88*n*62

androcentrism, 62*n*5

anthropocentric position, 277, 279*n*15

anthropological study, 37, 253; anthropologists, 157, 169, 183*n*42, 185*n*50, 185*n*52, 186*n*55, 400

antimonies, 327–29, 331

archeological study, 78

Aristotle, 193, 199, 204, 212, 377; Aristotelian rationalists, 123; Aristotelian space and time, 367–69, 377, 379

artificial intelligence, 268–69, 413, 418–19

Āryadeva, 325

Asaṅga, 218

astronomy, 10; astronomers, 272, 342

astrophysics, 3; astrophysicists, 406

Atiśa, 220–21

ātman, 104, 147, 381; *see also* self

Talbot, M., 326

tantra, *see* Vajrayāna

Taoism, 51, 134, 400–1; Taoists, 400, 405

Tarthang Tulku, 249, 251, 254

Taylor, C., 360*n*90

technology, 23–24, 292, 421; Tibetan
views of, 41–43, 69, 73–76, 89, 388;
and Buddhism, 48, 49, 95; *see also
under* measurement

Teller, P., 345–46

teleological explanations, 48

temperature, 291, 315–16

temples, 24

theistic religion, 63*n*16; theists, 14;
monotheism, 39; pantheism, 39, 40;
polytheism, 40, 62*n*8

theology, 8, 36, 400; theologians, 2;
Christian theology, 335, 393;
theological premises and assertions,
17, 22

Theravāda, 216, 298

thermodynamics, 291, 313, 317

Tholey, P., 250–7

Thompson, E., 54–56

Three Jewels, 318

three marks of existence, 172*n*4

three poisons, 148, 153–56, 169, 173*n*8,
174*n*11, 178*n*29

Thurman, R. A., xv–xvi; 52, 63*n*9,
64*n*28

Tilopa, 235–36

time, 13–14, 51, 113, 303–21, 326, 382–83;
atomicity of, 381; Buddhist analyses
of, 99; dimensions of, 375; points of,
372; as linear or cyclic, 39–40, 62*n*6,
62*n*8; and evolution, 152, 185*n*50;
and self, 118; absolute time, 368;
clock time, 122, 128, 135; path time,
378–79; temporality, 201, 211; temporal pulses, 201–3; temporal relations, 120; time interval, 297, 376; *see
also under* Aristotle; Einstein, A.;
Galileo; Planck, M.; space

Tomasello, M., 185*n*50

tonglen, 220–3, 225

Tooby, J., 182*n*41

totemic ancestors, 187*n*60

Training the Mind research project,
61*n*1, 63*n*12

transcendentalism, 8, 12, 18; *see also
under* realism

transcendental deduction, 350–51, 353,
360*n*90, 360*n*94, 361*n*103

transcranial magnetic stimulation
(rTMS), 197

Trivers, R., 158

Trungpa, C., Rinpoche, 217, 220, 227*n*6

truth, 3, 72, 186*n*54, 209, 310, 406, 411;
closed set of, 334; pursuit of, 15, 26;
the root of, 382; absolute truths, 22,
165–66, 283, 342; objective or
scientific truth, 12, 19, 26, 46, 326;
inner truths, 270; Buddhist truths,
26, 45; *see also* Four Noble Truths;
Two Truths

truth claims, 21–23, 25–27, 333

Tsongkhapa, 77, 217–8, 227*n*7, 236, 246

Tu W., 387

Tweed, T., 43

Two Truths, 289–90; conventional
truth, 289, 292, 299, 329–33, 340,
357*n*48, 357*n*49; ultimate truth, 289,
302*n*11, 329, 333, 340–41

vacuum state, 346

Vaibhāṣika, 94, 109

Vajrayāna, 96–97, 216, 219–20, 223, 231;
Kālacakra tradition, 377; *see also
under* physiology

Vajra-Yogini, 241

value, 3, 9, 17; values, 2, 51, 114, 127–28,
335, 411; value judgments, 130

Varela, F. J., xii–xiii, xviii, 54–56, 64*n*25,
193–94, 195–230, 237, 267–68, 335, 347,
356*n*38, 363, 417–20; Varela's autopoietic theory of cognition, 356*n*38